Computational Fluid Dynamics in Renewable Energy Technologies

This book is focused on combining the concepts of computational fluid dynamics (CFD) and renewable energy technologies. Besides introducing the fundamentals, the core of this book contains a series of practical examples providing useful information about the methods and smart solutions for CFD modeling of selected Renewable Energy Sources (RES) - based technologies. Each chapter includes a theoretical introduction to the discussed topic, descriptions of factors determining efficiency and other important parameters, followed by practical information concerning the CFD modeling methodology. A summary of the relevant recommendations and exemplary results with comments is also included.

Features:

- provides practical examples on the application of numerical methods in the analysis of renewable energy processes,
- includes an introduction to CFD for practitioners,
- explores selected aspects of the methodology used in CFD simulations of renewable energy technologies,
- discusses tips and hints for efficient use of CFD codes functionalities,
- contains additional exercise devoted to the geothermal systems.

This book is aimed at professionals and graduate students in energy engineering, renewable energy, CFD, energy systems, fluid mechanics and applied mathematics.

Computational Fluid Dynamics in Renewable Energy Technologies
Theory, Fundamentals and Exercises

Mateusz Szubel, Mariusz Filipowicz,
Karolina Papis-Frączek, and Maciej Kryś

CRC Press is an imprint of the
Taylor & Francis Group, an **informa** business

Designed cover image: Ms. Klaudia Duda

First edition published in 2024
by CRC Press
6000 Broken Sound Parkway NW, Suite 300, Boca Raton, FL 33487-2742

and by CRC Press
4 Park Square, Milton Park, Abingdon, Oxon, OX14 4RN

CRC Press is an imprint of Taylor & Francis Group, LLC

© 2024 Mateusz Szubel, Mariusz Filipowicz, Karolina Papis-Frączek and Maciej Kryś

Reasonable efforts have been made to publish reliable data and information, but the author and publisher cannot assume responsibility for the validity of all materials or the consequences of their use. The authors and publishers have attempted to trace the copyright holders of all material reproduced in this publication and apologize to copyright holders if permission to publish in this form has not been obtained. If any copyright material has not been acknowledged please write and let us know so we may rectify in any future reprint.

Except as permitted under U.S. Copyright Law, no part of this book may be reprinted, reproduced, transmitted, or utilized in any form by any electronic, mechanical, or other means, now known or hereafter invented, including photocopying, microfilming, and recording, or in any information storage or retrieval system, without written permission from the publishers.

For permission to photocopy or use material electronically from this work, access www.copyright.com or contact the Copyright Clearance Center, Inc. (CCC), 222 Rosewood Drive, Danvers, MA 01923, 978-750-8400. For works that are not available on CCC please contact mpkbookspermissions@tandf.co.uk

Trademark notice: Product or corporate names may be trademarks or registered trademarks and are used only for identification and explanation without intent to infringe.

ISBN: 9781032064185 (hbk)
ISBN: 9781032064192 (pbk)
ISBN: 9781003202226 (ebk)
ISBN: 9781032575155 (eBook+)

DOI: 10.1201/9781003202226

Typeset in Times
by codeMantra

To Magda and my students.
Mateusz Szubel
To my husband and parents.
Karolina Papis-Frączek

Contents

Preface xiii
Acknowledgments xv
Authors xvii

PART I Fundamentals of Computational Fluid Dynamics: Selected Issues

Chapter 1   Idea and Applications of CFD 3

    1.1   The Need for CFD 3
    1.2   The General Idea of CFD 3
    1.3   History 5
    1.4   Applications 6
    References 6

Chapter 2   A Practical Look at the Steps of CFD Model Development 9

    2.1   Preprocessing 9
        2.1.1   Geometry Design 9
        2.1.2   Meshing Process 10
        2.1.3   Preprocessing – Setting Solver 13
    2.2   Numerical Solution and Its Features 15
    2.3   Postprocessing 23
    References 23

Chapter 3   Transport Equations 25

    References 27

Chapter 4   Turbulent Flows in RANS Approach 29

    4.1   RANS-Based Zero-Equation and Non-Zero-Equation Turbulence Models ... 30
    4.2   One- and Multi-Equation Turbulence Models 33
    References 39

Chapter 5   Reactive and Multiphase Flows 43

    5.1   Classification of the Chemistry Modeling Approaches and Characteristics of the Selected Models 43
        5.1.1   Fast/Slow Chemistry Modeling 43
            5.1.1.1   Eddy Dissipation Model (EDM) 45
            5.1.1.2   Finite Rate Model (FRM) and Finite Rate Eddy Dissipation Model (FREDM) 45
            5.1.1.3   Eddy Dissipation Concept Model (EDCM) 46

	5.1.2	Premixed and Non-Premixed Combustion Models	46
5.2	Introduction to Homogeneous and Heterogeneous Combustion		47
5.3	Selected Types and Importance of Multiphase Transport in CFD		48
References			52

PART II Photothermal-Conversion-Based Technologies

Chapter 6 Theoretical Background 57

    6.1 Development of Technology 57
    6.2 Statistical Data 58
    6.3 Classifications and Characteristics 60
        6.3.1 Flat Plate Solar Collectors 61
        6.3.2 Evacuated Tube Solar Collectors 62
        6.3.3 Other Non-concentrating Technologies 63
        6.3.4 Concentrated Solar Power Systems 63
    6.4 Fundamentals of Energy Conversion and Balance 66
        6.4.1 Flat Plate Collector 66
        6.4.2 Evacuated Tube Collectors 71
        6.4.3 Concentrating Solar Collectors 72
    References 74

Chapter 7 Tutorial 1 – Flat-Plate Solar Collector 77

    7.1 Exercise Scope 77
    7.2 Preprocessing – Geometry 77
    7.3 Preprocessing – Meshing 86
    7.4 Preprocessing – Solver Settings 92
    7.5 Postprocessing 104

Chapter 8 Tutorial 2 – Evacuated Tube Solar Collector 111

    8.1 Exercise Scope 111
    8.2 Preprocessing – Geometry 112
    8.3 Preprocessing – Meshing 116
    8.4 Preprocessing – Solver Settings 122
    8.5 Postprocessing 129

Chapter 9 Tutorial 3 – Heat Receiver for a Solar Concentrating System 135

    9.1 Exercise Scope 135
    9.2 Preprocessing – Geometry 136
    9.3 Preprocessing – Meshing 142
    9.4 Preprocessing – Solver Settings 148
    9.5 Postprocessing 157
    9.6 Additional Exercise 159

PART III Photoelectric-Conversion-Based Technologies

Chapter 10 Theoretical Background 163

    10.1 Development of Technology 163
    10.2 Statistical Data 164
    10.3 Classifications and Characteristics 165
        10.3.1 Photovoltaic Modules and Systems 168
        10.3.2 Hybrid (Photovoltaic – Thermal) Collectors 168
    10.4 Fundamentals of Energy Conversion and Balance 170
        10.4.1 Photovoltaics 170
        10.4.2 Thermal Photovoltaics (PVT) 174
    References 178

Chapter 11 Tutorial 4 – Photovoltaic Farm 181

    11.1 Exercise Scope 181
    11.2 Preprocessing – Geometry 182
    11.3 Preprocessing – Meshing 186
    11.4 Preprocessing – Setting Solver 192
    11.5 Postprocessing 202
    11.6 Additional Exercise 205

PART IV Wind-Power-Based Technologies

Chapter 12 Theoretical Background 209

    12.1 Development of Technology 209
    12.2 Statistical Data 210
    12.3 Classifications and Characteristics 211
        12.3.1 Horizontal-Axis Wind Turbines 213
        12.3.2 Vertical-Axis Wind Turbines 214
    12.4 Fundamentals of Energy Conversion and Balance 216
        12.4.1 Horizontal-Axis Wind Turbines 218
        12.4.2 Vertical-Axis Wind Turbines 220
    References 221

Chapter 13 Tutorial 5 – Horizontal-Axis Wind Turbine 223

    13.1 Exercise Scope 223
    13.2 Preprocessing – Geometry 224
    13.3 Preprocessing – Meshing 236
    13.4 Preprocessor – Solver Settings 244
    13.5 Postprocessing 256

Chapter 14 Tutorial 6 – Vertical-Axis Wind Turbine 263

    14.1 Exercise Scope 263
    14.2 Preprocessing – Geometry – Part 1 264

14.3	Preprocessing – Meshing – Part 1	271
14.4	Preprocessing – Solver Settings – Part 1	279
14.5	Preprocessing – Geometry – Part 2	289
14.6	Preprocessing – Meshing – Part 2	293
14.7	Preprocessing – Solver Settings – Part 2	295
14.8	Postprocessing	297

PART V Biomass-Based Small-Scale Energy Applications

Chapter 15 Theoretical Background 305

15.1 Development of the Technology 305
15.2 Statistical Data 305
15.3 Classifications and Characteristics 308
  15.3.1 Direct-Combustion-Based Technologies 308
  15.3.2 Technologies for the Reduction of Environmental Impact of Particulate Matter (PM) Emissions 311
  15.3.3 Gasification-Based Technologies 313
  15.3.4 Heat Accumulation Systems for RES-Based Technologies 317
15.4 Fundamentals of Energy Conversion and Balance 318
  15.4.1 Fundamentals of Combustion, Pyrolysis, and Gasification Processes 318
  15.4.2 Heat Balance of the Biomass-Fired Heating Units 320
References 323

Chapter 16 Tutorial 7 – Syngas Burner 325

16.1 Exercise Scope 325
16.2 Preprocessing – Geometry 325
16.3 Preprocessing – Meshing 327
16.4 Preprocessing – Solver Settings 330
16.5 Postprocessing 341

Chapter 17 Tutorial 8 – Particulate Matter Separation in Cyclone 347

17.1 Exercise Scope 347
17.2 Preprocessing – Geometry 348
17.3 Preprocessing – Meshing 353
17.4 Preprocessor – Solver Settings 359
17.5 Postprocessing 365

Chapter 18 Tutorial 9 – Accumulation Heat Exchanger 373

18.1 Exercise Scope 373
18.2 Preprocessing – Geometry 374
18.3 Preprocessing – Meshing 380
18.4 Reprocessing – Solver Settings 384
18.5 Post-processing 393
18.6 Additional Exercise 396

Contents

PART VI Geothermal-Energy-Based Systems

Chapter 19 Theoretical Background 399
        19.1   Development of Technology 399
        19.2   Statistical Data 399
        19.3   Classifications and Characteristics 400
        19.4   Fundamentals of Energy Conversion and Balance 403
        References 409

Chapter 20 Tutorial 10 – Borehole Heat Exchanger 411
        20.1   Exercise Scope 411
        20.2   Preprocessing – Geometry 411
        20.3   Preprocessing – Meshing 418
        20.4   Preprocessing – Solver Settings 425
        20.5   Variant Analysis 436
        20.6   Post-processing 437

Index 443

Preface

Dear Reader,

This enhanced version includes a range of additional features to provide a greater understanding of the text. Key features include multiple movies by the author explaining about this book and trying to present the practical aspects of project execution using live examples, illustrations and videos. The quizzes have also been included for some chapters providing a deep study of those chapters. Both movies and quizzes help the fresh engineers to change the way they look at chemical engineering, and for professionals, it would be helpful in getting some quick tools and guidelines for a better understanding of computational fluid dynamics (CFD).

Rapid technological advances in the field of renewable energy significantly increase the importance of computer simulation as an essential tool supporting the conventional analytical and experimental procedures. It has become indispensable for further development of any industry.

Apart from the modeling of whole renewable-energy-based networks or energy systems, there is a huge need for numerical investigations of individual system elements – the devices and their components that are essential for drawing energy from renewable sources.

As long as it is possible to describe the above-mentioned devices using mathematical terms, it is possible to express the studied phenomena by appropriate mass, momentum and energy transport equations that can be discretized and solved, to get the desirable answer.

A powerful tool allowing to implement this methodology in any fluid and/or heat flow studies is CFD which is employed nowadays in many freeware and commercial codes. Due to this fact, it is crucial to train qualified simulation engineers.

This book presents the usefulness of CFD in studying the renewable energy technologies on the academic course level, and it is dedicated both to students and teachers (especially those who are introducing new courses in the field of renewable energy-based technologies, or running such courses for the first time).

Being part of academic courses devoted to CFD is a tall order both for teachers and for students. Teachers have to generate students' interest through the preparation of interesting exercises while being constrained by the limitations of academic hardware capacities. Students have to keep practicing software operation and find motivation to study the theory, which will allow them to design and solve models knowingly. No matter which of these two groups you belong to, the key is inspiration. If you are a student, this book will be for you a prompt to start your first real, practical projects. Something more than just basic models that you used to consider during your heat and mass transfer classes. If you are a teacher, we hope that the content will suggest you how to organize your academic course and provide you with useful exercises, or at least preliminary hints on how to create your own ones.

Without doubt, there are at least a dozen brilliant textbooks presenting theoretical side of the CFD. Each teacher has their favorite one and recommends it during classes. In the case of this book, the authors find it pointless to repeat the same, extended theoretical CFD background, while it can be found in the already-existing literature. On the other hand, the above-mentioned books, if at all, cover mainly basic exercises, allowing us to interpret the conservation principles or transport phenomena. Due to these facts, only basic theoretical issues which, based on the authors' practical experience, are recognized as the most problematic are briefly reminded in this book. The center of gravity is moved here from theory to practice (to 'learning by doing'). Thus, the tutorials covering RES-based technologies of significant practical application can be found here.

Apart from the above-mentioned things, the authors' additional goal was to go beyond the usual pattern of the textbooks devoted to CFD. All of the included technologies are considered taking into account the big picture of the issue. Apart from the practical CFD tutorials, each part covers the historical background, classifications, construction characteristics and the physics of energy conversion for the specific technology. This is a holistic point of view on the didactic requirements and the approach allowing to fulfill the gap in teaching the engineers in the RES field. Technical data regarding the considered technologies were carefully selected, which allows to save time that is usually required to collect theoretical information and physics description related to the system that is going to be analyzed using the CFD methods. This way the book constitutes a complete and integral source of essential knowledge.

The practical examples of the CFD applications that were presented in this book are dedicated to beginners and intermediate users of the CFD codes. The tutorials were created based on the example of Ansys Workbench software, which is currently one of the most powerful worldwide tools allowing running fluid flow simulations. The technical side of the tutorials was completely adapted to allow running the exercises using the student version (license) of the above-mentioned code.

This book introduces the reader to the theoretical background and practical aspects of the CFD analyses in the field of the RES-based technologies. The tutorials do not demand from the user any experience in using CFD codes. Each exercise guides the reader carefully through the procedure of developing and solving the model. On the other hand, full understanding of the transferred content requires basic knowledge of the heat transport and fluid mechanics theory.

We are a specific group of authors – we are of different ages, and we have very different experience and scientific backgrounds. Our common feature is our dedication to the fields of renewable energy technologies and computer simulations. We all have wide experience in running different kinds of professional trainings and practical, academic classes. Furthermore, after years of contribution in many commercial projects, we are familiar with the problems and specifics of the CFD applications in the industry.

We know and understand very well students' and teachers' needs and problems. We believe that this fact has allowed us to present different points of view on the considered topics and methods, and provide you with an excellent material that will introduce you to the issues of CFD in renewable energy technologies.

Acknowledgments

The authors wish to thank Ms. Elżbieta Kania for the proofreading of the text and Ms. Klaudia Duda for the cover design.

Authors

Mateusz Szubel (PhD, Eng.), is an active industry engineer and an assistant professor at the AGH University of Science and Technology, Faculty of Energy and Fuels, Krakow, Poland. Professor Szubel is a member of the research group focused on the issues of renewable energy technologies at the Department of Sustainable Energy Development. He is a research specialist on the conditions for the sustainable energy development. Furthermore, he is involved in the investigation of the possibilities of employing computational fluid dynamics (CFD) in the optimization of renewable energy technologies and increasing the energy efficiency of energy systems. He is focused on numerical modeling of energy storage and biomass thermochemical treatment. For 12 years, professor Szubel has been running over 20 academic courses related to the practical aspects of renewable energy technologies' applications and the use of commercial CFD codes to analyze the elements of these systems. He is the author and coauthor of 120 scientific and popular science papers.

Mariusz Filipowicz (DSc, Eng.) is an associate professor at the AGH University of Science and Technology, Faculty of Energy and Fuels, and former head of the Department of Sustainable Energy Development, Krakow, Poland. Professor Filipowicz graduated with an MSc in Technical Physics in 1991, obtained a PhD in 1998, and did habilitation in 2010 in the field of nuclear physics–nuclear fusion catalyzed by negative muons. Between 1994 and 1995, he completed a postgraduate course in Energy and Environment, the Tempus-Joint European Project (JEP) Postgraduate Course on Energy and Environment. Professor Filipowicz's main research areas are related to renewable energy technologies (mainly biomass, solar, and wind energy applications), energy efficiency, and nuclear physics (nuclear fusion reactions in the range of ultra-low energy). He is the author and coauthor of more than 250 scientific papers and leader of scientific projects at AGH, such as "BioEcoMatic: Construction of small-to-medium capacity boilers for clean and efficient combustion of biomass for heating" and "BioORC: Construction of cogeneration system with small to medium size biomass boilers" with the support of KIC InnoEnergy.

Karolina Papis-Frączek (MSc., PhD student) is an academic teacher and research assistant at the AGH University of Science and Technology, Faculty of Energy and Fuels, Krakow, Poland. She is a member of a research group focused on the issues of renewable energy technologies at the Department of Sustainable Energy Development. Ms. Papis' research interests include experimental and numerical studies on the efficiency of renewable energy technologies. Preparation of her doctoral thesis involves research into the use of concentrated solar radiation for the purpose of heat and electricity generation. She is the author and coauthor of 30 scientific and popular science papers.

Maciej Kryś (MSc) is an active industry engineer. He is a graduate of the Power Engineering Faculty at the Silesian University of Technology in Gliwice. For nearly a decade, he has been working as an Analysis Engineer, specializing in using Computational Fluid Dynamics tools. As part of his duties, he has assisted dozens of industrial companies and academic researchers in performing fluid dynamics simulations. His work did not only include executing simulations but also training engineers, specialists and researchers. Currently, he is working in Oil and Gas industry optimizing the performance of traditional industry equipment, but also actively supporting implementation of renewable energy technologies in this field.

Part I

Fundamentals of Computational Fluid Dynamics
Selected Issues

1 Idea and Applications of CFD

1.1 THE NEED FOR CFD

There are numerous limitations to finding solutions for complex flow problems, using analytical and experimental methods. Generally, the motion of a Newtonian fluid is described by the Navier-Stokes equation. Even though it was formulated nearly two centuries ago (1822–1850), only a few analytical solutions for trivial or strongly simplified cases are known today. Moreover, fluid flow is often accompanied by various physical phenomena like heat transfer, phase changes, chemical reactions, mechanical movement, *etc.*, which makes the analysis more complex [1]. Experimental methods may be used to solve these problems, but it should be highlighted that detailed experimental investigations not only require a significant amount of time and money but also carry a high risk of failure. Summarizing, there is clear evidence that finding a solution for complex flow problems requires another approach.

Computational fluid dynamics, known as CFD, is an interdisciplinary subject which combines three science branches: fluid mechanics, numerical analysis and computer science.

A specific fluid flow problem is described by a set of mathematical equations. These equations are known as governing ones, and they mainly describe the transport of mass, momentum, and energy in moving fluids. CFD uses numerical methods to convert these differential equations to the algebraic form, and then computers solve them (usually iteratively) [2]. The obtained results may be visualized in numerous ways.

CFD methods allow to significantly reduce the number of experimental tests, but never bring them down to zero, because they are still required to validate the numerical solution. This means that reliable results of CFD simulations should be consistent with experimental data. Obviously, the total cost of an accurate CFD model for a complex flow is money- and time-consuming, but the total cost of the additional experiments during the prototyping process is still greater. CFD gives an opportunity to prepare simulations on large scales and in nearly any conditions, even the ones impossible to create in a laboratory. Moreover, numerical simulations provide detailed information about the analyzed parameters at all points of the considered experimental system, not only in the places where sensors are installed.

There is a great advantage of combining an analytical approach, experimental measurements and numerical calculations because the mixed approach provides highly reliable solutions in the most cost-effective and time-saving way [1]. Nowadays, CFD is an essential tool in design and optimization processes [3]. CFD is also gaining significance as a substantial research tool in physical sciences [4].

1.2 THE GENERAL IDEA OF CFD

To solve any problem with CFD methods, it is necessary to describe the phenomena occurring in reality by physical laws (like mass, momentum, and energy conservation), described by mathematical equations. Analytical solutions of these equations are considered as continuous, *i.e.* valid at each point of the solution domain [2]. In contrast to analytical solutions, numerical results are obtained only at chosen points within the domain, so they are called discrete solutions (see Figure 1.1). Values at other locations are determined by approximation [3]. In order to specify the computation points, the solution domain is divided into smaller parts, known as computational cells, in a process called meshing or mesh/grid generation.

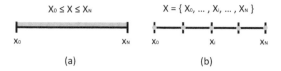

FIGURE 1.1 Graphical representation of (a) continuous domain and (b) discrete domain.

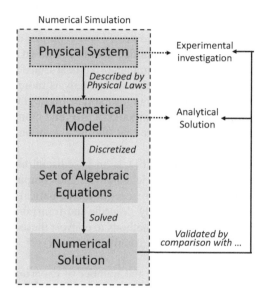

FIGURE 1.2 Flow chart presenting the general idea of a CFD simulation.

In the next step, the space- and time-dependent mathematical equations are approximated to the algebraic form with one of the numerical methods, for example, Finite Difference Method (FDM), Finite Element Method (FEM), or Finite Volume Method (FVM). One of the most popular methods in CFD is FVM, used in ANSYS CFX and ANSYS Fluent solvers. FVM is based on an integral form of the conservation laws [2], so it results in a set of equations describing flux conservation over each cell.

After the discretization process, all equations are iteratively solved for each cell by a computer, based on initial and boundary conditions, and using complex computer algorithms embedded within the CFD software. Apart from number values, numerical results could be displayed graphically in the form of vector fields, contours, plots or as numerical data. Validation cannot be forgotten – credibility of the prepared simulation has to be confirmed, *e.g.* by a comparison with experimental data. The scheme presenting a general idea of a CFD simulation is shown in Figure 1.2.

The most difficult phenomenon to describe with numerical methods is a fluid motion, especially in the transition and turbulent regime. It is due to the strong chaotic motion of the fluid, intensive mixing and increased momentum and energy exchange in such cases [4]. There are three approaches to model turbulence, with different degrees of numerical complexity and, therefore different accuracy:

- Direct Numerical Simulation – DNS
  The turbulence is simulated directly from the Navier-Stokes equations. This approach is possible only for simple flow cases characterized by low Reynolds numbers since the number of cells increases to the cube of the calculated Reynolds number [4]. Therefore,

DNS is extremely demanding on computer resources, which are unreachable for industrial applications [2]. Nevertheless, this method plays an important role in science, because it helps to understand the flow behavior in turbulent and transition flows and contributes to the development and validation of new and existing turbulence models [4].

- Large-Eddy Simulation – LES

  This approach uses a partial approximation in turbulence simulation, depending on the vortices size scale: large eddies are directly simulated, whereas the small ones are modeled [2]. This simplification is possible due to a more universal character of the small vortices. This assumption results in less strict requirements regarding the number of computational cells, and the flows at higher Reynolds numbers may be investigated. Nevertheless, LES still demands significant computer resources, and therefore it is still far from becoming an engineering tool. This approach finds common application in science, where detailed studies into fluid motion are required. This includes the following fields: separated flows, rotating flows, aerodynamic noise and mixing in combustion chambers or engines [4]. Interesting applications of LES can be found in [5–7].

- Reynolds – Averaged Navier-Stokes – RANS

  RANS represents the highest level of approximation in the turbulence simulation: the fluid motion is fully modeled. The instantaneous values of flow variables are decomposed into mean and fluctuating parts and inserted into the Navier-Stokes equations. This operation results in averaged Navier-Stokes equations with an additional term (in momentum equations): a Reynolds-stress tensor. To solve this set of equations, additional information about Reynolds stresses has to be provided [2]. This condition is fulfilled with the use of turbulence models [2]. RANS approach allows to use significantly coarser meshes than in the DNS and LES methods, which significantly reduces the computational requirements. However, this averaging process also results in the loss of all the detailed information about turbulent flow structures [4]. Nevertheless, the accuracy of RANS approach is sufficient for most engineering applications. More information about RANS is presented later in this book.

1.3 HISTORY

The history of CFD started in the 1922, when an Englishman Lewis Fry Richardson developed a weather forecasting system. His weather prediction scheme was based on differential equations approximated to the algebraic form with the finite differences method. Instead of computers, this "forecast factory" included 64,000 humans. Each person did one part of the flow calculations for a specified map location, using only a mechanical calculator. Due to the system imperfection, the attempt to numerically forecast weather for an eight-hour period consumed almost six weeks of real time, but failed [4].

Nevertheless, the idea of equation discretization was constantly developed. In 1933, Alex Thom published a paper describing the flow past a cylinder [8]. Similar results were obtained by a Japanese scientist Kawaguti in 1953, who worked on the solution for 20 hours per week for 18 months, using only a mechanical calculator [9].

In 1950, 28 years after Richardson's "forecast factory", the first successful weather forecast for a 24-hour period was performed by the ENIAC modern computer. A significant development of numerical methods known today as CFD was made in the 1960s, due to the increased availability of more powerful computers. The main contribution was made by scientists from Los Alamos National Lab and the aircraft industry. In that time, many numerical methods, which are still in use today, like k-epsilon turbulence model [3], were implemented. In the next decade, the existing concepts were improved and new ideas were carried through.

The crucial event in the development of CFD codes was in 1980 when Suhas V. Patankar published the book *Numerical Heat Transfer and Fluid Flow*, which is considered the most influential book on CFD to date. After this, many commercial CFD codes were created and they started being commonly implemented in numerous fields of industry [4].

Constant progress in CFD methods is still observable, but it would be misleading to consider CFD as a mature technology [1]. The issues related to turbulence and combustion modeling, efficient solution techniques for viscous flows, optimization techniques, discretization methods or connection of CFD with structural mechanics still require further development [4].

1.4 APPLICATIONS

Nowadays, CFD is applied to solve various engineering problems and it is also considered as an indispensable tool for science. Numerical methods are complementary to experiments and analytical investigations. Generally, the main applications of CFD simulations are in the fields of:

- laminar and turbulent flow [10],
- aircrafts, car and ship aerodynamics – reduction of the drag and maximization of lift force in order to: increase range and speed, decrease size and costs, minimize fuel usage [11],
- turbomachinery: pumps, fans, compressors and turbines – blade design, cavitation prevention [12],
- incompressible and compressible flow [13],
- multiphase flows – analysis of phase changes: condensation, evaporation; free surface flows; bursting or agglomeration of gas bubbles/liquid drops [14],
- heat transfer and thermal management – conduction, convection and radiation; heat storage systems [15],
- reactive flows [16],
- heating, air conditioning, and ventilation (HVAC) [17],
- rocket engines [18],
- battery simulation [19],
- electronics cooling [20].

What is more, CFD is applied in meteorology [5], oceanography [21], astrophysics [16], and also in architecture [4]. Many numerical techniques developed for CFD are used in the solution of Maxwell equations as well [4].

REFERENCES

1. T. Cebeci et al., *Computational Fluid Dynamics for Engineers*, From Panel to Navier-Stokes Methods with Computer Programs, Springer (2005), ISBN: 3-540-24451.
2. C. Hirsch, *Numerical Computation of Internal and External Flows Volume 1: Fundamentals of Computational Fluid Dynamics*, Elsevier (2007), ISBN: 978-0-7506-6594-0.
3. A. W. Date, *Introduction to Computational Fluid Dynamics*, Cambridge University Press, Cambridge, 2005.
4. J. Blazek, *Computational Fluid Dynamics: Principles and Applications*, Elsevier (2001), ISBN: 0-08-043009-0.
5. B. Lu, Q.-S. Li, Large eddy simulation of the atmospheric boundary layer to investigate the Coriolis effect on wind and turbulence characteristics over different terrains, *Journal of Wind Engineering and Industrial Aerodynamics*, 220 (2022), 1–10.
6. K. Zheng et al., Application of large eddy simulation in methane-air explosion prediction using thickening flame approach, *Process Safety and Environmental Protection*, 159 (2022), 662–673.
7. Y. Zhiyin, Large-eddy simulation: Past, present and the future, *Chinese Journal of Aeronautics*, 28 (1) (2015), 11–24.
8. A. Thom, The flow past circular cylinders at low speeds, *Proceedings of the Royal Society of London. Series A*, 141 (845) (1933), 651–666.

9. M. Kawaguti, Numerical solution of the Navier-Stokes equations for the flow around a circular cylinder at Reynolds number 40, *Journal of the Physical Society in Japan*, 8 (1953), 747–757.
10. M. Sorguna et al., CFD modeling of turbulent flow for Non-Newtonian fluids in rough pipes, *Ocean Engineering*, 247 (2022), 110777.
11. F. T. Johnson, E. N. Tinoco, N. J. Yu, Thirty years of development and application of CFD at Boeing Commercial Airplanes, Seattle, *Computers & Fluids*, 34 (10) (2005), 1115–1151.
12. P. Bradshaw, Turbulence modeling with application to turbomachinery, *Progress in Aerospace Sciences*, 32 (6) (1996), 575–624.
13. S. A. Khan, O. M. Ibrahim, A. Aabid, CFD analysis of compressible flows in a convergent-divergent nozzle, *Materials Today*, 46 (Part 7) (2021), 2835–2842.
14. M. Bracconi, CFD modeling of multiphase flows with detailed microkinetic description of the surface reactivity, *Chemical Engineering Research and Design*, 179 (2022), 564–579.
15. B. Pandey, R. Banerjee, A. Sharma, Coupled Energy Plus and CFD analysis of PCM for thermal management of buildings, *Energy and Buildings*, 231 (2021), 110598.
16. *Biomass in Small-Scale Energy Applications, Theory and Practice*, Edited by M. Szubel, M. Filipowicz, CRC Press, Taylor & Francis Group, Boca Raton, FL, 2020.
17. L. Zhang, Study of transient indoor temperature for a HVAC room using a modified CFD method, *Energy Procedia*, 160 (2019), 420–427.
18. V. Zubanov, V. Egorychev, L. Shabliy, Design of rocket engine for spacecraft using CFD-modeling, *Procedia Engineering*, 104 (2015), 29–35.
19. A. R. Sharma et al., Three-dimensional CFD study on heat dissipation in cylindrical lithium-ion battery module, *Materials Today*, 46 (Part 20) (2021), 10964–10968.
20. B. Ramos-Alvaradoa et al., CFD study of liquid-cooled heat sinks with microchannel flow field configurations for electronics, fuel cells, and concentrated solar cells, *Applied Thermal Engineering*, 31 (14–15) (2011), 2494–2507.

2 A Practical Look at the Steps of CFD Model Development

From the engineering point of view, the numerical solution carried out with CFD methods is done in three stages [1]:

- preprocessing, in general, all tasks that take place before the solution computation process: problem formulation, geometry designing/editing, mesh generation and solver setup (applying models, materials, cell zone and boundary conditions);
- solving (processing) all tasks directly regarding the numerical computation process: equation discretization, setting residual targets and monitors of selected solution data, initialization and iterative solution of mathematical equations;
- postprocessing all tasks regarding the solution data evaluation process: numerical and graphical presentation of generated results, analysis of obtained results and reporting of findings.

According to Figure 2.1, preprocessing is the most work-demanding stage of a CFD model development. More information about each stage can be found later in this book.

2.1 PREPROCESSING

2.1.1 Geometry Design

Depending on the point of interest, certain parts of the physical domain should be defined in the form of geometry [2]. Its final properties depend on the system's physical parameters (like dimensions) and on the parameters which will be analyzed (for example, in adiabatic flows, there is no

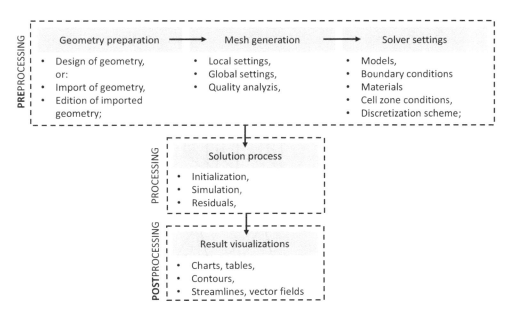

FIGURE 2.1 Three stages of a CFD model preparation.

need to define the wall thickness). It should be noted that the geometry prepared for CFD simulation should be simplified to shorten the solution time, for example, all unnecessary details (like roundings, chamfers, holes and construction elements, such as screws, welds, *etc.*) have to be removed. Another possibility to simplify the geometry is to apply a symmetry or periodic boundary condition. However, it is applicable only in the case when symmetry is observed not only in geometry but also in the occurring phenomena.

Generally, geometry may be created directly in the built-in software or imported from an external file [1]. Some embedded programs are usually available in commercial codes; *e.g.* in Ansys Workbench, there are two basic tools: DesignModeler and SpaceClaim DirectModeler. It is important to prepare a clean, watertight domain without any defects, such as missing or overlapping faces, sharp angles, highly-curved surfaces, *etc.*

2.1.2 Meshing Process

The second step of preprocessing is to discretize the previously prepared continuous geometry in a process called mesh/grid generation [1]. All fluid and solid bodies have to be divided into a finite number of isolated volumes. These control cells replace the continuity of the geometry, and they are a base for determining finite volume nodes (most often in CFD) or finite elements (in structural analyzes). In these points, the variables of governing equations are calculated and stored [3,4].

In the case of a two-dimensional domain, the most common cell shapes are: triangles and quadrilaterals, both shown in Figure 2.2. Triangles are easy and quick to generate. Moreover, they easily cover the whole, even complicated geometry. Quadrilateral cells are harder to generate, but they usually provide better numerical stability and significantly lower total cell number than triangular meshes. In some cases, such as supersonic flows, the quadrilateral mesh is a must.

Three-dimensional meshes contain cells shaped as tetrahedrons, hexahedrons, quadrilateral pyramids, triangular prisms and polyhedrons, as shown in Figure 2.3. After the extrusion of

FIGURE 2.2 Examples of (a) triangular and (b) quadrilateral cells.

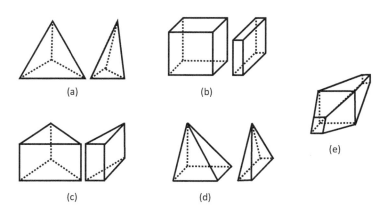

FIGURE 2.3 Examples of (a) tetrahedrons, (b) hexahedrons, (c) prisms, (d) pyramids, and (e) polyhedrons.

two-dimensional triangles and quadrilaterals, the prisms and hexahedrons are created, respectively. Prisms consist of two triangular and three quadrilateral faces, and they are commonly localized in the fluid boundary layer. Hexahedrons are bounded by six quadrilateral faces, and they provide the highest accuracy of the numerical solution among all cell shapes. Tetrahedral cells are bounded by four triangular faces, and a mesh with these elements can be generated automatically. Pyramid cells consist of four triangular and one quadrilateral faces. Due to this, pyramids are commonly used as transition elements between cells with square and triangular faces. The most complex shape belongs to polyhedral cells. They may be bounded by any number of faces in different shapes.

The stage of preprocessing requires dedicated software tools and is commonly considered the most complex and labor-intensive part of the CFD simulation, especially in the case of three-dimensional domains with different scale objects (for example, large diameter channels with small detail, like thermocouple). The quantitive and qualitative parameters of the mesh are crucial for the accuracy of a numerical simulation [3].

The number of mesh elements is directly related to their size. Smaller elements better imitate the continuum, especially in detailed geometries. Moreover, the distance between solution points is shorter, so the numerical approximation is more accurate. However, extensive meshes lead to unreasonable computation times. It should be noted that there is a limit of mesh elements, beyond which the solution does not change significantly. A general relation between the number of mesh elements, solution accuracy and calculation time is shown in Figure 2.4.

Usually, the suitability of a particular mesh is not known a priori. It is necessary to investigate the effect of mesh refinement on the level of the solution accuracy until no significant changes in simulation results are observed during further mesh modifications. This process is known as mesh independence analysis.

Even a mesh optimized for quantity may be inadequate for the solution process if its quality is poor. Mesh quality refers to the influence of cell shapes on the solution results. Generally, the more regular the shape is, the better conditions for the discretization process. Highly skewed (with sharp internal angles) or extremely oblong mesh elements lead to unreliable results and convergence errors. When comparing two meshes with the same number of control cells, the one with higher quality provides a more accurate solution in a shorter time.

The quality of the generated mesh can be assessed on the basis of numerous indicators. One of them, which is often monitored in CFD models, is called skewness. It can be expressed as a measure of the proximity of the cell shape to the ideal shape, *i.e.* a figure with equal edges or angles. Skewness takes values from 0 to 1; the smaller the skewness value, the better the mesh quality. The

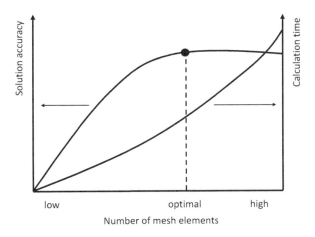

FIGURE 2.4 General relation between the number of mesh elements, solution accuracy and calculation time.

highest acceptable value is 0.95, with the average not being higher than 0.8. The method of interpreting the quality of the mesh cells depending on the skewness is presented in Table 2.1.

Orthogonal quality (Table 2.2) is another indicator commonly used in CFD models. It determines the degree of similarity of the cell shape to a regular solid. The maximum quality of cells occurs when their orthogonal quality is 1. Cells with the worst, unacceptable quality are characterized by an orthogonal quality close to 0. In practice, the minimum value of orthogonal quality should not be lower than 0.05 and the average should be higher than 0.6.

Information about another quality parameters such as: aspect ratio, smoothness, warpage, Jacobian, *etc.* can be found in [5].

Meshes can be classified into two categories: structured and unstructured (Figure 2.5). The first one is characterized by regularity: there is a family of lines, and mesh points are located only at the intersection of these lines. Due to this, a two- or three-dimensional array is created, filled with quadrilateral or hexahedral cells. Their regular shape and relatively high surface area/volume provide good convergence as well as space- and memory-efficiency [1,4]. According to the orientation between grid lines and the coordinate axes, the structured grids are divided into Cartesian and body-fitting categories, according to the orientation between the grid lines and the coordinate axes. The first category lines are parallel to the coordinate axes, and the others are adjusted to fit the geometry shape.

On the contrary, the unstructured grids include meshes with irregular connectivity. This means these meshes do not follow any regular pattern and they may be built of elements in any shape. This results in an inefficient use of space: an unstructured grid needs more cells than a structured one to fill the same domain. Nevertheless, they are commonly used in industrial applications due to relatively easy conforming to the shape of complex geometries [3].

TABLE 2.1
Mesh Quality Depending on Skewness

			Skewness		
Perfect	**Very Good**	**Good**	**Acceptable**	**Bad**	**Not Allowed**
0.00–0.25	0.25–0.50	0.50–0.80	0.80–0.94	0.95–0.97	0.98–1.00

TABLE 2.2
Grid Quality Depending on the Orthogonal Quality

			Orthogonal Quality		
Not Allowed	**Bad**	**Acceptable**	**Good**	**Very Good**	**Perfect**
0.000–0.001	0.001–0.50	0.50–0.20	0.20–0.69	0.70–0.95	0.95–1.00

(a)   (b)

FIGURE 2.5 Examples of (a) structured and (b) unstructured meshes.

2.1.3 Preprocessing – Setting Solver

The final step of preprocessing includes all actions related to setting physical properties of the analyzed case. In Solver, the time dependency of the model has to be specified. The steady-state approach is recommended for problems of a stationary nature (in the balanced state of its physics), or for cases characterized by relatively small dynamics. On the contrary, when time significantly influences the occurring phenomena, the case should be treated as a transient one. Dynamic simulations lead to a great increase in the solution time because the computation procedure has to be repeated in each time step.

Popular CFD solvers offer work in two modes: pressure- or density-based. In the first approach, density in the domain is determined from the equation of state, based on the pressure field. In the second approach, the computation process starts with the determination of density from the continuity equation, with the assumption of constant pressure in the domain. Then, the pressure is computed from the equation of state [6]. The general criterion which decides the usability of these modes is a Mach number. For low fluid velocity ($Ma < 0.3$) and when the fluid flow can be treated as uncompressible, the pressure-based solver is usually applied. Otherwise, the density-based approach is recommended. In a major part of CFD applications for renewable energy technologies, the pressure-based solver is applied. In this case, the pressure is computed based on the so-called pressure (Poisson) equation which is derived from the momentum and continuity equations. Then density can be computed based on the state equation.

In preprocessing, mathematical models describing physical phenomena are selected. Modern solvers allow activating built-in detailed models, including:

- Turbulence models,
- Multiphase flows,
- Radiation models,
- Species transport,
- Discrete phase,
- Phase changes,
- Acoustics,
- Battery and potential models,
- Other.

The selection of transport equations and models should be carried out with great attention, because, depending on circumstances, some models work better than others. It is necessary to check which model possesses the features allowing to describe the investigated case reliably. Certain models have got strong limitations in their application and their inappropriate use leads to great discrepancies between the CFD results and the experimental data [3]. Generally, the more detailed the model is, the higher number of equations and the longer time required for the computation. In each simulation, the balance between the computation time and the required accuracy should be stated in engineering applications a relatively simple model can often give an equally good solution as the more complex one [2] Thus, the key issue is to decide what model output information is most wanted. This approach significantly simplifies the model and the number of potential mistakes. Never try to complicate the CFD analysis just because this way it looks more serious or impressive.

To obtain a solution for the set of equations describing the behavior of the system, it is necessary to define the values of the physical quantities on the domain boundaries. The most popular boundary condition types are:

- Inlet,
- Outlet,
- Wall,
- Symmetry.

Generally, the closer to the real conditions the solver settings are, the lower the risk of failure.

Moreover, in solver, the material properties are defined and assigned to domains [1]. Preprocessing includes also settings concerning the discretization of mathematical equations in differential or integral form into an algebraic set of equations for each grid point [3].

The above-mentioned types of boundary conditions which are defined applying a series of solver windows and drop-down lists (when the model is set up in the solver with a graphical user interface) correspond directly to different boundary condition types defined by the fundamentals of the numerical method theory.

In the case of Dirichlet, or first-type boundary condition, an ordinary second-order differential equation can be considered, where:

$$f'' + f = 0 \tag{2.1}$$

if the domain of $f(\varphi)$ function is $[a, b]$, then its values at the system boundaries are numbers equal to y_1 and y_2, according to the equations below:

$$f(a) = y_1 \tag{2.2a}$$

$$f(b) = y_2 \tag{2.2b}$$

It has to be noted that for the partial differential equation with Laplacian as below:

$$\nabla^2 f + f = 0 \tag{2.3}$$

Dirichlet boundary conditions in domain $\Omega \subset R^n$ are defined as follows:

$$f(\varphi) = g(\varphi), \forall \varphi \in \partial\Omega \tag{2.4}$$

where g is a function defined at the limit $\partial\Omega$.

So, it can be simply concluded that Dirichlet boundary condition specifies the value of the function at the system boundary. Its practical interpretation can be, for example, a heater which is installed inside the electric kettle. When the kettle is working, water temperature around the heater is changing, heat, mass and momentum transport occurs. However, assuming a certain level of case simplification, it can be claimed that the heater surface has the same temperature (constant temperature at the wall). Furthermore, water velocity is always equal to zero (zero velocity at the wall) on this surface which is another good example.

For equation 2.1 and with the assumption that the domain of $f(\varphi)$ function is the same, Neumann (or second-type) boundary conditions can be written as:

$$f'(a) = y_1 \tag{2.5a}$$

$$f'(b) = y_2 \tag{2.5b}$$

On the contrary, for equation 2.3 in the domain $\Omega \subset R^n$ they are described as follows:

$$\frac{\partial f}{\partial n}(\varphi) = g(\varphi), \forall \varphi \in \partial\Omega \tag{2.6}$$

where n is normal vector, external at the limit $\partial\Omega$.

A Practical Look at the Steps of CFD Model Development

So contrary to the first-type boundary condition, the Neumann does not correspond to the constant value at the boundary. Referring to the previously given examples, it can be said that the second-type boundary condition describes the heat flux across the boundary instead of its fixed temperature.

Another useful boundary condition type is Robin boundary condition. Among others, it is applicable for a description of the convection at the system boundary. Actually, it can be said that it is a combination of the two previously discussed types. Thus, it includes a combination of the linear value of a function and its ordinary derivative at the boundary of the area defined for this function [7,8]. Taking into account this definition, Robin boundary condition, in the same domain and limit as before, can be written as:

$$au + b\frac{\partial u}{\partial n} = h \tag{2.7}$$

wherein a and b are non-zero constants or functions here, whereas h is the function defined at $\partial\Omega$ limit.

2.2 NUMERICAL SOLUTION AND ITS FEATURES

In the case of CFD modeling of renewable energy technology elements, 3D cases are of specially great importance. Furthermore, many models have to be considered as dynamic, due to the characteristics of certain device operation (no balance conditions or presence of moving elements). Contrary to classic fluid mechanics, flows that are considered in the case of renewable energy technologies are often unstable, non-isothermal, reactive, or even multiphase. This fact seriously impedes the modeling process and justifies the necessity to modify general numerical methods (applied in CFD) in a specific way.

There are three essential components defining the numerical solution method:

- mathematical model discretization method,
- computational grid (mesh),
- method of solving a very large algebraic equation system.

Equations creating a mathematical model of the considered problem are expressed in a differential or integro-differential form. Discretization is the process of converting these primary transport equations into their algebraic approximations. The role of these final equations is to describe relations between dependent variables in the selected discretization points, in space and time.

Discretization methods for CFD are described in many literature sources, like [9,10]. Currently, their general classification including two main method groups (Figure 2.6) can be assumed. The first one is grid discretization method group. It can be concluded that it includes classic CFD methods that are applied in popular commercial CFD codes. These methods apply the approximation of Navier-Stokes (N-S) equations inside particular areas of the numerical grid (mesh). The other group includes particle discretization methods. They use molecular static mechanics equations. In general, the above-listed groups are usually treated as two completely different; however, the lattice Boltzmann equation can be treated as an explicit, discrete Lagrange approximation of Navier-Stokes equations. Although Finite Difference Methods (FDM) have certain advantages, like simplicity and easy analysis of approximation accuracy [3,11], they have been found less numerically efficient in contemporary applications, compared to Finite Element Method (FEM) and Finite Volume Method (FVM).

These two methods (apart from the collocation method, least squares method and others) are classified as so-called weighted residual methods [12,13]. They differ in the way of selecting weight (or test) functions.

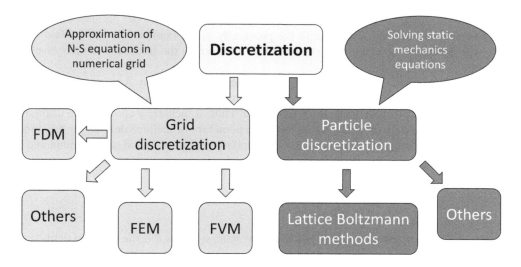

FIGURE 2.6 Simplified classification of numerical discretization methods for CFD.

To explain the operation principle of these methods, an L operator meaning transport equation can be applied auxiliary here:

$$L[\phi(x,t)] = 0 \tag{2.8}$$

Where ϕ is a dependent variable determined in an independent variable set (x, t).

In numerical methods an accurate solution of transport equations (ϕ) is approximated by $\hat{\phi}$ in discrete time and space points. These points are nodes coming from the generated mesh. The already-mentioned residual R is defined as the difference between the L operator for approximated solution, $\hat{\phi}$, and for an accurate solution ϕ (equation 2.9):

$$R = L[\hat{\phi}(x,t)] - L[\phi(x,t)] = L[\hat{\phi}(x,t)] \tag{2.9}$$

Usually, $\hat{\phi}$ is approximated in the form of a sum of the intercept $\hat{\phi}_0$, so-called approximation functions (or test functions) $\varphi_j(x)$ and their coefficients $a_j(t)$. Both intercept and approximation functions are known [12,13]:

$$\hat{\phi} = \hat{\phi}_0 + \sum_{j=1}^{M} a_j(t)\varphi_j(x) \tag{2.10}$$

Coefficients $a_j(t)$ are calculated from the condition that the integral from the weighted residual $W \cdot R$ should be zero in the domain of integral (equation 2.11).

$$\iint_{tV} W(x) \cdot R(x,t) dV \, dt = 0 \tag{2.11}$$

Weights W can be constant or depend on the space coordinate. This is the specific feature of each discretization method from the group of weighted residual method group. The two most popular methods with significant practical meaning are Finite Element Method and Finite Volume Method. Currently, in commercial CFD solvers, the latter one is usually applied.

A Practical Look at the Steps of CFD Model Development 17

Although the discretization process leads to obtaining algebraic equations, the number of these equations is great. Of course, it depends on the number of discrete locations in space in which the equations are discretized and solved. The number of the algebraic equations can be estimated as the product of the numbers of mesh nodes and transport equations (it is equal to or even greater than the product result). Thus, the number of final equations to solve in the case of 2D laminar, isothermal flow (continuity equation and two momentum equations) in the computational domain that includes 512,000 nodes (limit for student license in the case of Ansys Fluent solver) exceeds 1.5 million equations. Of course, commercial simulations with appropriate result quality usually require consideration of many more nodes and they are not laminar or/and isothermal flows.

In such a case an equation system in matrix notation is usually presented as follows:

$$A \cdot \Phi = B \qquad (2.12)$$

where:

Φ vector of searched dependent variable values in the approximation nodes,
A matrix of known coefficients obtained according to the applied discretization method,
B vector of coefficients which contains information about boundary values (conditions).

Numerical mesh is created by its nodes – the points in space and time in which approximate values of dependent variables ϕ are computed. A detailed description of selected basic mesh features was presented in Section 2.1.2.

Methods of solving very large equation systems, as presented in equation 2.12, depend on the specifics of the considered problem. One of the factors of great importance is whether the studied phenomena are dynamic (fields of flow-variables change in time). In the case of transient models, the so-called time-marching method is usually applied (which is typical for parabolic issues). In this method, the result of time discretization is a set of time steps. Depending on the observed phenomena time scale, time steps have to be respectively longer or shorter. Appropriate adjustment of a time step is a crucial issue from the point of view of capturing the dynamics of the system behavior. During model solution an individual time step is treated as pseudo-steady state. It means that for the solution purposes, the time step takes the qualities of a steady-state issue. Then solution can be carried out iteratively. When convergence criteria for all equations (or iteration limit) are reached, solver can pass to the next time step. It can be concluded that convergence criteria are acceptable limits of the variable imbalance in the control volume (in Finite Volume Method!), in transport equations considered in a certain model. If the case is strongly nonlinear, the pseudo-time-marching (pseudo-transient) method can be applied. Then, in each time step, solver starts solution from the selected initial state and gradually reaches the steady state (and convergence).

Because CFD solvers available in Ansys Workbench environment, which was used to present execution of the tutorials in Part II of the book are based on FVM, this discretization method was described in detail.

To keep the notation clarity and make the interpretation of the main method assumptions easier, only 1D diffusion (for example heat transfer by conduction) was considered in the example (Figure 2.7).

Try to imagine that x dimension is much smaller than the other two. It is like very long and very high wall which allows to treat the case as one-dimensional (linear).

Heat transfer (or diffusion in the general case) in the system occurs between the walls, from the left one with temperature T_b to the right one with temperature T_a. Additionally, there is heat source q in the system.

As a result of contractual spatial discretization of the system, three nodes have been obtained, as in Figure 2.8. The reference point for discretization is the control cell located in the system center (gray cell limited by rectangular frame).

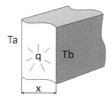

FIGURE 2.7 Model 1 dimensional model of the heat transfer in a solid.

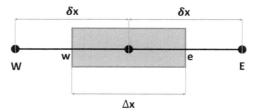

FIGURE 2.8 System of nodal points in the computational grid of the considered 1D diffusion heat transfer case.

Of course "cell" term is contractual too, due to the linear nature of the system. In this example the considered central P node is located in the cell centroid and it is not on the system boundary. To make the analysis easier, the popular naming convention based on world directions was adopted (E, e – east, W, w – west).

The distance between P point and two adjacent nodes (W on the left side and E on the right one) is δx. Other two important points are w and e, which are located at the boundaries of the control cell which includes P node.

Taking into account all the listed assumptions, mathematical description of this case can be done applying transport equation that includes the diffusion term and the source term:

$$\frac{\partial}{\partial x}\left(\Gamma \frac{\partial \phi}{\partial x}\right) + S_\phi = 0 \qquad (2.13)$$

where:
Γ – diffusion coefficient,
ϕ – general variable,
S_ϕ – source term.

By using the divergence theorem, it is possible to express the derivative from equation 2.13 for transport in x direction as follows:

$$\frac{\partial \phi}{\partial x} = \frac{1}{\Delta V}\int_V \frac{\partial \phi}{\partial x}dV = \frac{1}{\Delta V}\int_A \phi_i A_i^x \qquad (2.14)$$

wherein A_i^x should be actually understood as $A_i \cdot n'_x$ (n'_x – unit vector), because the above-presented transformation by definition requires the application of a vector variable. The application of the operation from equation 2.14 to approximate the diffusion term was presented in equation 2.15:

$$\frac{\partial}{\partial x}\left(\Gamma \frac{\partial \phi}{\partial x}\right) = \frac{1}{\Delta V}\int_V \frac{\partial}{\partial x}\left(\Gamma \frac{\partial \phi}{\partial x}\right)dV = \frac{1}{\Delta V}\int_A \left(\Gamma \frac{\partial \phi}{\partial x}\right)dA_i^x \approx \frac{1}{\Delta V}\sum_{i=1}^{2}\left(\Gamma \frac{\partial \phi}{\partial x}\right)A_i^x \qquad (2.15)$$

A Practical Look at the Steps of CFD Model Development

where A_i^x – predicted, projected, contractual boundary area.

It is worth noting that the number of the sum components at the end (rhs) of the expression 2.15 comes from the fact that in the case of linear problem, variable ϕ transport is possible only in x direction, to W and E points.

If, in agreement with the initial assumption, the considered variable gradient in the computational domain is from E to W, the projected areas for the considered control cell can be expressed with appropriate signs:

$$A_1^x = -A_W \tag{2.16a}$$

$$A_2^x = A_E \tag{2.16b}$$

Using equations 2.16, equation 2.15 can be written as follows:

$$\sum_{i=1}^{2}\left(\Gamma\frac{\partial\phi}{\partial x}\right)A_i^x = \frac{1}{\Delta V}\left(\Gamma\frac{\partial\phi}{\partial x}\right)_e A_E - \frac{1}{\Delta V}\left(\Gamma\frac{\partial\phi}{\partial x}\right)_w A_W \tag{2.17}$$

An approximation of the source term (in equation 2.13) that has so far been omitted in the considerations can be realized according to expression 2.18.

$$\frac{1}{\Delta V}\int_V S_\phi dV = S_\phi \tag{2.18}$$

So combining expressions 2.17 and 2.18, the final form of discretized diffusion transport equation of ϕ variable can be obtained:

$$\frac{1}{\Delta V}\left(\Gamma\frac{\partial\phi}{\partial x}\right)_e A_E - \frac{1}{\Delta V}\left(\Gamma\frac{\partial\phi}{\partial x}\right)_w A_W + S_\phi = 0 \tag{2.19}$$

To write expression 2.19 in the algebraic form, with respect to the nodal points E, P and W, it is required to approximate the ϕ variable gradient at the boundaries of the control cell (e and w points). For this purpose piecewise linear gradient change can be assumed. However, subsequent linear segments refer to the point pairs: E-P and P-W. This approach allows to effectively approximate the first derivative for e and w:

$$\left(\Gamma\frac{\partial\phi}{\partial x}\right)_e A_E = \Gamma A_E\left(\frac{\phi_E - \phi_P}{\delta x_E}\right) \tag{2.20a}$$

$$\left(\Gamma\frac{\partial\phi}{\partial x}\right)_w A_W = \Gamma A_W\left(\frac{\phi_P - \phi_W}{\delta x_W}\right) \tag{2.20b}$$

Then transformed diffusion terms from equations 2.20 can be substituted into equation 2.19 to obtain:

$$\frac{1}{\Delta V}\Gamma_e A_E\left(\frac{\phi_E - \phi_P}{\delta x_E}\right) - \frac{1}{\Delta V}\Gamma_w A_W\left(\frac{\phi_P - \phi_W}{\delta x_W}\right) + S_\phi = 0 \tag{2.21}$$

It is worth noting that the difference of the diffusion fluxes in E and W directions is balanced by a source term S_ϕ. Actually, equation 2.21 is discretized balance of the general variable in the control cell surrounding P node.

Conversion of equation 2.21 allows to obtain its algebraic form. Then it is possible to read its subsequent coefficients:

$$\frac{1}{\Delta V}\left(\frac{\Gamma_e A_E}{\delta x_E}+\frac{\Gamma_w A_W}{\delta x_W}\right)\phi_P = \frac{1}{\Delta V}\left(\frac{\Gamma_e A_E}{\delta x_E}\right)\phi_E + \frac{1}{\Delta V}\left(\frac{\Gamma_w A_W}{\delta x_W}\right)\phi_W + S_\phi \tag{2.22}$$

Thus, the algebraic form of the discretized transport equation with appropriate coefficients can be finally written as:

$$a_p \phi_P = a_E \phi_E + a_W \phi_W + b \tag{2.23}$$

where:

$$a_p = a_E + a_W \tag{2.24a}$$

$$a_E = \frac{\Gamma_e A_E}{\Delta V \delta x_E} \tag{2.24b}$$

$$a_W = \frac{\Gamma_w A_W}{\Delta V \delta x_W} \tag{2.24c}$$

$$b = S_\phi \tag{2.24d}$$

Special attention has to be paid to the fact that in the case of a homogeneous, regular mesh (equal distances between nodal point and cell faces in each cell):

$$\delta x_i = \text{const} \tag{2.25}$$

and constant diffusion coefficient (in this case it is heat conduction coefficient), independent from the nodal point:

$$\Gamma_e = \Gamma_w = \Gamma \tag{2.26}$$

following relations apply:

$$a_E = \frac{\Gamma A}{\Delta V \delta_x} = \frac{\Gamma A}{\Delta x \Delta y \Delta z \delta_x} = \frac{\Gamma A}{\Delta x A \delta_x} = \frac{\Gamma}{\delta_x \cdot \delta_x} = \frac{\Gamma}{\delta_x^2} \tag{2.27a}$$

$$a_W = \frac{\Gamma}{\delta_x^2} \tag{2.27b}$$

$$a_p = a_E + a_W \tag{2.27c}$$

The above-presented explanation leads to right conclusion that the FVM is a discretization method which can be successfully applied for control cells characterized by different shapes.

A Practical Look at the Steps of CFD Model Development

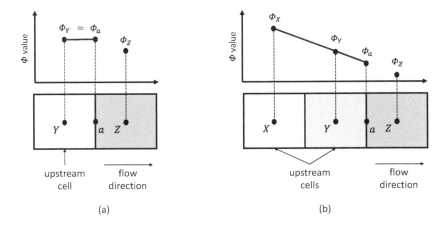

FIGURE 2.9 Visual representation of advection schemes: (a) first order and (b) second order.

The example covers a description of the central differencing scheme as an efficient method to approximate the variable value at the face between adjacent control volumes, when the diffusion term is being discretized. In case of the convection terms in transport equations, so-called upwind schemes are applied by CFD solvers like Ansys Fluent and Ansys CFX.

Advection schemes determine the numerical mechanism of transporting a quantity (like velocity and turbulent kinetic energy) through the solution domain. The data from one cell is used to approximate variable values in adjacent cells: firstly, only at the boundary and then at the center of the cell. Among others, there are two most popular advection schemes: first and second orders, which differ in the number of points used for approximation of data at the face between adjacent cells (see Figure 2.9). In the first order schemes, it is assumed that the value at the cell boundary (cell face) is the same as in the upstream cell. Due to this, the solution is easy to converge, but less accurate: especially gradients tend to be smeared out. In the second-order schemes, the value at the cell face is approximated based on the values from two upstream cells. This scheme results in a more accurate solution, but it is harder to obtain [2].

During the stage of the CFD simulation (solution process – processing), the algebraic system of equations is solved iteratively to compute the mesh nodal values of the variables in the considered transport equations. Therefore, the calculation time strongly depends on the number of mesh cells, number of discretized equations and their complexity, the discretization schemes used and the computing power [1,3].

The solution continues until the convergence (residual target) is reached for each transport equation. Otherwise, the computation process stops when the given iteration limit is achieved. This term means that the output from the numerical simulation is a valid approximation of the exact solution, which is generally not known [3]. So how does it work? There are also two other terms that describe the reliability of numerical solution: consistency and stability. The consistency requirement says that the discretization of a mathematical model should result in a good representation of the original equation. Stability means that the errors of discretization do not increase during the progress of the computation. The Peter Lax theorem (Figure 2.10) connects the consistency, stability and convergence. It states that a consistent and stable algorithm will automatically satisfy the convergence requirement [3].

In each iteration, the level of the results accuracy is growing. This may be observed in solver in the form of a chart which displays the residual level for each transport equation depending on the iteration number. Residuum describes the maximum imbalance in the transport of a variable in mesh elements. More information can be found in Ref. [2].

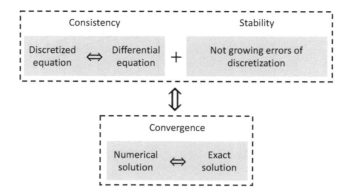

FIGURE 2.10 Graphical representation of Lax's theorem.

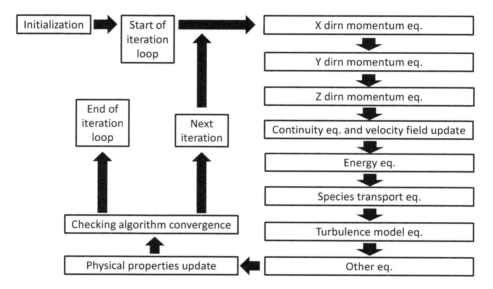

FIGURE 2.11 Schematic of the iteration loop in the case of a solution applying the segregated pressure-velocity coupling algorithm. (Based on Ref. [14].)

Residuum relates directly to the degree of balance in the transport of a given variable within the control volume. In other words, the appearance of a source or associated sink in the transport equation with errors of the numerical method or discretization, and not a mathematical description of the problem, is observed in a given iterative step of the solution as the remainder from the balance of the convective and diffusion terms of a given equation. The maximum value resulting from this transport imbalance in the cell is represented by the solver as residuum.

The solver achieves convergence when the imbalances for all equations are relatively small. For example, in engineering applications these values for velocity are 10^{-3}, whereas in scientific applications 10^{-5}. Converged solution means that computation is completed. Solution process can apply different algorithms to solve transport equations in certain order.

For example, there are two basic groups of algorithms applied by Ansys Fluent (so-called pressure-velocity coupling algorithms).

Figures 2.11 and 2.12 present flowcharts referring to the solution order of an individual transport equation, depending on the selected pressure-velocity coupling algorithm.

A Practical Look at the Steps of CFD Model Development

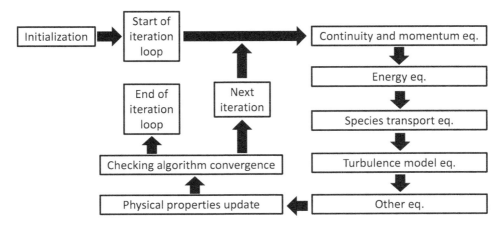

FIGURE 2.12 Schematic of the iteration loop in the case of a solution applying the coupled pressure-velocity coupling algorithm. (Based on Refs. [14,15].)

It is worth recalling that in the case of segregated algorithms, continuity equation is solved based on a determined velocity field (from momentum equations), which then is updated (to satisfy the principle of momentum conservation) thanks to the pressure equation and pressure correction. The pressure equation is derived applying both the continuity and momentum equation [14].

2.3 POSTPROCESSING

The CFD simulation generates lots of data, which have to be properly displayed and analyzed. Postprocessors are powerful visualization systems. They allow to investigate qualitatively and quantitatively obtained results. Data can be plotted in the form of graphs, plots, tables or displayed as selected variables field, known as contours [1]. The flow field may also be represented in the form of vector plots, streamlines. There are also various calculators, which, among others, allow to determine the average or maximum values of the selected parameters in certain regions of the computational domain [3]. The wide range of postprocessor tools is presented in the practical part of the book. An interpretation of the obtained results is crucial to answer the questions of interest.

REFERENCES

1. A. W. Date, *Introduction to Computational Fluid Dynamics*, Cambridge University Press, Cambridge, 2005.
2. J. Blazek, *Computational Fluid Dynamics: Principles and Applications*, Elsevier (2001), ISBN: 0-08-043009-0.
3. C. Hirsch, *Numerical Computation of Internal and External Flows Volume 1: Fundamentals of Computational Fluid Dynamics*, Elsevier (2007), ISBN: 978-0-7506-6594-0.
4. J. F. Thompson, Z. U. A. Warsi, C. Wayne Mastin, *Numerical Grid Generation: Foundations and Applications*, North Holland (1985), ISBN: 9780444009852.
5. https://ansyshelp.ansys.com. (access: 11.06.2023)
6. T. Jiyuan, Y. Guan-Heng, L. Chaoqun, *Computational Fluid Dynamics. A Practical Approach*, Butterworth-Heinemann, Oxford, 2013.
7. E. Isaacson, *Analysis of Numerical Methods*, Dover Publication Inc., New York, 1994.
8. G. Dahlquist, Å. Björck, *Numerical Methods in Scientific Computing: Volume 1*, Society for Industrial and Applied Mathematics, Philadelphia, 2008.
9. I. Huismann, J. Stiller, J. Fröhlich, Efficient high-order spectral element discretizations for building block operators of CFD, *Computers and Fluids*, 197 (2020), 104386.
10. K. Takabatake, M. Sakai, Flexible discretization technique for DEM-CFD simulations including thin walls, *Advanced Powder Technology*, 31 (2020), 1825–1837.

11. T. Cebeci et al., *Computational Fluid Dynamics for Engineers, from Panel to Navier-Stokes Methods with Computer Programs*, Springer (2005), ISBN: 3-540-24451.
12. L. Lapidus, G. F. Pinder, *Numerical Solution of Partial Differential Equations in Science and Engineering*, Wiley, Canada, 1999.
13. H. Mohammad, *Weighted Residual Methods. Principles, Modifications and Applications*, Elsevier Science Publishing Co Inc., 2017.
14. H. Wang, et al., Literature review on pressure–velocity decoupling algorithms applied to built-environment CFD simulation, *Building and Environment*, 143 (2018), 671–678.
15. T. Uroića, H. Jasak, Block-selective algebraic multigrid for implicitly coupled pressure-velocity system, *Computers & Fluids*, 167 (2018), 100–110.

3 Transport Equations

A general, integral form of the transport equation for a variable in the finite volume method, with the assumption of certain control volume *CV*, limited by the control surface *CS*, for a dynamic (time dependent) system can be written as follows [1,2]:

$$\int_t^{t+\Delta t}\left(\frac{\partial}{\partial t}\int_{CV}\rho\phi dV\right)dt + \int_t^{t+\Delta t}\left(\int_{CS}\rho\phi u * n dS\right)dt$$
$$= \int_t^{t+\Delta t}\left(\int_{CS}\Gamma\,\text{grad}\,\phi * n dS\right)dt + \int_t^{t+\Delta t}\left(\int_{CV}S_\phi dV\right)dt \quad (3.1)$$

where:
- t – time $[s]$,
- ρ – density $\left[\dfrac{\text{kg}}{\text{m}^3}\right]$,
- ϕ – general variable,
- V – volume $[\text{m}^3]$,
- u – fluid element velocity vector $\left[\dfrac{\text{m}}{\text{s}}\right]$,
- n – unit vector normal to the S surface $[-]$,
- Γ – diffusion coefficient $\left[\dfrac{\text{mm}^2}{\text{s}}\right]$,

wherein expression 3.2 is an equivalent form [1,2]:

$$\frac{\partial(\rho\phi)}{\partial t} + (\rho\phi u) = (\Gamma\,\text{grad}\,\phi) + S_\phi \quad (3.2)$$

Expression 3.1 can be treated as the basic one used to compute discretized transport equations. However, it has to be noted that usually certain simplifications are applied. They allow to obtain a relatively uncomplicated system of differential equations. This form was used below to briefly discuss basic transport equations for the CFD model of a laminar, non-isothermal flow.

In this case, it is necessary to take into account at least three equations: continuity, momentum and energy equations.

The continuity equation that was presented below in the conservation form (3.3) is a direct consequence of the validity of the mass conservation principle in the system. According to this principle, mass balance within the fluid element has to be equal zero. Equation 3.3 expresses the sum of density change rates and is often called the advection term [3]:

$$\frac{\partial\rho}{\partial t} + \left(\frac{\partial\rho u}{\partial x} + \frac{\partial\rho v}{\partial y} + \frac{\partial\rho w}{\partial z}\right) = 0 \quad (3.3)$$

where u, v and w components of the local velocity vector, are location and time(t) functions.

One basic, exemplary and practical, consequence of the continuity equation solution is the possibility to observe formation of a boundary layer in the fluid flow (in the case of its zero velocity on the wall).

Depending on the number of dimensions in the considered system case, the model has to take into account an appropriate number of momentum equations, which are a natural consequence of the momentum conservation principle: inertia forces acting on the fluid element are balanced by the sum of any other forces that exert an influence on this element. Equations describing rates of momentum changes in a three-dimensional coordinate frame (for X, Y and Z directions) are presented below:

$$\rho\frac{\partial u}{\partial t} + \rho u\frac{\partial u}{\partial x} + \rho v\frac{\partial u}{\partial y} + \rho w\frac{\partial u}{\partial z} = \frac{-\partial p}{\partial x} + \mu\left(\frac{\partial^2 u}{\partial x^2} + \frac{\partial^2 u}{\partial y^2} + \frac{\partial^2 u}{\partial z^2}\right) + \Sigma f_i^x \quad (3.4a)$$

$$\rho\frac{\partial v}{\partial t} + \rho u\frac{\partial v}{\partial x} + \rho v\frac{\partial v}{\partial y} + \rho w\frac{\partial v}{\partial z} = \frac{-\partial p}{\partial x} + \mu\left(\frac{\partial^2 v}{\partial x^2} + \frac{\partial^2 v}{\partial y^2} + \frac{\partial^2 v}{\partial z^2}\right) + \Sigma f_i^y \quad (3.4b)$$

$$\rho\frac{\partial w}{\partial t} + \rho u\frac{\partial w}{\partial x} + \rho v\frac{\partial w}{\partial y} + \rho w\frac{\partial w}{\partial z} = \frac{-\partial p}{\partial x} + \mu\left(\frac{\partial^2 w}{\partial x^2} + \frac{\partial^2 w}{\partial y^2} + \frac{\partial^2 w}{\partial z^2}\right) + \Sigma f_i^z \quad (3.4c)$$

where:
p – pressure [Pa],
f_i – body forces $\left[\frac{N}{m^3}\right]$.

Basic physical consequence of the momentum conservation principle taking effect in a simple fluid flow model (with constant density) is the fact that hydrodynamic entrance length is the function of a fluid velocity (determining fluid inertia), viscosity and pressure (that represent contribution of surface forces – molecular interactions). Of course the above conclusion applies only in the simplified fluid flow case. Furthermore, mass forces related to the acceleration cannot be overlooked. The simplest example here is gravity acting on the fluid of certain density, which is especially significant in the natural convection issues.

Referring to equation 3.4 it can be concluded that the approach to the description of the energy transport can be analogous to the momentum transport [3–5]:

$$q\frac{\partial T}{\partial t} + qu\frac{\partial T}{\partial x} + qv\frac{\partial T}{\partial y} + qw\frac{\partial T}{\partial z} = \frac{\lambda}{C_p}\left(\frac{\partial^2 T}{\partial x^2} + \frac{\partial^2 T}{\partial y^2} + \frac{\partial^2 T}{\partial z^2}\right) + \Sigma S_i \quad (3.5)$$

where:
T – absolute temperature [K],
λ – thermal conductivity coefficient $\left[\frac{W}{mK}\right]$,
C_p – specific heat $\left[\frac{J}{kgK}\right]$,
S_i – source term (additional heat source or sink) $\left[\frac{W}{m^3 s}\right]$.

Energy equation is a direct consequence of the first law of thermodynamics, which concerns the conservation of energy in any considered system. The overall message behind this principle is that the energy cannot be created or destroyed, but only converted from one form to another or transferred (in the form of heat and/or work). In the case of fluid mechanics, usually in a simplified case of the energy transport equation, total internal energy of a fluid is balanced.

It is worth noting that the transport equations which are the subject of this chapter create only one subset among equations describing fluid flow physics. Good examples of the completing ones are state equations [6–9], like the ones allowing to compute local density, including the equation of the incompressible ideal gas model below:

$$\rho = \frac{P_r}{\frac{R}{M_w}T} \qquad (3.6)$$

where:
P_r – operation pressure in the system [Pa],

R – universal gas constant $\left[\dfrac{J}{mol \cdot K}\right]$,

M_w – gas molar mass $\left[\dfrac{kg}{mol}\right]$,

T – absolute temperature [K].

Of course the above state equation is just an example representing a broad group of equations which are applied in the model to describe its thermodynamic state.

REFERENCES

1. C. Hirsch, *Numerical Computation of Internal and External Flows Volume 1: Fundamentals of Computational Fluid Dynamics*, Elsevier (2007), ISBN: 978-0-7506-6594-0.
2. J. Blazek, *Computational Fluid Dynamics: Principles and Applications*, Elsevier (2001), ISBN: 0-08-043009-0.
3. T. Jiyuan, Y. Guan-Heng, L. Chaoqun, *Computational Fluid Dynamics. A practical Approach*, Butterworth-Heinemann, Oxford, 2013.
4. L. Lapidus, G. F. Pinder, *Numerical Solution of Partial Differential Equations in Science and Engineering*, Wiley, Canada, 1999.
5. H. K. Versteeg, W. Malalasekera, *An Introduction to Computational Fluid Dynamics. The Finite Volume Method*, Pearson Education Limited, Harlow, 2007.
6. B. Liu et al., Decompression of hydrogen—natural gas mixtures in high-pressure pipelines: CFD modelling using different equations of state, *International Journal of Hydrogen Energy*, 44 (14) (2019), 7428–7437.
7. A. Elshahomi et al., Decompression wave speed in CO_2 mixtures: CFD modelling with the GERG-2008 equation of state, *Applied Energy*, 140 (2015), 20–32.
8. X. Liu et al., Source strength and dispersion of CO_2 releases from high-pressure pipelines: CFD model using real gas equation of state, *Applied Energy*, 126 (2014), 56–68.
9. J. R. Travis et al., Real-gas equations-of-state for the GASFLOW CFD code, *International Journal of Hydrogen Energy*, 38 (19) (2013), 8132–8140.

4 Turbulent Flows in RANS Approach

In the case of renewable energy technologies, many (or presumably even most) transport phenomena of practical significance are related to the flows of fluids characterized by relatively low viscosity. One of the consequences of this fact is that these flow phenomena are unstable. This results in a phenomenon called turbulence (or turbulent flow). The significance of the above-mentioned issue in fields like, among others, wind-energy-based technologies [1], solar heat exchangers [2], biomass combustion [3] and particulate matter separation systems is confirmed by the tutorials included in the second part of this book, where different turbulence models are applied. Due to the great importance of turbulent flows in the CFD modeling of renewable energy technologies, this matter has been described below.

Among the approaches to the modeling of turbulent flows which were characterized in Section 1.2, the Reynolds-Averaged Navier-Stokes (RANS) method is of the greatest importance when it comes to engineering applications, including studies of renewable energy technologies. Although RANS describes turbulence in a flow rather from statistical perspective, it provides answers to the most significant questions referring to the widely understood operation efficiency of the modeled RES-based system elements.

At the heart of this thesis, there is an assumption that at a certain time and location, the general variable (it can be for example velocity U, as in Figure 4.1), in the flow, can be expressed as the sum of the time-averaged component (location-dependent) and the fluctuating component (which depends on time and location in the system).

For any general variable, the above-described idea of the so-called Reynolds decomposition can be expressed by the equation:

$$\phi(x,t) = \bar{\phi}(x) + \phi'(x,t) \tag{4.1}$$

where:
$\bar{\phi}$ – time-averaged component of the variable,
ϕ' – fluctuating component of the variable.

Averaging process average component in $t+1$ period has to be understood as:

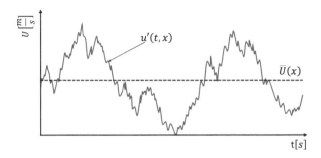

FIGURE 4.1 Graphical interpretation of a turbulent flow in the RANS approach, on the example of the U velocity, where $\acute{U}$ is the average component and u' is the fluctuating term, characterized by a time-dependent amplitude.

DOI: 10.1201/9781003202226-5

$$\bar{u}_t(x) = \frac{1}{T}\int_t^{t+1} u_i(t;x)dt \qquad (4.2)$$

wherein another averaging of already averaged velocity u_i leads to the same result:

$$\bar{\bar{u}}_t(x) = \frac{1}{T}\int_t^{t+1} \bar{u}_l(x)dt = \frac{\bar{u}_l(x)}{T}\int_t^{t+1} dt = \bar{u}_l(x) \qquad (4.3)$$

By definition, the time-averaged value of the individual fluctuating component is:

$$\overline{u'_l}(x) = \frac{1}{T}\int_t^{t+1}\left[u_i(t;x) - \bar{u}_t(x)\right]dt = \bar{u}_t(x) - \bar{u}_t(x) = 0 \qquad (4.4)$$

The previously presented (equation 4.1) decomposition operation has to be applied in the same way for the transport equation components. In the momentum equation, this procedure refers to velocity and pressure (equations 4.5a and 4.5b).

$$U = \bar{U} + u' \qquad (4.5a)$$

$$P = \bar{P} + p' \qquad (4.5b)$$

On the other hand, this decomposition does not apply to fluid properties, for example in the case of density (equation 4.6a) and viscosity (equation 4.6b).

$$\rho = \rho' \qquad (4.6a)$$

$$v = \text{constant} \qquad (4.6b)$$

Turbulence models based on RANS allow modeling turbulences in the whole length scale range. For this reason, no eddies are directly solved (applying classic Navier-Stokes equations). Although this fact makes RANS most applicable in engineering (industrial) issues, it results in certain serious limitations of the approach. One of the most significant is, undoubtedly, treating the turbulence phenomenon as an isotropic one [4–7], which makes analysis of certain classes of fluid flow issues impossible [8,9].

4.1 RANS-BASED ZERO-EQUATION AND NON-ZERO-EQUATION TURBULENCE MODELS

Based on the decomposition and averaging rules which were presented in the previous point, transport equations are converted to forms that correspond to turbulent flow conditions. Special attention has to be paid to the conversion of the momentum equation (equations 3.4a–3.4c). For example, in the case of an x-direction flow, after Reynolds decomposition and averaging, the following expression can be obtained:

$$\rho\frac{D\overline{U_x}}{Dt} = \frac{-\partial(\bar{P})}{\partial x} + \mu\left(\frac{\partial^2(\overline{U_x})}{\partial x^2} + \frac{\partial^2(\overline{U_x})}{\partial y^2} + \frac{\partial^2(\overline{U_x})}{\partial z^2}\right) - \rho\left[\frac{\partial}{\partial x}\left(\overline{u_x}^2\right) + \frac{\partial}{\partial y}\left(\overline{u_x u_y}\right) + \frac{\partial}{\partial z}\left(\overline{u_x u_z}\right)\right] \qquad (4.7)$$

Turbulent Flows in RANS Approach

If a 3D flow case in the *XYZ* coordinate frame is considered, based on the above-presented equation and an additional two for the remaining flow directions (*XY, Z*), Reynolds stress in equation 4.7 can be expressed in total as follows:

$$[\sigma;\tau] = \begin{bmatrix} -p & 0 & 0 \\ 0 & -p & 0 \\ 0 & 0 & -p \end{bmatrix} + 2\mu \begin{bmatrix} \frac{\partial \overline{U_x}}{\partial x} & \frac{1}{2}\left(\frac{\partial \overline{U_x}}{\partial y} + \frac{\partial \overline{U_y}}{\partial x}\right) & \frac{1}{2}\left(\frac{\partial \overline{U_x}}{\partial z} + \frac{\partial \overline{U_z}}{\partial x}\right) \\ \frac{1}{2}\left(\frac{\partial \overline{U_y}}{\partial x} + \frac{\partial \overline{U_x}}{\partial y}\right) & \frac{\partial \overline{U_y}}{\partial y} & \frac{1}{2}\left(\frac{\partial \overline{U_y}}{\partial z} + \frac{\partial \overline{U_z}}{\partial y}\right) \\ \frac{1}{2}\left(\frac{\partial \overline{U_z}}{\partial x} + \frac{\partial \overline{U_x}}{\partial z}\right) & \frac{1}{2}\left(\frac{\partial \overline{U_z}}{\partial y} + \frac{\partial \overline{U_y}}{\partial z}\right) & \frac{\partial \overline{U_z}}{\partial z} \end{bmatrix}$$

$$+ \rho \begin{bmatrix} \overline{-u'^2_x} & \overline{-u'_x u'_y} & \overline{-u'_x u'_z} \\ \overline{-u'_y u'_x} & \overline{-u'^2_y} & \overline{-u'_y u'_z} \\ \overline{-u'_z u'_x} & \overline{-u'_z u'_y} & \overline{-u'^2_z} \end{bmatrix} \tag{4.8}$$

The second term on RHS expresses the surface stress tensor which is related to viscosity. It contains only average velocity components which are determinable. Thus, this tensor is not a problem from the mathematical point of view. However, decomposition and averaging cause occurrence of additional stresses, which are expressed in the form of the products of averaged fluctuating components. The stress tensor coming from this fact (the last one on the RHS of equation 4.8) is called the Reynolds stress tensor [10–13]. The occurrence of this additional element of the Reynolds equations causes the requirement for additional operations to close the system of equations. The form of such an equation written in the summation convention is:

$$\rho\left(\frac{\partial \overline{U_i}}{\partial t} + \overline{U}_j \frac{\partial \overline{U_i}}{\partial x_j}\right) = \frac{\partial}{\partial x_i}\left(\sigma_{ij}\right) + \overline{F}_i \tag{4.9}$$

where: σ_{ij} is total stress including problematic tensor R_{ij}:

$$\sigma_{ij} = -\overline{P} \cdot \delta_{ij} + \mu \overline{D_{ij}} - \rho \cdot \overline{u'_i u'_j} \tag{4.10}$$

$$R_{ij} = -\rho \overline{u'_i u'_j} \tag{4.11}$$

To close the equation system which is extended with new variables, it is required to apply a new, synthetic parameter – so-called eddy viscosity (or in other words turbulent viscosity) [14–17], which allows to treat Reynolds stress analogously to viscous stress (equation 4.12). Justification of such an approach is the fact that both stress types cause similar results in the fluid flow. Due to the above, the conclusion that Reynolds stress should fulfill a relation analogous to Newton's law of viscosity looks reasonable. It is so-called Boussinesq hypothesis (or Boussinesq eddy viscosity hypothesis) [18,19].

$$-\rho \overline{u'_i u'_j} = \rho v_T \overline{D_{ij}} \tag{4.12}$$

Application of this hypothesis in a description of the Reynolds equation total stresses allows to eliminate stresses coming from the fluctuating component products. This tensor is then substituted by the product of viscous stress tensor and one new variable – eddy (turbulent) viscosity:

$$[\sigma;\tau] = \begin{bmatrix} -p & 0 & 0 \\ 0 & -p & 0 \\ 0 & 0 & -p \end{bmatrix} + 2\mu \begin{bmatrix} \frac{\partial \overline{U_x}}{\partial x} & \frac{1}{2}\left(\frac{\partial \overline{U_x}}{\partial y} + \frac{\partial \overline{U_y}}{\partial x}\right) & \frac{1}{2}\left(\frac{\partial \overline{U_x}}{\partial z} + \frac{\partial \overline{U_z}}{\partial x}\right) \\ \frac{1}{2}\left(\frac{\partial \overline{U_y}}{\partial x} + \frac{\partial \overline{U_x}}{\partial y}\right) & \frac{\partial \overline{U_y}}{\partial y} & \frac{1}{2}\left(\frac{\partial \overline{U_y}}{\partial z} + \frac{\partial \overline{U_z}}{\partial y}\right) \\ \frac{1}{2}\left(\frac{\partial \overline{U_z}}{\partial x} + \frac{\partial \overline{U_x}}{\partial z}\right) & \frac{1}{2}\left(\frac{\partial \overline{U_z}}{\partial y} + \frac{\partial \overline{U_y}}{\partial z}\right) & \frac{\partial \overline{U_z}}{\partial z} \end{bmatrix}$$

$$+ 2\mu_T \begin{bmatrix} \frac{\partial \overline{U_x}}{\partial x} & \frac{1}{2}\left(\frac{\partial \overline{U_x}}{\partial y} + \frac{\partial \overline{U_y}}{\partial x}\right) & \frac{1}{2}\left(\frac{\partial \overline{U_x}}{\partial z} + \frac{\partial \overline{U_z}}{\partial x}\right) \\ \frac{1}{2}\left(\frac{\partial \overline{U_y}}{\partial x} + \frac{\partial \overline{U_x}}{\partial y}\right) & \frac{\partial \overline{U_y}}{\partial y} & \frac{1}{2}\left(\frac{\partial \overline{U_y}}{\partial z} + \frac{\partial \overline{U_z}}{\partial y}\right) \\ \frac{1}{2}\left(\frac{\partial \overline{U_z}}{\partial x} + \frac{\partial \overline{U_x}}{\partial z}\right) & \frac{1}{2}\left(\frac{\partial \overline{U_z}}{\partial y} + \frac{\partial \overline{U_y}}{\partial z}\right) & \frac{\partial \overline{U_z}}{\partial z} \end{bmatrix}$$

(4.13)

The final closure of the equation system is possible, among others, thanks to the Prandtl's hypothesis, which defines eddy viscosity as a certain function of the so-called mixing length (equation 4.14). Prandtl's mixing-length hypothesis is based on the postulate that turbulent fluid flow structures collide with each other – like fluid molecules. Mixing length can be understood as a distance passed by a fluid portion before it mixes with the surrounding fluid. Knowledge of this value provides a solution to the N-S equation system for a turbulent fluid flow. It is possible to determine the mixing length by applying algebraic methods [20]. For instance, in a near-wall region, the characteristic turbulence length scale (mixing length) l_m can be approximated by equation 4.15:

$$\mu_T = \rho l_m^2 \left|\frac{\partial \overline{U_x}}{\partial y}\right| \quad (4.14)$$

$$l_m = \kappa y \quad (4.15)$$

where κ is the von Kármán constant, $\kappa = 0.41$.

Due to the fact that the above-presented method of describing turbulent flow does not require any additional transport equations (for variables describing turbulence), models based on such an approach are called zero-equation turbulence models. Although these models can be successfully applied in numerous fluid flow cases, they fail when it is required to model very accurately phenomena occurring in the viscous sublayer or when the boundary layer separation is considered.

4.2 ONE- AND MULTI-EQUATION TURBULENCE MODELS

In the cases when zero-equation models fail, it is needed to expand the model with a new equation or often an equation describing transport of the variables related to the turbulence characteristics. In such a situation, we consider one- or multi-equation models.

Theoretical background of this approach comes from the description of the phenomenon that was observed around 1920 by L. F. Richardson. This phenomenon was gradual disintegration of eddies which were previously produced thanks to the kinetic energy of the mean flow. Disintegration processes result in the creation of smaller and smaller eddies, which are finally dissipated due to the balancing of their inertia by the fluid molecular viscosity forces. In this last stage, the remaining whirl energy is converted to heat [21–23]. The chart presented in Figure 4.2 allows to imagine the characteristics of these turbulence energy changes during the above-discussed process.

Figure 4.3 is a kind of idealized visualization of the same phenomenon from the perspective presented by Richardson, with a big whirl being split into two smaller ones, and so on, to the viscosity balancing kinetic energy of the smallest whirls at the dissipation rate of the length scales l_s.

When eddies reach appropriate length scales (so-called Kolmogorov length scales), they pass from the integral range to the range in which the inertial energy cascade takes place.

Energy change in this range can be treated as a linear one, thus there is a certain coefficient determining turbulence dissipation rate. Its symbol is ε. This coefficient allows to define turbulence transport in fluid flow modeled applying RANS method, taking into account turbulence models other than the zero-equation ones [9,20,21,24]. The energy that is dissipated should be understood as turbulent kinetic energy, whose transport equation can be derived starting with the assumption (simplification) that in the turbulence phenomenon in the considered fluid flow, the influence of normal stresses (relative to the fluid element surfaces) is dominant.

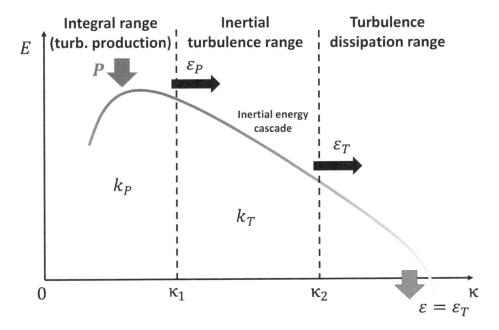

FIGURE 4.2 Chart presenting changes of the eddy energy in a turbulent fluid flow including the range of inertial energy cascade.

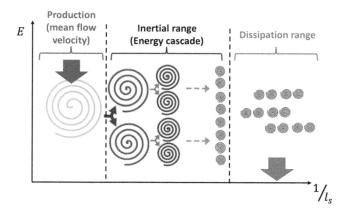

FIGURE 4.3 Transfer of eddies energy during their splitting and dissipation.

In such a situation the turbulent stress tensor has the form of a diagonal matrix and the equation of the turbulent kinetic energy transport (being at the same time equation of tensor τ) can be written as follows:

$$\operatorname{tr}\left(\left[\tau_{ij}\right]\right) = \tau_{ii} = -2\overline{u'_i u'_i} = -2\left(\overline{u'_1 u'_1} + \overline{u'_2 u'_2} + \overline{u'_3 u'_3}\right) = 2k \tag{4.16}$$

The evolutionary equation of turbulent kinetic energy can be obtained as a result of contraction of the transport equations of Reynolds stress tensors [9]. Then this equation is:

$$\frac{\partial k}{\partial t} + \overline{U}_k \frac{\partial k}{\partial x_k} = \tau_{ik}\frac{\partial \overline{u_l}}{\partial x_k} - \nu \overline{\frac{\partial u'_l}{\partial x_k}\frac{\partial u'_l}{\partial x_k}} + \frac{\partial}{\partial x_k}\left[\nu\frac{\partial k}{\partial x_k} - \frac{1}{2}\overline{u'_l u'_l u'_k} - \frac{1}{\rho}\overline{p' u'_k}\right] \tag{4.17}$$

In equation 4.17, it is possible to separate out the so-called "production term", or in other words turbulences production term P_k which is modeled using Boussinesq's hypothesis:

$$P_k = \tau_{ik}\frac{\partial \overline{U}_i}{\partial x_k} \tag{4.18a}$$

$$P_k \approx 2\nu_t \overline{S}_{ik}\frac{\partial \overline{U}_i}{\partial x_k} = 2\nu_t \overline{S}_{ik}\overline{S}_{ik} = \nu_t S^2 \tag{4.18b}$$

where S_{ik} is the first invariant of stress, assuming a diagonal matrix of the turbulent stress tensor. Local generation of fluctuating eddy structures in a fluid flow is proportional to the product of the covariance of fluctuation velocities (Reynolds stress) and the gradient of mean velocity. The second term on RHS of the evolutionary equation 4.17 corresponds approximately to the dissipation of turbulent kinetic energy ε:

$$\varepsilon \approx \nu \overline{\frac{\partial u'_l}{\partial x_k}\frac{\partial u'_l}{\partial x_k}} \tag{4.19}$$

In the case of two-equation turbulence models, ε is computed from the second-order partial derivative. The last two terms on RHS of the evolutionary equation 4.17 describe turbulence diffusive transport. Their modeling is carried out based on the gradient-diffusion hypothesis:

$$\frac{-1}{2}\overline{u'_i u'_i u'_k} - \frac{1}{\rho}\overline{p' u'_k} \approx \frac{v_t}{\sigma_k}\frac{\partial k}{\partial x_k} \qquad (4.20)$$

where σ_k is turbulent Prandtl number, which takes a constant value, often equal 1 [9].

Thus, if all the above-presented assumptions (4.18–4.20) are implemented in the evolutionary equation 4.17, it takes the form including production P_k and dissipation ε terms:

$$\frac{\partial k}{\partial t} + \overline{U}_k \frac{\partial k}{\partial x_k} = P_k - \varepsilon + \frac{\partial}{\partial x_k}\left[\left(v + \frac{v_t}{\sigma_k}\right)\frac{\partial k}{\partial x_k}\right] \qquad (4.21)$$

or even simpler, taking into account general form of the diffusion term defined in equation 4.20:

$$\frac{Dk}{Dt} = P_k - \varepsilon + \left(\frac{v_t}{\sigma_k}\operatorname{grad} k\right) \qquad (4.22a)$$

$$\frac{Dk}{Dt} = P_k - \varepsilon + \operatorname{Diff}(k) \qquad (4.22b)$$

An analysis of the above-presented expression leads to the correct conclusion that computation of the turbulence dissipation rate scalar field requires to apply an individual transport equation:

$$\frac{D\varepsilon}{Dt} = C_{\varepsilon 1}\frac{\varepsilon}{k}P_k - C_{\varepsilon 2}\frac{\varepsilon^2}{k} + \left(\frac{v_t}{\sigma_\varepsilon}\operatorname{grad}\varepsilon\right) \qquad (4.23)$$

where σ_ε is another turbulent Prandtl number and $C_{\varepsilon 1}$, $C_{\varepsilon 2}$ are empirical constants of models or functions of characteristic turbulence length scale [9].

Extending a fluid flow RANS model by two additional transport equations is currently the most popular approach in modeling full-scale renewable energy technologies, due to relative simplicity. Although in each popular two-equation turbulence model, in commercial solvers like Ansys Fluent, one of these equations describes transport of turbulent kinetic energy k, the second one does not have to refer directly to the turbulence dissipation rate - ε. An example of another approach is a two-equation "$k - \omega$" model, in which the second transported variable is similar to the ε, but not the same – it is the specific dissipation rate [9,25–27]:

$$\frac{D\omega}{Dt} = C_{\omega 1}\frac{\omega}{k}P_k - C_{\omega 2}\omega^2 + \operatorname{Diff}(\omega) \qquad (4.24)$$

where $\omega \propto \frac{\varepsilon}{k}$, and $C_{\omega 1}$, $C_{\omega 2}$ are empirical constants from models.

Interestingly, "$k - \omega$" model, compared with "$k - \varepsilon$", is more accurate in determining shear stress in the boundary layer region. At the same time, it is unfortunately more sensitive to the boundary transport parameters. Furthermore, it requires good mesh quality with very fine cells in the near-wall region, which comes from direct solving of the flow model in the viscous layer. To assess the reasonability of the "$k - \omega$" model application, the below described so-called "dimensionless wall distance" (y^+ parameter) can be applied [28]:

$$y^+ = \frac{\Delta y_1 \cdot \rho \cdot U_\tau}{\mu} \qquad (4.25)$$

where:
Δy_1 – wall distance from the first cell node [m],
μ – dynamic viscosity [Pa s],
y^+ – dimensionless wall distance [-],
ρ – fluid density $\left[\frac{kg}{m^3}\right]$,
U_τ – so-called dynamic viscosity depending on shear stress $\left[\frac{m}{s}\right]$.

In the case of application of the "$k - \omega$" model, the above-presented indicator must not exceed five; however, it is recommended to keep it rather around 1 [28].

Regardless of which two-equation RANS-based turbulence model is implemented in the analysis, the awareness of their limitations is crucial. Apart from the isotropic behavior of the modeled turbulence phenomena, an important disadvantage is the approximated form of the turbulent transport equations, which affects the whole model quality. Particularly inaccurate is the approximation of the source term in the turbulence dissipation rate transport equation. The fundamental assumption regarding eddy viscosity, according to the Launder and Spalding theory from 1974 [20], says that it is only k and ε function:

$$\mu_t = \rho C_\mu \cdot \frac{k^2}{\varepsilon} \qquad (4.26)$$

where C_μ coefficient in the mean flow (away from walls) takes the value of 0.09 but is also strongly wall-distance-dependent. Due to this fact, it increases significantly close to the walls, which in practice makes it impossible to implement such turbulence models in the boundary layer. A solution to this problem is so-called wall functions [29–31], which can be applied in different variants, depending on the modeled fluid flow case. Despite the implementation of the wall functions, in the case of the so-called "Standard $k - \varepsilon$" model, certain problems related to the eddy viscosity overestimation in the boundary layer and turbulent length scale in the flow stagnation regions (so-called stagnation point anomaly) may occur [9,18,32]. To avoid this situation, it is required to apply the so-called damping function f_μ, to modify the classic definition of eddy viscosity:

$$v_t = f_\mu C_\mu \frac{k^2}{\varepsilon} \qquad (4.27)$$

$$f_\mu = \exp\left[\frac{-3.4}{\left(\frac{1-R_T}{50}\right)^2}\right] \qquad (4.28)$$

$$R_T = \frac{k^2}{v\varepsilon} \qquad (4.29)$$

Turbulent Flows in RANS Approach

Another solution allowing to reduce efficiently the computation errors in the boundary layer is to implement there an additional one-equation model instead of a simple wall function. This model is activated in the near-wall region if certain strictly determined criterion is fulfilled there. This approach, known as "Enhanced Wall Treatment" is a so-called two-layer model (or blended-wall model) [33–35]. The above-mentioned criterion, allowing to switch between standard "$k - \varepsilon$" and additional one-equation model, is the turbulent Reynolds number Re_y:

$$\mathrm{Re}_y = \frac{y\sqrt{k}}{\nu} \tag{4.30}$$

$$l_\mu = f(\mathrm{Re}_y) \tag{4.31a}$$

$$\lambda_{Ry} = f(\mathrm{Re}_y) \tag{4.31b}$$

$$\nu_{t,\mathrm{2layer}} = C_\mu l_\mu \sqrt{k} \tag{4.31c}$$

$$\nu_{t,\mathrm{EnhWall}} = \lambda_{Ry} \nu_{t,k\varepsilon} + (1 - \lambda_{Ry}) \nu_{t,\mathrm{2layer}} \tag{4.31d}$$

where y of course refers to the wall distance.

The switch from the one-equation model of the boundary layer to the classic "$k - \varepsilon$" takes place when the turbulent Reynolds number exceeds 200.

A significant reduction of the stagnation point anomaly can be achieved by application of the modified "$k - \varepsilon$" model - "$k - \varepsilon$ Realizable" [36]. Then a reduction of the synthetic stress causing turbulence overproduction is realized thanks to the modified transport equation for the turbulence dissipation rate and a different method of computing C_μ coefficient in eddy viscosity ν_t equation (equation 4.27) which is actually substituted by $C_{\mu,\mathrm{modif}}$. The turbulent kinetic energy equation remains unchanged, thus it is not discussed below:

$$\frac{\partial \varepsilon}{\partial t} + U_j \frac{\partial \varepsilon}{\partial x_j} = C_1 S\varepsilon - C_2 \frac{\varepsilon^2}{k + \sqrt{\nu\varepsilon}} + \frac{\partial}{\partial x_j}\left[\left(\nu + \frac{\nu_t}{\sigma_\varepsilon}\right)\frac{\partial \varepsilon}{\partial x_j}\right] \tag{4.32}$$

$$C_1 = \mathrm{MAX}\left[0.43 \frac{\eta}{(\eta + 5)}\right] \tag{4.33a}$$

$$\eta = \frac{Sk}{\varepsilon} \tag{4.33b}$$

$$S = \sqrt{2S_{ij}S_{ij}} \tag{4.33c}$$

$$\nu_t = C_{\mu,\mathrm{modif}} \frac{k^2}{\varepsilon} \tag{4.34}$$

$$C_{\mu,\mathrm{modif}} = \frac{1}{4.04 + A_1 \frac{kS}{\varepsilon}} \tag{4.35a}$$

$$S = \sqrt{S_{ij}S_{ij} + \tilde{\Omega}_{ij}\tilde{\Omega}_{ij}} \tag{4.35b}$$

$$\tilde{\Omega}_{ij} = \Omega_{ij} - 2\varepsilon_{ijk}\omega_k \tag{4.35c}$$

$$A_1 = \sqrt{6}\cos(W) \tag{4.35d}$$

$$W = \frac{S_{ij}S_{jk}S_{ki}}{S^3} \tag{4.35e}$$

where:

ω_k – angular velocity $\left[\dfrac{\text{rad}}{\text{s}}\right]$,

Ω_{ij} – average tensor of rotation speed in the reference frame rotating with ω_k $\left[\dfrac{\text{rad}}{\text{s}}\right]$,

S_{ij}, S_{jk}, and S_{ki} – appropriate components of turbulent stress tensor [Pa].

Likewise, the standard "$k - \omega$" model has its modification which, in certain degree, makes computation results independent from the mean flow conditions. In the basic model, there was great probability of errors in the mean flow region. The model improved by Mentner was called "$k - \omega$ SST" (Shear Stress Transport) and its newest version proposed by Wilcox [37] implements the function included in the eddy viscosity expression which limits stress. k and ω transport is described by equations as follows:

$$\frac{Dk}{Dt} = P_k - \beta^* \omega k + \frac{\partial}{\partial x_j}\left[\left(v + \sigma_k \frac{k}{\omega}\right)\frac{\partial k}{\partial x_j}\right] \tag{4.36}$$

$$\frac{D\omega}{Dt} = \alpha \frac{\omega}{k} P_k - \beta\omega^2 + \frac{\partial}{\partial x_j}\left[\left(v + \sigma_\omega \frac{k}{\omega}\right)\frac{\partial \omega}{\partial x_j}\right] + \frac{\sigma_d}{\omega}\frac{\partial k}{\partial x_j}\frac{\partial \omega}{\partial x_j} \tag{4.37}$$

$$P_k = v_t S^2 \tag{4.38}$$

$$\beta = \beta_0 f_\beta \tag{4.39a}$$

$$f_\beta = \frac{1+85\chi_\omega}{1+100\chi_\omega} \tag{4.39b}$$

$$\chi_\omega = \left|\frac{\Omega_{ij}\Omega_{jk}\hat{S}_{ki}}{(\beta^*\omega)^3}\right| \tag{4.39c}$$

$$\hat{S}_{ki} = S_{ki} - \frac{1}{2}\frac{\partial u_m}{\partial x_m}\delta_{ki} \qquad (4.39d)$$

$$\Omega_{ij} = \frac{1}{2}\left(\frac{\partial u_i}{\partial x_i} - \frac{\partial u_j}{\partial x_i}\right) \qquad (4.39e)$$

where:
$\alpha = \frac{13}{25}, \beta_0 = 0.0708, \beta^* = 0.09, \sigma_k = 0.6, \sigma_\omega = 0.5$

$\sigma_d = 0$, for $\left(\frac{\partial k}{\partial x_j}\frac{\partial \omega}{\partial x_j} \leq 0\right)$

$\sigma_d = \frac{1}{8}$, for $\left(\frac{\partial k}{\partial x_j}\frac{\partial \omega}{\partial x_j} > 0\right)$

Eddy viscosity is computed using relation:

$$v_t = \frac{k}{\tilde{\omega}} \qquad (4.40a)$$

$$\tilde{\omega} = \mathrm{MAX}\left(\omega, C_{\lim}\sqrt{\frac{2S_{ij}S_{ij}}{\beta^*}}\right) \qquad (4.40b)$$

where $C_{\lim} = \frac{7}{8}$, and the second term in brackets in equation 4.40b is the so-called "stress limiter"

Based on the discussed examples of popular two-equation turbulence models it can be concluded that their development concerns reduction of their sensitivity under the conditions occurring in the boundary layer or under the mean flow velocity influence. Quite efficient from this point of view are "$k - \varepsilon$ Realizable" and "$k - \omega$ SST". On the other hand, it still cannot be said that either of these approaches is a universal one. Thus, all the attempts to combine the functionalities of these two approaches are reasonable. An interesting example of the model that incorporates the above-mentioned idea is "Shear Stress Transport (SST)" for Ansys CFX solver, where, thanks to the "blending function", a smooth transition between "$k - \omega$" (at the wall) and "$k - \varepsilon$" (in the bulk flow) is implemented. Even two "blending functions" are applied in the model, wherein the second one influences the eddy viscosity value [38–40].

REFERENCES

1. P.-L. Delafin et al., Comparison of low-order aerodynamic models and RANS CFD for full scale 3D vertical axis wind turbines, *Renewable Energy*, 109 (2017), 564–575.
2. A. S. T. Tan et al., Performance analysis of a solar heat collector through experimental and CFD investigation, *Materials Today: Proceedings*, 52 (2022), 1338–1344.
3. R. Laubscher, S. van der Merwe, Heat transfer modelling of semi-suspension biomass fired industrial watertube boiler at full- and part-load using CFD, *Thermal Science and Engineering Progress*, 25 (2021), 100969.

4. J. Yu, M. Li, T. Stathopoulos, Strategies for modeling homogeneous isotropic turbulence and investigation of spatially correlated aerodynamic forces on a stationary model, *Journal of Fluids and Structures*, 90 (2019), 43–56.
5. G. Cao, L. Pan, K. Xu, Three dimensional high-order gas-kinetic scheme for supersonic isotropic turbulence I: Criterion for direct numerical simulation, *Computers & Fluids*, 192 (2019), 1–14 (104273).
6. H. Shen et al., Stochastic modeling of subgrid-scale effects on particle motion in forced isotropic turbulence, *Chinese Journal of Chemical Engineering*, 27 (12) (2019), 2884–2891.
7. G. Wang et al., LBM study of aggregation of monosized spherical particles in homogeneous isotropic turbulence, *Chemical Engineering Science*, 201 (2019), 201–211.
8. H. K. Versteeg, W. Malalasekera, An introduction to computational fluid dynamics. *The Finite Volume Method*, Pearson Education Limited, Harlow, 2007.
9. S. Kubacki, B. Górecki, W. Sadowski, Introduction to Large Eddy Simulation, training materials: "Selected issues of turbulence modeling", Quickersim, Warsaw (19–20.09.2019).
10. M. L. A. Kaandorp, R. P. Dwight, Data-driven modelling of the Reynolds stress tensor using random forests with invariance, *Computers & Fluids*, 202 (2020), 1–16 (104497).
11. D. Lengani, D. Simoni, S. Kubacki, E. Dick, Analysis and modelling of the relation between the shear rate and Reynolds stress tensors in transitional boundary layers, *International Journal of Heat and Fluid Flow*, 84 (2020), 108615.
12. G. Busco, E. Merzari, Y. A. Hassan, Invariant analysis of the Reynolds stress tensor for a nuclear fuel assembly with spacer grid and split type vanes, *International Journal of Heat and Fluid Flow*, 77 (2019), 144–156.
13. S. Gavrilyuk, H. Gouin, Geometric evolution of the Reynolds stress tensor, *International Journal of Engineering Science*, 59 (2012), 65–73.
14. K. Hanjalić, M. Popovac, M. Hadžiabdić, A robust near-wall elliptic-relaxation eddy-viscosity turbulence model for CFD, *International Journal of Heat and Fluid Flow*, 25 (6) (2004), 1047–1051.
15. X. Li, J. Tu, Evaluation of the eddy viscosity turbulence models for the simulation of convection–radiation coupled heat transfer in indoor environment, *Energy and Buildings*, 184 (2019), 8–18.
16. M. Z. I. Qureshi, A. L. S. Chan, Influence of eddy viscosity parameterisation on the characteristics of turbulence and wind flow: Assessment of steady RANS turbulence model, *Journal of Building Engineering*, 27 (2020), 1–9 (100934).
17. Y. Duan, C. Jackson, M. D. Eaton, M. J. Bluck, An assessment of eddy viscosity models on predicting performance parameters of valves, *Nuclear Engineering and Design*, 342 (2019), 60–77.
18. J. Blazek, *Computational Fluid Dynamics: Principles and Applications*, Elsevier (2001), ISBN: 0-08-043009-0.
19. J. K. Lai, E. Merzari, Y. A. Hassan, Sensitivity analyses in a buoyancy-driven closed system with high resolution CFD using Boussinesq approximation and variable density models, *International Journal of Heat and Fluid Flow*, 75 (2019), 1–13.
20. B. E. Launder, D. B. Spalding, The numerical computation of turbulent flows, *Computer Methods in Applied Mechanics and Engineering*, 78 (2) (1974), 269–289.
21. T. Jiyuan, Y. Guan-Heng, L. Chaoqun, *Computational Fluid Dynamics. A Practical Approach*, Butterworth-Heinemann, Oxford, 2013.
22. T. Cebeci et al., *Computational Fluid Dynamics for Engineers, From Panel to Navier-Stokes Methods with Computer Programs*, Springer (2005), ISBN 3-540-24451.
23. M. A. R. Sadiq Al-Baghdadi, *Computational Fluid Dynamics Modeling in Development of Renewable Energy Applications*, M. A. R. Sadiq Al-Baghdadi, Computational Fluid Dynamics Modeling in Development of Renewable Energy Applications, International Energy and Environment Foundation, Scotts Valley, 2011.
24. L. Lapidus, G. F. Pinder, *Numerical Solution of Partial Differential Equations in Science and Engineering*, Wiley, Canada, 1999.
25. C. F. Panagiotou, F. S. Stylianou, S. C. Kassinos, Structure-based transient models for scalar dissipation rate in homogeneous turbulence, *International Journal of Heat and Fluid Flow*, 82 (2020), 1–14 (108557).
26. R. M. Horwitz, A. E. Hay, Turbulence dissipation rates from horizontal velocity profiles at mid-depth in fast tidal flows, *Renewable Energy*, 114 (A) (2017), 283–296.
27. G. A. Degrazia, Autocorrelation function formulations and the turbulence dissipation rate: Application to dispersion models, *Physica A: Statistical Mechanics and Its Applications*, 389 (24) (2010), 5808–5813.
28. https://ansyshelp.ansys.com (last access: 11.06.2023).

29. F. Berni, S. Fontanesi, A 3D-CFD methodology to investigate boundary layers and assess the applicability of wall functions in actual industrial problems: A focus on in-cylinder simulations, *Applied Thermal Engineering*, 174 (2020), 115320.
30. B. Blocken, J. Carmeliet, T. Stathopoulos, CFD evaluation of wind speed conditions in passages between parallel buildings—Effect of wall-function roughness modifications for the atmospheric boundary layer flow, *Journal of Wind Engineering and Industrial Aerodynamics*, 95 (9–11) (2007), 941–962.
31. B. Blocken, T. Stathopoulos, J. Carmeliet, CFD simulation of the atmospheric boundary layer: wall function problems, *Atmospheric Environment*, 41 (2) (2007), 238–252.
32. M. Sarwar et al., *Large Eddy Simulation of Fire Behaviour in Landscapes (POSTER)*, Bushfire CRC, Melbourne, 2012.
33. W. Xu, Q. Chen, A two-layer turbulence model for simulating indoor airflow: Part I. Model development, *Energy and Buildings*, 33 (6) (2001), 613–625.
34. W. Xu, Q. Chen, A two-layer turbulence model for simulating indoor airflow: Part II. Applications, *Energy and Buildings*, 33 (6) (2001), 627–639.
35. T. Jongen, Y. P. Marx, Design of an unconditionally stable, positive scheme for the k–ϵ and two-layer turbulence models, *Computers & Fluids*, 26 (5) (1997), 469–487.
36. M. Lateb et al., Comparison of various types of k–ε models for pollutant emissions around a two-building configuration, *Journal of Wind Engineering and Industrial Aerodynamics*, 115 (2013), 9–21.
37. Wilcox, D. C., Formulation of the k-omega turbulence model revisited, *AIAA Journal*, 46 (11) (2008), 2823–2838.
38. P. A. Costa Rocha et al., A case study on the calibration of the k–ω SST (shear stress transport) turbulence model for small scale wind turbines designed with cambered and symmetrical airfoils, *Energy*, 97 (2016), 144–150.
39. P. A. Costa Rocha et al., k–ω SST (shear stress transport) turbulence model calibration: A case study on a small scale horizontal axis wind turbine, *Energy*, 65 (2014), 412–418.
40. X. L. Yang, Y. Liu, L. Yang, A shear stress transport incorporated elliptic blending turbulence model applied to near-wall, separated and impinging jet flows and heat transfer, *Computers & Mathematics with Applications*, 79 (12) (2020), 3257–3271.

5 Reactive and Multiphase Flows

For flows involving more complex phenomena, such as combustion, multiphase and multi-species flows with possible other effects (chemical reactions), as in fire simulations, we need to model the physical laws describing these phenomena and provide the best possible approximations.

A valid definition of the interaction types between turbulences and chemistry is the essential stage of the development of the CFD modeling of the reactive flows. The problem is that a large source term (for example representing production or consumption of certain species) changes the flow variables rapidly in space and in time. The changes due to a strong source term happen at much smaller time scales than those of the flow equations. This increases the *stiffness* of the governing equations significantly. The stiffness is defined as the ratio of the largest to the smallest eigenvalue of the Jacobian matrix $\frac{\delta R}{\delta W}$. The stiffness can also be viewed as the ratio of the largest to the smallest time scale, equation 5.1 [1].

$$\frac{dW_{I,J,K}}{dt} = \frac{-1}{\Omega_{I,J,K}} \left[\sum_{m=1}^{N_F} (F_c - F_v)_m \Delta S_m - (Q \cdot \Omega)_{I,J,K} \right] \quad (5.1)$$

where:
  $W_{I,J,K}$ – vector of conservative variables,
  $\Omega_{I,J,K}$ – a particular volume,
  N_F – number of control volume faces (four in 2D and six in 3D),
  ΔS_m – area of the face m,
  Q – source term (comprises all volume sources due to body forces and volumetric heating),
  F_c and F_v – flux vector, related to the convective transport of quantities in the fluid and vector related to the viscosity, containing the viscous stresses, as well as the heat diffusion, respectively.

More details can be found in Ref. [1] and references therein. A detailed overview of the equations governing a chemically reacting flow, together with the Jacobian matrices of the fluxes and their eigenvalues, can also be found in Ref. [1].

5.1 CLASSIFICATION OF THE CHEMISTRY MODELING APPROACHES AND CHARACTERISTICS OF THE SELECTED MODELS

Certain possible reacting flow models are given in Figure 5.1, based on the Ansys Fluent solver case [2].

According to the classification presented in the figure, first of all, two features of the reactive flow have to be considered: chemistry and flow configuration. These issues were briefly discussed in the next two subsections.

5.1.1 Fast/Slow Chemistry Modeling

Two essential criteria can be applied for consideration in relation to chemistry modeling: Reynolds number and Damköhler number defined in equation (5.2):

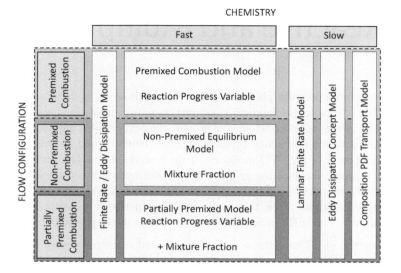

FIGURE 5.1 Ansys Fluent reacting flow models.

$$Re = \frac{\rho U L}{\mu} \quad (5.2a)$$

$$Da = \frac{\text{mixing time scale}}{\text{chemistry time scale}} = \frac{\dfrac{L}{U}}{\dfrac{\rho_{pd}}{R_{\text{slow}}}} = \frac{\dfrac{k}{\epsilon}}{\dfrac{\rho_{pd}}{R_{\text{slow}}}} \quad (5.2b)$$

where:

ρ – density (kg/m³),
U – characteristic velocity (m/s),
L – characteristic length for flow (m),
μ – dynamic viscosity (Pa S)
ρ_{pd} – density of the adiabatic flame surface (1/mm)
R_{slow} – rate of the slowest stoichiometric reaction (mol/dm³/s)

If the Damköhler number is much larger than 1 ($Da \gg 1$), it means that subtracts are reacting at the moment of contact. This situation describes the so-called *fast chemistry*. The opposite case ($Da \sim 1$) describes "*slow chemistry*".

From the CFD point of view, various modeling approaches have to be applied to those two cases. For $Da \gg 1$ kinetics of chemical reaction may be omitted. Then turbulent mixing can be assumed as the predominant factor of the chemical reaction occurs. Therefore, a proper selection of the turbulence model is essential in this case.

The four most important models presented in Figure 5.1 (based on the example of Ansys Fluent [3]) are those encompassing fast, slow, and transition chemistry cases:

Eddy dissipation model (EDM)
Finite rate model (FRM)
Finite rate/Eddy dissipation model (FREDM)
Eddy dissipation concept (EDC)

First of all, the selection of the appropriate approach needs to be based on the chemistry – the turbulence relation and the assessment of the significance of detailed reaction kinetics.

Of course, the evaluation of the significance of the kinetics is based on the Damköhler number (equation 5.2).

A brief description of subsequent models from the list above was included in the following subsections.

5.1.1.1 Eddy Dissipation Model (EDM)

For a highly turbulent reacting flow, where the mixing rate is much higher than the chemical rate, it may be assumed that the reaction time is negligible. In that case, the reaction is limited by mixing intensity. Once the reagents reach contact, the reaction is completed. In this situation $Da \gg 1$ and the EDM approach is applicable.

However, as in other approaches, EDM has some drawbacks that need to be considered before using it in a given case. The most important feature of the EDM is that the kinetics of the reactions is excluded from the considerations. Therefore, the rate of chemical reactions is controlled by the presence of large eddies in flows. It means that, for example, there is no combustion in certain zones of laminar flows.

Formulas for the expression of the rate of chemical reactions can be found, *e.g.*, in Ref. [3].

5.1.1.2 Finite Rate Model (FRM) and Finite Rate Eddy Dissipation Model (FREDM)

The FREDM allows the reaction rates between kinetics-dependent and mixing-dependent to be switched. The evaluation of conditions is based on the calculation of both the turbulence timescale and the Arrhenius equation (or Arrhenius chemical rate coefficient) that is related to the activation energy and temperature and thus, the kinetics.

The rates of production and consumption of reagents calculated using these two methods are compared and the smaller value is selected as the one that controls the reaction in a given reactor area. In dynamic simulations, of course, besides the region of the computational domain, the chemistry type may vary in time owing to the changes in the flow characteristics and the temperature field.

Every time a classic case of slow chemistry is considered (such as gasification in laminar flow), the FRM can be used separately. Then, only the detailed chemistry approach is applied. This approach is useful in each case where not just the result, but – most of all – the chemical process is considered. It is possible to create sets of transition reactions and study advanced chemistry – for example, related to the formation of pollutants.

This advantage, however, may also pose a problem. A reliable representation of complex chemistry in certain processes requires the setting of kinetic parameters and rate exponents.

A widely used method, which allows to describe the kinetics of pyrolysis and thus, delivers the required, above mentioned parameters, is thermogravimetric analysis (TGA). The method is based on precise measurement of the mass loss of the fuel sample in the conditions of increasing temperature and inert atmosphere (see, *e.g.*, Ref. [4]).

For example, most of the commercially available biomass-based small-scale energy devices can be considered fast chemistry systems. In some of the reactors with a relatively large reaction zone (straw-fired batch boilers, gasifiers, and incinerators), or a simplified air distribution system (as in fireplaces), it is possible to distinguish flow-stagnation regions besides the turbulent-flow area, where the temperature is high enough to allow certain chemical reactions to proceed owing to radiation (such as decomposition), even if no efficient mixing occurs. For example, such an approach was described in a paper [5], where combustion in a relatively huge combustion chamber was modeled.

In this case, the application of a model based on fast chemistry, such as the classic EDM, leads to seriously disturbed simulation results. In these complex cases, a combination of two different approaches that control the process in specific reaction conditions may be implemented. It is also common to use one global reaction mechanism for the kinetic reaction when no detailed chemistry is required [1].

5.1.1.3 Eddy Dissipation Concept Model (EDCM)

It is always a challenge to represent actual chemical phenomena in a turbulent, reacting mixture flow. In CFD simulations, the representation of actual chemical mechanisms proceeding under turbulent combustion conditions entails a high computational cost. However, this approach is also widely used in studies that require a detailed analysis of the phenomena that take place in the reactor. In these instances, it is necessary to employ methods fully encompassing the interactions between the chemistry and the turbulences in the considered system.

In 1996, Gran and Magnussen developed an offshoot of the EDM extended by finite rate chemistry. However, the new method has significantly differed from the original model, owing to, among other things, the scales of reactions.

To understand the idea behind the eddy dissipation concept model (EDCM), one may imagine that the reaction area is somewhat discretized to a certain number of very small, individual reactors where the chemistry occurs. It means that the analyses relate to small-scale phenomena. The processes in these "small-scale reactors" take place under constant pressure enthalpy conditions. Based on the above, it can be concluded that a certain length of the scale is certainly important, but the timescale of the reaction occurring in a given reactor additionally needs to be considered.

Equations 5.3 and 5.4 describe the length fraction of small individual reactors (γ) and the timescale of the reaction (τ) that occurs inside, respectively:

$$\gamma = 2.1377 \left(\frac{\nu \epsilon}{k^2} \right)^2 \quad (5.3)$$

$$\tau = 0.4082 \sqrt{\frac{\nu}{\epsilon}} \quad (5.4)$$

Based on these two parameters and species, the mean reaction rate can be expressed as provided below (equation 5.5):

$$S_k = \frac{\rho \gamma^2}{\tau (1 - \gamma^3)} \left(Y_k^* - Y_k \right) \quad (5.5)$$

where Y_k is the mass fraction of the "k" species and Y_k^* is its mass fraction in small scale.

The EDCM provides the possibility to decide on how complicated the chemistry in a considered CFD simulation should be. Furthermore, it has to be noted that in the case of the EDCM, there is no requirement to apply certain unintuitive constants (as in the EDM). However, it should always be remembered that, even without these constants, the more complex the case, the more complicated the reaction description will have to be considered.

5.1.2 PREMIXED AND NON-PREMIXED COMBUSTION MODELS

Generally, two combustion types may be distinguished:

- non-premixed combustion, which is related to the so-called diffusion flame that is formed in the conditions of individual inlets of fuel and oxidizer to the system; reagents can be transported to the reaction area from each side of the flame.
- premixed combustion, which causes the formation of a kinetic flame, where both reagents are completely mixed and the reaction front is directly related to the temperature front in the flame [1].

Reactive and Multiphase Flows

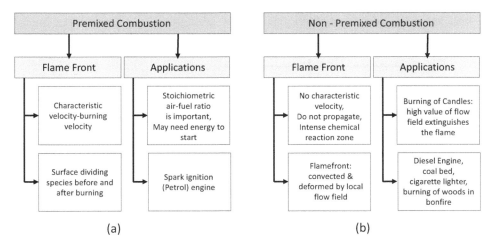

FIGURE 5.2 Flame characteristics and applications of: (a) premixed and (b) non-premixed combustion.

The idea of this type of combustion is presented in Figure 5.2, where the main characteristics and applications are indicated. For the premixed combustion, the flame front propagates spontaneously with a characteristic velocity, called the burning velocity, in a direction normal to itself, to consume the available reactant mixture. The flame front is the surface of discontinuity between the burnt and unburnt regions. For the non-premixed case, flame fronts do not have a characteristic velocity and do not propagate, the combustion zone has a distinct flame front where fuel and oxidizers are in the stoichiometric ratio [2].

Premixed combustion is much more difficult to model than non-premixed combustion. The reason for this is that premixed combustion usually occurs as a thin, propagating flame that is stretched and contorted by turbulence. To capture the laminar flame speed, the internal flame structure would need to be resolved, as well as the detailed chemical kinetics and molecular diffusion processes. Since practical laminar flame thicknesses are of the order of millimeters or smaller, resolution requirements are usually unaffordable. The essence of premixed combustion modeling lies in capturing the turbulent flame speed, which is influenced by both the laminar flame speed and the turbulence. Non-premixed combustions are simplified to a mixing problem, and equations for individual species are not solved. Instead, species concentrations are derived from the predicted mixture fraction fields [2].

Probability Density Function (PDF) Flamelet (PFM) Combustion model (in Ansys CFX solver) is used to simulate such systems. Here, the combustion phenomena are assumed to occur on thin surface of zero thickness called flamelets (or the flame front). The turbulent flame is treated as a combination of laminar flamelets and the individual species transport equations are not solved. The effect of a turbulent-flow field on flames and reaction is incorporated statistically utilizing a probability density function.

Excerpts from ANSYS Tutorials: "When you use the non-premixed combustion model, you need to create a PDF table. This table contains information on the thermo-chemistry and its interaction with turbulence. Ansys Fluent interpolates the PDF during the solution of the non-premixed combustion model." Examples include jets, jets in cross- or co- or swirl-flows [2].

5.2 INTRODUCTION TO HOMOGENEOUS AND HETEROGENEOUS COMBUSTION

Homogeneous combustion is determined by the flow of gases and by interdiffusive phenomena characteristic, *e.g.*, of mixtures of flammable gases and liquid vapors mixed with air. It is worth

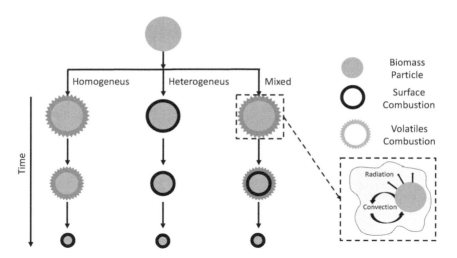

FIGURE 5.3 Outlook view of the main ideas of homo and hetero combustion.

mentioning that the conditions for a continuous reaction are: fuel, heat source with a given intensity and temperature, required concentration of flammable substance related to oxygen, and access of oxygen to the flame zone (thin external layer of a flame).

Heterogeneous combustion (otherwise known as combustion in porous media) is mainly characterized by the following properties:

- occurs for substances that during combustion are not transferred into gaseous form, there is glowing but no flame, *e.g.*, charcoal glowing,
- a high surface area solid is immersed into a gaseous reacting flow,
- its characteristics are determined by the exchange of heat and material with the surface of the solid,
- a solid and gas phase interacts to promote the complete transfer of reactants to their lower energy potential products.

In this type of combustion, additional fluid phases may or may not be present. Chemical reactions and heat transfer occur locally in each phase and between both phases. In this regime of combustion, thermal heat released from the combustion by-products is transferred into the solid phase by convection. Both conduction and radiation result then in upstream heat transfer (along with adverse convection within the gas phase). Heat is then convectively transferred to the unburnt reactants [6].

The main differences between homo- and heterogeneous combustions are presented in an outlook form in Figure 5.3.

Such two types of combustion are observed simultaneously during biomass combustion, where we have the gaseous form from degasification and solid – charcoal, as presented in Figure 5.4.

5.3 SELECTED TYPES AND IMPORTANCE OF MULTIPHASE TRANSPORT IN CFD

A multiphase flow is defined as a flow in which more than one phase (*i.e.*, gas, solid, and liquid) exist. Such flows are ubiquitous in the industry, examples being gas-liquid flows in evaporators and condensers, gas-liquid-solid flows in chemical reactors, solid-gas flows in pneumatic conveying, *etc*. Some examples will be given later in this section. This introductory section attempts to give an overview, with more detailed material appearing on each type of multiphase flow in separate entries. In gas-solid and liquid-solid flows, the dispersed phase is always in the solid phase because

Reactive and Multiphase Flows

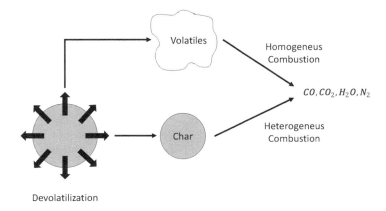

FIGURE 5.4 Homo and hetero combustions during biomass combustion.

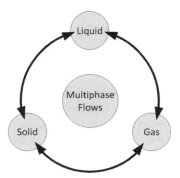

FIGURE 5.5 An example of a multiphase flow and four phases is presented: oil, air, water and solid particles.

solid particles never coalesce with each other. On the other hand, in gas-liquid and liquid-liquid flows, the dispersed phase is determined mainly by the flow rates of both phases since the interface between them is deformable, and the dispersed phase coalesces and finally becomes the continuous phase as the flow rate increases. For example, in gas-liquid flows, the gas phase is dispersed when the gas flow rate is low compared with the liquid flow rate [7,8]. An example of the multiphase flow idea is presented in Figure 5.5.

Multiphase flows have wide applications in various processes, refrigeration, air conditioning, petroleum, oil and gas, food processing, automotive, power generation and metal industries including phenomena like mixing, particle-laden flows, CSTR – Continuously Stirred Tanks Reactor [gas dispersion and floating particles suspension in an agitated or stirred vessel], Water Gas Shift Reaction (WGSR), fluidized bed, fuel injection in engines, bubble columns, mixer vessels, Lagrangian Particle Tracking (LPT) [8]. Figure 5.6 presents multiphase flow classification based on phase type [9].

Another practical example of real gas is the simulation of steam or, which is more demanding, of wet steam in turbomachinery applications. In this case, the steam is mixed with water droplets, it is either possible to solve an additional set of transport equations or to trace the water droplets along a number of streamlines. These simulations have very important applications in the design of modern steam turbine cascades. Similarly, in ORC turbines simulation methods are extended to two-phase turbines under the assumption of a two-phase homogeneous fluid under thermal equilibrium. Dry steam and liquid with vapor bubbles are treated as two phases [10]. Also, heat exchangers with

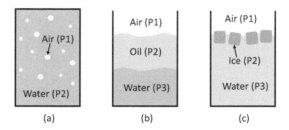

FIGURE 5.6 Multiphase flow examples. For (a) two-phase flow with two different types of matter, heat transfer and immiscible fluids, (b) three-phases (three different types of matter), (c) as (b) but different phases of the same matter (ice and water) exists and mass transfer – ice to water and melting occurs.

phase change materials can work in a regime of two-phase and superheated regions formed inside a tube carrying the ORC working fluid refrigerant [11].

The analysis of the flow past turbine blades can, for instance, help to understand the occurrence of supercritical shocks by condensation and flow instabilities, responsible for an additional dynamic load on the blades, resulting in the loss of efficiency.

The application of renewable energies technology (RET) is wide and concerns practically each field. For example:

- solar energy devices, like flat-plate collectors, where organic fluid or refrigerant is applied, can be used, *e.g.*, as evaporators for a heat pump cycle and have many advantages over conventional collectors. In this case, working fluid experiences a phase change. One important factor is that the multiphase heat transfer coefficient is larger than for a single-phase flow, resulting in increasing heat gain [12]. Another example can be found, *e.g.*, in Ref. [13] where a multiphase flow was considered in the construction of a multiphase heat exchanger. Compared to a single-phase flow, a multiphase flow has much better heat transfer behavior than single-phase flow. In the cited example, a phase change material (PCM) layer surrounds the MgH_2 used for hydrogen desorption reaction to achieve thermal energy storage.
- water energy – multiphase flow phenomena can be also observed during the operation of hydraulic turbines. The flow field of this turbine is extremely complex because it has a free surface, and its internal and external flows are mixed [14]. Besides that, water splashing and spray formation should be analyzed [15]. Likewise, the effect of cavitation can be treated as a multiphase process, involving the formation of bubbles inside a liquid medium. Cavitation is an important factor because creates noise, vibration, surface erosion, and efficiency loss in hydro turbines [16]. Application of wave energy system also requires a multiphase approach, because the models used assume the existence of two phases: the denser fluid (usually water) and empty space representing the lighter fluid (usually air) [17].
- geothermal systems can be characterized by multiphase phenomena, but the fundamental multiphase flow and phase transition behavior during geothermal exploitation is not well understood. Geothermal resources (hot media – hot water and steam) can exist in such phases as: gas-liquid and supercritical liquid and gas in porous media [18].
- biomass system – multiphase flow occurs during biomass gasification – *i.e.*, gas phase and the solid phase – solid biomass particles. Also multiple solid phase (biomass and inert sand) should be taken into account for such an analysis. In the presented study [19] an Eulerian fluid model for the gas phase with a Lagrangian discrete element model (DEM) for the solid phase were employed. Modeling and simulation of biomass gasification in a bubbling fluidized bed gasifier require analysis of such phases as biomass (saw dust), sand and gases. Eulerian- Eulerian approach can be used for such modeling. The gas mixture

phase is then considered as the primary continuous phase and the solid phase is the secondary dispersed phase [20]. Many biomass devices require the application of a multiphase model for simulation, *e.g.*, complete-mix digesters and anaerobic biohydrogen fermenters – gas-liquid two-phase models, for anaerobic biofilm reactors gas-liquid-solid three-phase CFD simulations are applied. In the case of photobioreactors, except for single-phase models, Gas-liquid two-phase are used [21]. The fast pyrolysis process is a multiphase reacting flow field that includes a complex gas-solid mixture. The gas-solid multiphase flow characteristics greatly influence the consequent pyrolysis reaction field. Applying a proper simulation method allows for the investigation of such characteristics as: effects of reaction temperature and gas velocity on hydrodynamics, heat transfer, and the consequent pyrolysis reaction [22].

- energy storage (mostly applications for RET) – for compressed air storage in aquifers where two phases: compressed air and water exist, gas saturation and bubble formation during air injection should be considered to obtain valid values of energy recovery efficiency [23,24]. Phase change materials (PCMs) can be an interesting option for thermal energy storage. For such materials latent thermal energy is stored and released during the material's phase transitions. In those cases such phases are solid-liquid, liquid-gas and both solid-liquid/liquid-gas phases. For wider application, such factors as phase separation, deterioration of thermal properties, mechanical properties, *etc.* are essential [25].

For multiphase flow, several models can be adopted according to Figure 5.7 [26]:

The use of the conventional Eulerian model for granular flows and Lagrangian approach incorporated with the Discrete Element Method (CFD-DEM) are quite well proven; however, some

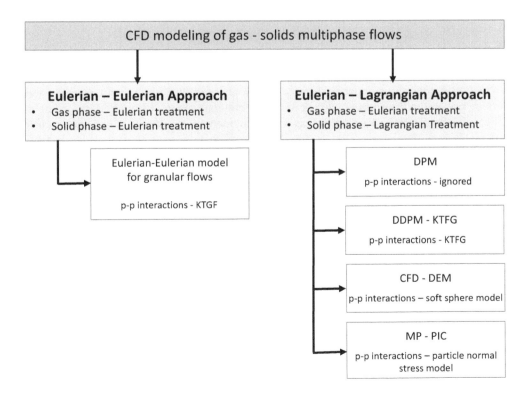

FIGURE 5.7 Summary of model approaches to gas-solids multiphase flow modeling.

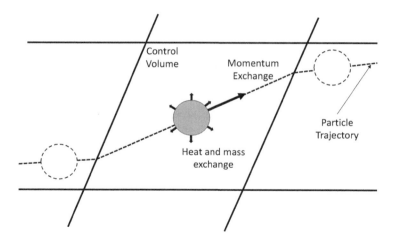

FIGURE 5.8 Diagram of a discrete phase model.

limitations restrict the use of these models in a wide range of applications. Therefore, some new models have been introduced to model gas-solids flows, for example:

- Dense Discrete Phase Model incorporated with Kinetic Theory of Granular Flow (DDPM-KTGF),
- Dense Discrete Phase Model incorporated with Discrete Element Method (DDPM-DEM) and
- Computational Particle Fluid Dynamics (CPFD) numerical scheme incorporated with the Multiphase-Particle-In-Cell (MP-PIC) method.

These models have been validated for certain applications under certain conditions, however, further validation of these models is still a necessity [26].

As an example, some consideration relating to the DPM (Discrete Phase Model) model will be presented. DPM allows to simulate a discrete second phase in the Lagrangian frame of reference, in addition to solving transport equations for the continuous phase. Also, DPM can compute the trajectories of the discrete phase entities, as well as momentum transfer, heat and mass transfer to/from them (Figure 5.8).

The discrete phase model or DPM, is a classic method of simulating discrete phase transport in the Lagrangian reference frame in the case of fluid droplets and solid particles. This approach is limited to a maximum of 10% of discrete phase volume in relation to the continuous phase. This is due to the assumption of the lack of interaction between the particles. Many examples related to such an approach can be found in the literature. For example, in biomass applications for gasification of large biomass particles at high temperatures for hydrogen generation [27] and wood-chips gasification [28]. Other examples can be related to, *e.g.*, solar energy – [29] where particle tracking and heat transfer enhancement in a solar cavity receiver is analyzed.

In principle, including multiphase flow phenomena in a CFD model is always a challenge. Although considering multiphase flows is not recommended for the beginning users of CFD codes, certain simplified example related to this issue was presented in the book practical part.

REFERENCES

1. J. Blazek, *Computational Fluid Dynamics: Principles and Applications*, Elsevier (2001), ISBN: 0-08-043009-0.
2. Combustion simulation, cfdyna.com (last access: 01.03.2022).

3. *Biomass in Small-Scale Energy Applications, Theory and Practice*, Edited by M. Szubel, M. Filipowicz, CRC Press, Taylor & Francis Group, Boca Raton, FL, 2020.
4. M Szubel, A Dernbecher, T Dziok, Determination of kinetic parameters of pyrolysis of wheat straw using thermogravimetry and mathematical models, *IOP Conference Series: Earth and Environmental Science*, 214 (2019), 012131.
5. M. Szubel et al., Homogenous and heterogeneous combustion in the secondary chamber of a straw-fired batch boiler, *EPJ Web of Conferences*, 143 (2017), 02125.
6. T. Tadao, A theoretical study on an excess enthalpy flame. *Symposium (International) on Combustion*, 18 (1) (1981), 465–472. doi:10.1016/j.apenergy.2016.06.128.
7. Multiphase flow - MR CFD - CFD analysis, consultation, training, simulation, mr-cfd.com (last access: 15.04.2022).
8. Multi-phase flow simulations in OpenFOAM, cfdyna.com (last access: 15.04.2022).
9. Multiphase flow in CFD: Basics and modeling | SimScale Blog, https://www.simscale.com/blog/2017/12/multiphase-flow/ (last access: 15.04.2022).
10. M. T. White, Cycle and turbine optimisation for an ORC operating with two-phase expansion, *Applied Thermal Engineering*, 192 (2021), 116852.
11. R. Majumdar, S. Singh, S. K. Saha, Quasi-steady state moving boundary reduced order model of two phase flow for ORC refrigerant in solar-thermal heat exchanger, *Renewable Energy*, 126 (2018), 830e843.
12. H. U. Helvaci, Z. A. Khan, Mathematical modelling and simulation of multiphase flow in a flat plate solar energy collector, *Energy Conversion and Management*, 106 (2015), 139–150.
13. J. Yao, P. Zhu, L. Guo, F. Yang, Z. Zhang, J. Ren, Z. Wu, Study of a metal hydride based thermal energy storage system using multi-phase heat exchange for the application of concentrated solar power system, *International Journal of Hydrogen Energy* 46 (2021), 29332e29347.
14. Y. Nishi et al., A study of the flow field of an axial flow hydraulic turbine with a collection device in an open channel, *Renewable Energy*, 130 (2019), 1036e1048.
15. S. Bhattarai et al., Novel trends in modelling techniques of Pelton Turbine bucket for increased renewable energy production, *Renewable and Sustainable Energy Reviews*, 112 (2019), 87–101.
16. K. Kumar, R. P. Saini, A review on operation and maintenance of hydropower plants, *Sustainable Energy Technologies and Assessments*, 49 (2022), 101704.
17. J. V. Ringwood, Wave energy control: status and perspectives, IFAC PapersOnLine 53-2 (2020), 12271–12282.
18. G. Feng et al., Multiphase flow modeling and energy extraction performance for supercritical geothermal systems, *Renewable Energy*, 173 (2021), 442e454.
19. A. Porcu et al., Experimental validation of a multiphase flow model of a lab-scale fluidized-bed gasification unit, *Applied Energy*, 293 (2021), 116933.
20. M. Anil et al., Performance evaluation of fluidised bed biomass gasifier using CFD, *Energy Procedia*, 90 (2016), 154–162.
21. B. Wu, Advances in the use of CFD to characterize, design and optimize bioenergy systems, *Computers and Electronics in Agriculture*, 93 (2013), 195–208.
22. H. C. Park, H. S. Choi, Fast pyrolysis of biomass in a spouted bed reactor: Hydrodynamics, heat transfer and chemical reaction, *Renewable Energy*, 143 (2019), 1268e1284.
23. L. Yang et al., Numerical investigation of cycle performance in compressed air energy storage in aquifers, *Applied Energy*, 269 (2020), 115044.
24. C. Guo et al., The promise and challenges of utility-scale compressed air energy storage in aquifers, *Applied Energy*, 286 (2021), 116513.
25. L. Yang et al., A comprehensive review on sub-zero temperature cold thermal energy storage materials, technologies, and applications: State of the art and recent developments, *Applied Energy*, 288 (2021), 116555.
26. W. K. Hiromi Ariyaratne et al., CFD approaches for modeling gas-solids multiphase flows – A review, *Proceedings of the 9th EUROSIM & the 57th SIMS*, Oulu, Finland, (12–16.09.2016), pp. 680–686.
27. H. Ansarifar, M. Shams, Numerical simulation of hydrogen production by gasification of large biomass particles in high temperature fluidized bed reactor, *International Journal of Hydrogen Energy*, 43 (2018), 5314–5330.
28. I. Janajreh, M. Al Shrah, Numerical and experimental investigation of downdraft gasification of wood chips, *Energy Conversion and Management*, 65 (2013), 783–792.
29. C. Ophoff, N. Ozalp, D. Moens, A numerical study on particle tracking and heat transfer enhancement in a solar cavity receiver, *Applied Thermal Engineering*, 180 (2020), 115785.

Part II

Photothermal-Conversion-Based Technologies

6 Theoretical Background

6.1 DEVELOPMENT OF TECHNOLOGY

The sun can be treated as the primary source of renewable energy, providing the possibility of using solar power, as well as that of wind and biomass. Its great potential was already recognized by ancient civilizations, which apart from treating it with reverence as something supernatural and even divine, also understood its beneficial power.

Centuries ago people learned how to use magnifying glass to concentrate solar radiation to ignite a stack of dry grass and twigs. Antique civilizations, like the Greeks or Romans, knew very well how to build houses and public buildings to utilize passive heating (large south-oriented windows) [1]. Literature even provides examples of whole metropolises designed and constructed in the way maximizing the usage of solar power to increase the indoor temperature, especially in winter [2,3].

Furthermore, ancient history provides examples of much more surprising applications of solar power, including concentrated radiation, which today is sometimes recognized as more advanced than classic solar thermal technologies. The idea of Archimedes was to use numerous polished metal plates as mirrors concentrating solar radiation on Roman warships [4].

As J. Perlin writes in his book [6], the 19th-century engineers were able to use solar thermal power to run the steam engine, and not much later (early 20s) solar power was used for irrigating plantations. The first modern-like commercial solar thermal collector ("Climax Solar-Water Heater"), invented by Clark Kemp, was introduced into the market in the 90s of the 19th century (USA) [5]. Figure 6.1 presents the commercial for this device. The beginning of the 20th century brought faster development of solar thermal collectors technology, including the incorporation of energy storage systems [6].

Increasing interest in solar thermal power generation technology led to its employment in big-scale systems, not just for heating, but primarily for electricity generation, which was accomplished using solar radiation concentration systems. The first important pilot plants based on solar thermal electric technologies were run during the early 1980s. Within ten years, their operational cost decreased by four-fifths [4].

FIGURE 6.1 Commercial for the climax solar thermal collector introduced to the market in 19th century [5].

6.2 STATISTICAL DATA

Figure 6.2 illustrates the installed capacity of solar thermal technologies in the top ten countries around the world in 2021. The largest installed capacity is observed in Spain (2304 MW) and in the USA (1496 MW). That is far more than for any other country. The third place belongs to China with 570 MW installed. Next come the countries with beneficial climate conditions, such as: Morocco (540 MW), South Africa (500 MW), India (343 MW), Israel (242 MW) and Chile (108 MW). The last two countries in this ranking are located on the Arabian Peninsula: the United Arab Emirates (110 MW) and Kuwait (50 MW). The most striking observation to emerge from the data is that the sum of the installed capacity of solar thermal technologies for Spain and the USA is much greater than the cumulative capacity for the rest of the world (2587 MW). That is due to the fact that these two countries house the world largest solar facilities, including both Concentrated Solar Thermal (CST) and Concentrated Solar Power (CSP) technologies.

Figure 6.3 compares the installed capacity of low-, medium-, or high-temperature solar thermal technologies in the world, in a period of ten years (between 2012 and 2021). Overall, the installed capacity increased over this time by a factor of 2.5: from 2.6 GW in 2012 to 6.4 GW in 2021. The most significant annual net addition was observed in 2013 and it was equal to 1.3 GW. The smallest change in the installed capacity was observed in 2017 and it was only 0.1 GW. In 2021, there was a slight decrease in cumulative installed capacity by 0.1 GW.

The most common type of solar thermal collector is the evacuated tube one, with a 71% world market share, followed by the flat plate collectors with a 23% share. Concentrated Solar Thermal (CST) and Concentrated Solar Power (CSP) technologies are still being developed and thus, these facilities are expensive. Nevertheless, a significant decline in the average global costs of concentrated solar thermal technologies was noted in the analyzed period: from 8,281 USD/kWh in 2012 to 4,581 USD/kWh in 2020.

Figure 6.4 shows the electricity generation of solar thermal technologies in 2019. The country with the largest generation in Spain (over 5600 GWh). The second place is occupied by the USA,

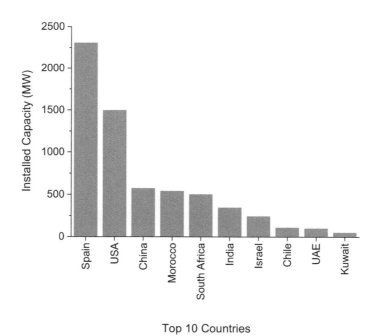

FIGURE 6.2 Top ten countries with the highest installed capacity of solar thermal technologies in 2021. (Based on Ref. [7].)

Theoretical Background 59

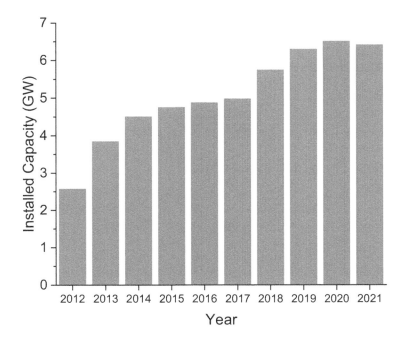

FIGURE 6.3 Cumulative installed capacity of solar thermal technologies in the world [7].

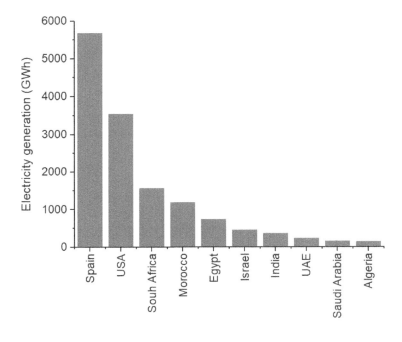

FIGURE 6.4 Top ten countries with the highest electricity generation from solar thermal technologies in 2019. (Based on Ref. [7].)

with 3,500 GWh of generated energy. The dominance of Spain and the USA is proved by their highly-developed installations and high installed capacity. Apart from these two countries, the largest energy generation is noted in South Africa, Morocco and Egypt, with values ranging from about 1,500 to 700 GWh. The remaining countries: Israel, India, the United Arab Emirates, Saudi Arabia and Algeria have the generation of 500 GWh or less. Together, these top ten countries account for 99% of the total worldwide energy generation from solar thermal technologies in 2019. Generally, high electricity generation from solar thermal installations is noted for African and Middle East countries, where climatic conditions are the most beneficial for that kind of power plants.

6.3 CLASSIFICATIONS AND CHARACTERISTICS

Regardless of technical characteristics (like non-concentrating or concentrating system) of the collector, the efficiency of conversion of solar radiation into heat depends on certain fundamental factors, one of them being a material able to absorb sunlight. The heat coming from the solar energy conversion is distributed over the absorbing object and the temperature increase rate depends on the specific heat of the absorbing material. Alternatively, the absorbed energy can be transferred to an operating medium, transporting it to a thermal storage. Of course, the absorber both absorbs and emits the energy. Although the heat release occurs by convection and radiation, contribution of these mechanisms to the heat balance varies. It depends on the device technology, ambient conditions and temperature of the heat dispersing object. The amount of solar energy potentially available for the absorbing object is a resultant of the external conditions, like geographical location, and local conditions (such as cloudiness or turbidity), month of the year, day and time, but also the absorber orientation and its shape [8,9]. The latter can be quite complicated in the case of modern solar-power-based devices. In general, heat loss can be reduced by keeping the temperature gradient between the absorber and its ambient environment small. It is partially possible due to the heat transfer to the operating medium (so the energy absorbing section cannot get significantly hot); however, as long as the absorber is in contact with the ambient environment, some undesirable heat release will still occur. To minimize this phenomenon, application of appropriate insulation techniques with transparent external covers is required.

The above-mentioned issues of the system operation change when it is required to obtain high temperature of the operating medium from the receiver (for example for solar thermal power generation using heat engines). In such a case there is no possibility to achieve satisfying thermal conditions without concentrating solar radiation on a smaller area. The challenge then is not only to construct a high temperature resistant, hermetic receiver [10–12] but also to apply an appropriate operating medium and efficient radiation-concentrating technology [4].

Comparing with non-concentrating thermal solar collectors, which can easily use glycol water as the medium (it starts to degrade when the temperature exceeds around 120°C), high-temperature concentrating ones are much more demanding in this field. Among others, hydrocarbon oils or molten salts are recommended here [9,10]. Solar radiation-concentrating systems can be based on mirrors or lenses [10–12], which allow for increasing the temperatures obtained by receivers [13,14]. It has to be noted here that both low and high-temperature solar radiation-concentrating technologies are available. The latter allows to warm up the operating medium even a couple of times more, compared with non-concentrating solar thermal collectors.

Based on the short characteristic presented above it can be concluded that there are many different criteria allowing to classify solar thermal collectors. Extensive discussion on that topic has been included in many literature sources, like Refs. [9] and [14].

Figure 6.5 presents quite an intuitive, simplified classification, assuming the possibility of concentration of solar radiation as the basic criteria of division. It has to be noted that many other features can be taken as the basis of the classification of collectors into different types. For example, we can adopt a division from the point of view of sun tracking ability (and then division into one-/two-axis tracking, different technologies allowing that) or operating medium (water, air and so on).

Theoretical Background

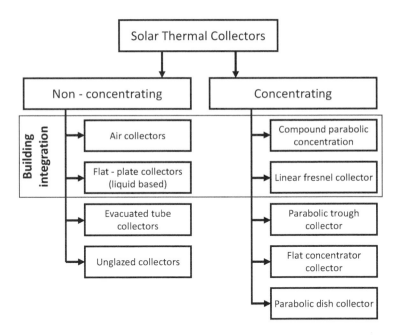

FIGURE 6.5 One of a couple of possible approaches to general classification of solar thermal collectors. (Based on Ref. [9].)

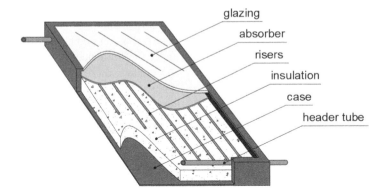

FIGURE 6.6 Construction of a flat plate solar collector.

6.3.1 FLAT PLATE SOLAR COLLECTORS

A flat plate solar collector (FPC) is a device which converts incident solar radiation into heat and is characterized by a flat, continuous absorbance area [15] (Figure 6.6). An FPC consists of a single or multiple layers transparent cover, located at the top of the collector. The glazing is covered with antireflective coating [16], which provides high absorption of visible light and low emittance in the infrared. The glazing fulfills two main roles: reduces thermal losses from the collector and simultaneously protects the absorber against the adverse weather conditions [17].

The absorber plate is an area that collects the solar energy transmitted through the transparent cover. To increase the efficiency of this process, a high conductivity absorber sheet (copper or aluminum) is additionally covered by a highly-absorptive, dark coating. Risers are welded to the bottom part of the absorber plate. The channels may be arranged in many ways, such as: harp, serpentine or winding, depending on the collector main function. The channels are filled with working fluid [18]. The two most widely used fluids are water (in hot climate) or water-glycol solution (in temperate climate where freezing is possible). In the FPC working fluid usually heats up to 80°C [19]. Then, the hot fluid is transmitted to a large diameter header tube, located at one side of the collector, which carries the heat to the storage system. The fluid, after being cooled down, is directed to the second header tube, placed at the opposite side of the collector, which delivers the cold fluid to the collector. The sides of the collector and the underside of the risers are covered with insulation. It minimizes the heat losses from the device to the ambient environment [20]. The back and the sides of the FPC are covered with a protective metal or plastic case.

FPCs are usually used for water and space heating. They also find applications in numerous industry sectors, like dairy, paper, and textile [17,21].

6.3.2 Evacuated Tube Solar Collectors

An evacuated tube solar collector (ETC) is a device which usually consists of 15–40 parallel glass tubes with a discontinuous absorber and converts solar radiation into heat [15]. Figure 6.7 shows the main components of a single tube. It consists of two concentric borosilicate glass tubes sealed at one end. The outer tube (1) is transparent to the incident solar radiation and has got antireflective coating, whereas the inner (2) is covered with both: antireflective and selective coating [16]. In some constructions, the inner tube is also transparent and the function of the absorber is performed by narrow metallic sheets.

Between the glass tubes the vacuum space (3) is created, providing excellent insulation. Therefore, the energy conversion efficiency is greater for an ETC than for an FPC [20, 22]. The heat is collected by the copper pipe (4) located inside the construction. The heat pipe is sealed at one end (similar to the glass tubes) and filled with a phase-changing material (5). This substance heats up, evaporates and moves up to the heat exchanger (6) placed inside a larger diameter header pipe, called manifold (7). Another working fluid (8) flows through the main channel and cools down the vapor. The condensation occurs and then the fluid flows down, under the influence of gravity, to the heat pipe [17]. The manifold is covered by an insulation layer (9) to reduce thermal losses and placed inside a plastic or metal case. Operation temperature of the ETC varies from 50°C to 200°C. [23]

ETCs are usually used for water and space heating, solar refrigeration and desalination. They are also applied in industrial sectors such as: dairy, chemical and oil refinery plastics which require medium-temperature heat [17,21].

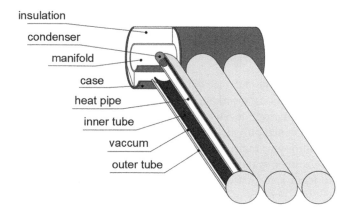

FIGURE 6.7 Construction of evacuated tube solar collector with heat pipe.

Theoretical Background

6.3.3 Other Non-concentrating Technologies

Of course FPC and ETC are the two most widespread technologies of non-concentrating solar thermal collectors. The main advantage of the two remaining representatives of this category (according to Figure 6.5: air collectors and unglazed collectors) is relative simplicity, which comes from construction and operation parameters, as well as the operating medium itself. However, for the same reasons the application of the unglazed collector or the air-based one in some systems can constitute a problem.

Unglazed collectors are simply made of plastic, usually the black one. Although such devices are capable of absorbing solar radiation efficiently (exactly due to the absence of glazing), the lack of thermal insulation results in potentially much higher heat losses, for example, compared with the flat plate and, especially, the evacuated tube collectors. As a consequence of the above-mentioned disadvantage, the achievable output temperatures in the case of this technology are quite low, allowing it to satisfy the required level at most around 30°C. However, it doesn't mean that this technology is practically useless – it can still cover the requirement for the heating of certain systems, especially outdoor ones, like swimming pools [9,24].

Solar air-based collectors can be an attractive solution for any application that doesn't require heat storage systems, analogous to the one typical for fluid-based collectors. The heated air from the collector is used directly, for example for space heating or, rather in case of industrial process, for drying. An interesting application, combining the usage of two different RES-based technologies, is using the system of air-based solar collectors for drying biomass before its thermo-chemical treatment (combustion, gasification, others) [25].

In a classic air solar collector, the air that is being heated up flows between two surfaces, one of which is an absorber. Gas flow can occur above or under the absorber face. In the first case it is necessary to provide a transparent layer separating the air from the environment. Because the air-based collectors can cover relatively big areas, like whole facades, air circulation forced by fans is introduced [9,26] instead of natural convection.

6.3.4 Concentrated Solar Power Systems

Concentrated solar power (CSP) systems are facilities which generate high-temperature heat or electricity by concentrating solar radiation onto a small area. For this purpose optical elements such as lenses or mirrors are used. Based on the reflector shape they may be divided into four categories: linear Fresnel reflectors (for linear Fresnel collectors – LFC), parabolic trough collectors (PTC), heliostat field reflectors (for heliostat field collectors – HFC) and parabolic dish reflectors (for parabolic dish collectors PDC). LFC and PTC are line-focus solutions that concentrate sunlight on a linear receiver, whereas HFC and PDL are known as point-focus solutions [27].

Regardless of the category, each concentrated solar system consists of similar elements but in different shapes and sizes. The optical element known as a concentrator follows the movement of the sun and reflects the beam of sunrays onto a receiver located at the focal point. The energetic potential of each technology is limited by the concentration ratio [27].

Parabolic dish reflectors for PDC concentrate solar radiation onto a receiver located at the focal point of the dish as shown in Figure 6.8. This construction tracks the sun position in two axes to reflect the beam directly into the thermal receiver. The absorbed energy is transferred to the working fluid circulating inside the thermal receiver. Finally, the heat is delivered to the power-conversion system [20]. PDR systems achieve a 100–1000 sun concentration ratio. Therefore, the heat receiver can reach temperatures above 1,500°C and the heat absorbed by the working medium is commonly used for electricity generation purposes. Due to its construction type, the PDC system is modular: a single facility may work standalone or be part of a larger power plant [15].

The Heliostat field collector (HFC) system consists of numerous flat mirrors arranged in a circular array (Figure 6.9). They focus radiation onto the common point where the receiver is located.

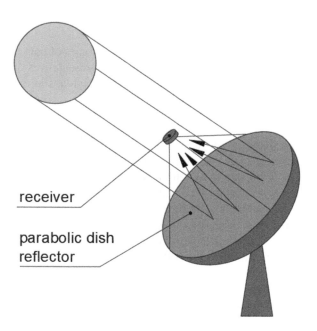

FIGURE 6.8 Construction of a parabolic dish collector system.

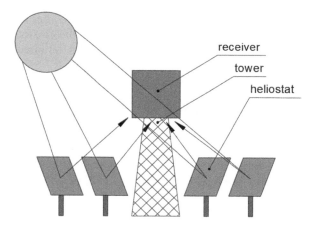

FIGURE 6.9 Construction of a heliostat field collector system.

The absorber is usually mounted at the top of a high tower, built in the center of the heliostat field [28]. High concentration ratios between 300 and 1,500 suns are achieved so the working fluid (water or molten salt) heats up to 2,000°C, which allows the generation of industrial thermal power or heat [29].

Parabolic trough collectors (PTC) are linear-focus systems. The optical element is a parabolic-shaped sheet of highly reflective material. Solar radiation is concentrated with the intensity of 15–45 suns onto a focal line, where a glass or metallic tubular receiver is placed (Figure 6.10). To increase the absorption efficiency, the casing is covered with a dark-colored, highly-absorptive coating. The heat transfer fluid flows along a pipe and gradually heats up. Because the construction is lengthwise, the sun is tracked on only one axis [30]. This type of solar-concentrating system reaches temperatures up to 400°C but also works efficiently in temperatures below 100°C.

Another type of solar-concentrating systems are linear Fresnel collectors (LFC) shown in Figure 6.11. These facilities consist of several linear optical elements located parallel and spaced

Theoretical Background 65

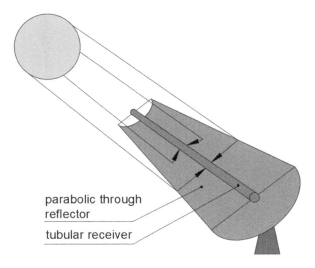

FIGURE 6.10 Construction of a parabolic trough collector system.

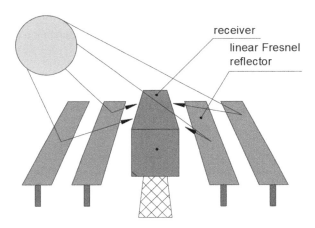

FIGURE 6.11 Construction of linear Fresnel collector system.

apart to avoid the shading effect. The Fresnel reflectors concentrate incident radiation onto a linear receiver located between them with a 10–40 suns ratio [31]. Because that only one receiver is used, the LFCs are more economical than PTCs. They reach the maximum temperature range of up to 250°C–300°C.

In fact, concentrating and non-concentrating solar thermal collector technologies can be combined into one integrated system. Figure 6.12 presents an example of a small-scale cogeneration system containing classic flat plate collectors and a high-temperature receiver, heated up by a parabolic dish reflector. The system has been constructed at the AGH University of Science and Technology (Krakow, Poland) at the Department of Sustainable Energy Development.

In the presented installation the operating medium is preheated by non-concentrating solar thermal collectors, then it is transported to a high-temperature receiver, where further heating of the medium takes place. The thermal output of the receiver provides heat which is enough to power the absorption heat pump. Additionally, the temperature of the working fluid at the output of the chiller is high enough to allow for storing the heat in an accumulation system [32].

Further subsections describe briefly the construction of the most widespread types of solar thermal collectors.

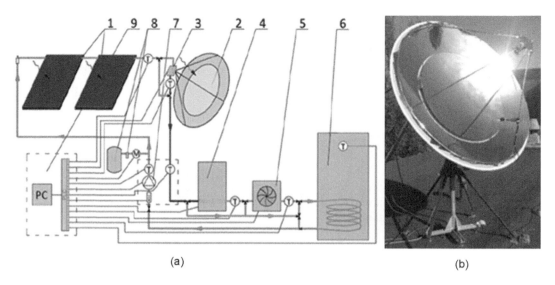

FIGURE 6.12 (a) Schematic of an exemplary cogeneration system integrating concentrating and non-concentrating thermal solar technologies: 1) flat plate solar collectors, 2) Parabolic dish concentrator, 3) high-temperature receiver, 4) absorption chiller, 5) fan-based cooler, 6) buffer tank, 7) medium pump, 8) pressurizer, 9) measurement – control system. (b) Picture of the collectors, high-temperature receiver and concentrator in the real installation.

6.4 FUNDAMENTALS OF ENERGY CONVERSION AND BALANCE

6.4.1 Flat Plate Collector

The goal of this section is to give general background necessary for the development of more advanced numerical models and to provide various relations required to determine the amount of useful energy collected, as well as the influence of different constructional parameters on the performance of a solar collector.

The theory of a flat plate solar collector (FPC) includes the relationship between solar radiation as an energy source, energy losses to the ambient environment and the energy available for the user. It is based on physical laws concerning solar radiation and heat transfer [15].

The first step is to calculate an absorber temperature profile – function $T(x)$ along the cross section of the profile as shown in Figure 6.13. The x-axes zero point is located at the same distance from two neighboring pipes, the thermal contact is represented by a "welding" (bond); however, the consideration is also valid for other types of thermal contact and can be represented by the coefficient of heat transfer of the contact and its thickness.

Let's consider a part of the absorber (shown in Figure 6.13 inside the circle) – a detailed diagram of the calculations is as follows (Figure 6.14).

Heat transfer inside the absorber is given by the thermal conductivity coefficient λ_p, heat losses by the overall coefficient of heat losses by U_L and solar radiation is given by S_a.

The energy balance is as follows:

- Amount of solar energy, according to formula: $S \cdot \Delta x$, where S is the intensity of solar radiation (W/m²)
- Energy losses to the ambient environment: $U_L \cdot (T_p - T_\infty)$
- Heat transfer along an absorber forced by the temperature difference in points x and $x + \Delta x$ is: $T(x) - T(x + \Delta x)$.

Theoretical Background

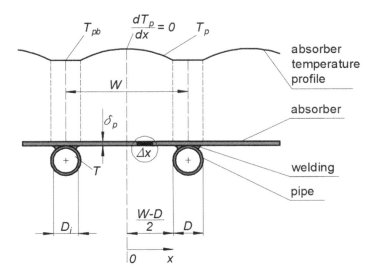

FIGURE 6.13 Diagram for the calculation of a flat plate collector. The following notation is used: δ_p-thickness of the absorber, W-distance between pipes, D-external pipe diameter, D_i – internal pipe diameter, δ_s – average thickness of the contact layer.

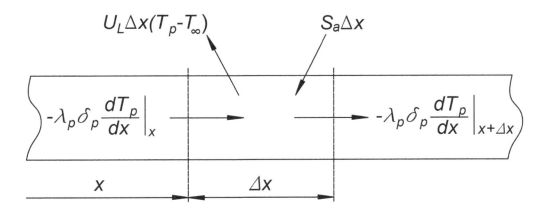

FIGURE 6.14 Heat balance scheme (it is assumed that the length of this element is equal 1 m, therefore the variable Δx represents a surface element).

Grouping all these terms we obtain the following equation for the energy balance:

$$S\Delta x - U_L \Delta x \cdot (T_p - T_\infty) + \left(-\lambda_p \delta_p \frac{dT_p}{dx}\right)_x - \left(-\lambda_p \delta_p \frac{dT_p}{dx}\right)_{x+\Delta x} = 0 \qquad (6.1)$$

Applying the definition of a derivative to the two last terms of (6.1) we can rewrite the formula in the following way:

$$\frac{d^2 T_p}{dx^2} = \frac{U_L}{\lambda_p \delta_p} \cdot \left(T_p - T_\infty - \frac{S}{U_L}\right) \qquad (6.2)$$

It is visible that the formula (6.2) is the second order differential equation. Its basic properties are:

- $T_p(x)$ – fin temperature is symmetrical to $x = 0$,
- The temperature of a fin base, T_{pb} is known.

To solve equation 6.2 it is first necessary to simplify its mathematical form. We insert new variables:

$$\chi = T_p - T_\infty - \frac{S}{U_L} \tag{6.3}$$

$$m^2 = \frac{U_L}{\lambda_p \delta_p} \tag{6.4}$$

Applying equations 6.3 and 6.4 to 6.2, we obtain a new form of equation 6.2:

$$\frac{d^2\chi}{dx^2} - m^2 \chi = 0 \tag{6.5}$$

The solution according to appropriate rules gives:

$$\frac{T_p - T_\infty - \dfrac{S}{U_L}}{T_{pb} - T_\infty - \dfrac{S}{U_L}} = \cosh\left[\frac{m(W-D)}{2}\right] \tag{6.6}$$

Using equation 6.6, several important parameters can be derived and the behavior of a flat plate collector (FPC) can be examined. Below, some selected parameters are briefly presented:

- Fin efficiency

  It is a relation of the heat taken by a fin for the real profile of the absorber temperature $T_p(x)$ to the heat which would be taken, if the thermal conductivity of the plate material was infinite $\lambda_p \to \infty$ (which is equivalent to the constant temperature of the absorber $T_p(x) = T_{pb} = \text{const}$):

$$F = \frac{tgh\left[m \cdot \dfrac{W-D}{2}\right]}{m \cdot \dfrac{W-D}{2}} \tag{6.7}$$

Function F has the following shape:

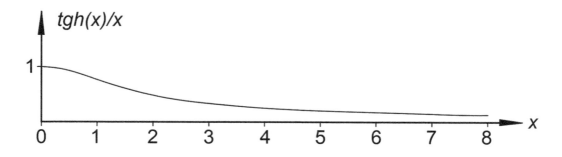

FIGURE 6.15 Function $tgh(x)/x$ for x from zero to eight.

Theoretical Background

It is visible that the efficiency tending to unity can be obtained for $x \to 0$, which means that $m \cdot (W - D) \to 0$. This is valid for $W=D$, and it means that the distance between pipes is equal 0 (all the absorber surface is densely covered by pipes) or $m \to 0$, which is equivalent to perfect thermal insulation ($U_L \to 0$), a very high heat conduction coefficient of the absorber material ($\lambda_p \to \infty$ and/or a large thickness of the absorber ($\delta_p \to \infty$). In practice it is not possible to fulfill all these conditions; however, it is possible to obtain the efficiency of 80% or more.

- Collector efficiency

A collector efficiency can be interpreted as a relation of heat losses from the absorber to heat losses from the working medium. The following factors have to be included – conduction of the thermal contact between the absorber materials and the external surface of the pipes, and the heat transfer to the working medium.

$$f = \frac{\dfrac{1}{U_L}}{W \left\{ \dfrac{1}{U_L [D + (W-D) \cdot F]} + \dfrac{\delta_s}{\lambda_s D} + \dfrac{1}{\pi D_i \alpha} \right\}} \quad (6.8)$$

where: λ_s represents the heat transfer coefficient of the thermal contact and α is the convective heat transfer coefficient (from the internal side of pipes to working liquid).

The maximal efficiency can be obtained when the following conditions are fulfilled:
- $F \to 1$
- $\lambda_s \to \infty$
- $\alpha \to \infty$

The first condition illustrates that fin efficiency should tend to unity, the second one that thermal contact between an absorber and pipes should be as high as possible, and the last one that the coefficient of the heat transfer from the internal surface of the pipe to working medium should be as high as possible. If the above-mentioned conditions are fulfilled, the collector efficiency tends to unity regardless of the value of the U_L coefficient.

- Heat removal factor

It is a relation of the actual efficient heat power of the collector to the power obtained if the absorber temperature was equal to the temperature of the working medium, and is given by the formula:

$$F_R = \frac{\dot{m} c_p}{A \cdot U_L} \cdot \left\{ 1 - \exp\left(\frac{-A U_L f}{\dot{m} c_p} \right) \right\} \quad (6.9)$$

where: $\dot{m}$ is the working medium flow, c_p is its specific heat and A is the active surface of the collector. It is visible that for the following cases:
a. Very large mass flow ($\dot{m} \to \infty$), $F_R = f$, which means that the heat removal efficiency is equal to the fin efficiency (in practice for large flows).
b. No flow ($\dot{m} = 0$), $F_R = 0$, which means that the medium is simply equal to 0.
Similar cases can be considered for small/large specific heat values, insulation efficiency, etc.

- Useful heat power

  It is a product of the mass flow, specific heat and the temperature outlet-inlet difference and in the considered case it takes the form of the following equation:

$$\dot{Q} = A \cdot F_R \left[S - U_L (T_i - T_a) \right] \tag{6.10}$$

This equation is often called Hottel-Whillier-Bliss (HWB) equation.

- Temperature of the outlet medium

  This temperature is given by a relation between useful heat power and working medium flow. It is described by the following formula:

$$\frac{T_o - T_\infty - \dfrac{S}{U_L}}{T_i - T_\infty - \dfrac{S}{U_L}} = \exp\left(\frac{-AU_L f}{\dot{m} c_p} \right) \tag{6.11}$$

where: T_i, T_o – inlet and outlet temperatures, respectively.

The following specific cases can be considered:
a. Very large flow $(\dot{m} \to \infty)$, then $T_o = T_i$ (no temperature increase)
b. No flow $(\dot{m} = 0)$, then

$$T_o = T_\infty + \frac{S}{U_L} \tag{6.12}$$

This case is called "stagnation", we can see that the temperature can reach high values, especially for good overall thermal efficiency of the collector (small U_L), e.g. for high solar radiation intensity equal to 1000 W/m² and thermal insulation of 3 W/(m² K) the temperature inside the solar collector can rise to 300°C! This may lead to collector damage.

It is often convenient to express the performance of solar collectors in terms of efficiency defined as:

$$\eta_{th} = \frac{\dot{Q}}{S \cdot A} = \frac{AF_R \left[S - U_L (T_p - T_\infty) \right]}{S} \tag{6.13}$$

Generally, this efficiency depends on several factors, like: the construction of the collector, its operation regime and ambient conditions. Some simplified considerations related to this efficiency will be presented further.

To measure the performance of a solar collector, it is easier to assess the absorber plate temperature T_p as the average of the inlet and outlet fluid temperature.

The solar collector efficiency can also be defined as a ratio of the useful energy gain to solar radiation entering the collector aperture area, therefore, solar collector efficiency can be written as:

$$\eta = \frac{\dot{Q}}{AS} = \frac{\dot{m} c_p (T_o - T_i)}{AS} \tag{6.14}$$

Equations 6.13 and 6.14 are equivalent; however, the application of (6.13) is somewhat problematic in practice as the determination of T_p is inconvenient.

Theoretical Background

In practice, the collector efficiency is often plotted against $(T_m - T_\infty)$, then:

$$\eta = \eta_0 - a_1 \frac{(T_m - T_\infty)}{S} - a_2 \frac{(T_m - T_\infty)^2}{S} \tag{6.15}$$

where T_m is the average temperature of the collector (medium), $\eta_0 = \frac{\dot{Q}}{AS}$ is called optical efficiency of a collector (for $\Delta T = 0$). The term $\frac{(T_m - T_\infty)}{S}$ is called reduced temperature difference. The factor η_0, coefficients a_1 and a_2 are determined experimentally using data fitting. These coefficients and the efficiency plots are often applied for the comparison of various types of collectors. Some examples will be given in Section 6.4.2.

6.4.2 Evacuated Tube Collectors

All the equations derived for flat plate collectors (FPC) are valid for evacuate tube collectors (ETC). Compared to FPCs, ETCs are better insulated as the vacuum insulation is used (smaller value of U_L), but optical efficiency η_0 is smaller for ETCs due to the thicker glass of the vacuum-resistant tubes.

The energy loss from the absorber plate to the glass tube is by radiation only, since conduction and convection heat transfer are inhibited by the vacuum inside the glass tube. The energy loss from the glass tube to the ambient environment is by radiation and convection.

The radiation loss from both sides of the absorber plate to the glass tube can be expressed as [33]:

$$Q_{loss} = \frac{2\sigma(T_p^4 - T_a^4)}{\frac{1-\epsilon_p}{\epsilon_p A_c} + \frac{1}{A_p} + \frac{1-\epsilon_p}{\epsilon_g A_g}} = h_g(T_g - T_a) + \epsilon_g \sigma(T_g^4 - T_s^4) \tag{6.16}$$

where ϵ_p and ϵ_g are the emissivity of the absorber plate and glass tube, respectively, h_g is the convection heat transfer coefficient from the glass tube to the ambient air, and A_p and A_g are the areas of the absorber plate and glass tube, respectively.

A comparison of FPC & ETC can be performed using equation 6.15 and comparing coefficients a_1, a_2 and the plot for parameter $(T_m - T_a)$. Figure 6.16 presents a comparison of typical collectors available in Polish market.

It is visible that an FPC has better optical efficiency, but due to poorer insulation, its efficiency decreases fast with the temperature difference increase.

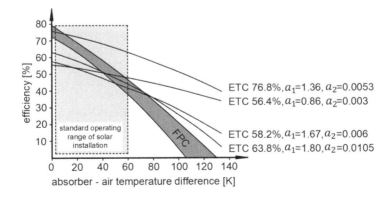

FIGURE 6.16 Comparison of PCS and ETC available on the Polish market, optical efficiencies and a_1, a_2 coefficients are given.

6.4.3 CONCENTRATING SOLAR COLLECTORS

The basic parameter describing concentrating systems is the gain of solar radiation due to concentration. The concentration level C can be defined as:

$$C = A_{ap}/A_{rec} \tag{6.17}$$

where A_{ap} is the surface of an optical aperture, A_{rec} is the surface of a receiver.

All the equations for a collector working in average radiation conditions can be adapted here with replacing intensity of solar radiation S by $G = C \cdot S$.

Due to a very large working medium temperature (big difference between inlet and outlet temperatures) some simplification of the equations can be done.

We can use the equations presented in the further part of the subchapter to carry out practical calculations with sufficient accuracy.

Thermal efficiency of concentrating solar collectors (CSC) can be calculated using the formula adopted from (6.13):

$$\eta_{th} = \frac{A_{rec}F_R\left[G - U(T - T_a)\right]}{GA_{rec}C} \tag{6.18}$$

The working medium temperature T_o at the outlet of the receiver is given by the following equation:

$$T_o = T_i + \frac{Q}{\dot{m}c_p} \tag{6.19}$$

where: T_i – inlet temperature, Q – amount of heat transferred to the medium, $\dot{m}$ – mass flow of the medium and c_p – specific heat of the medium.

Q is given by the formula (similar to the formula 6.10 with S replaced by G):

$$Q = A_{rec}F_R \cdot \left[G - U_L(T_i - T_o)\right] \tag{6.20}$$

The F_R coefficient is given by equation 6.9.

The normal to the concentrator surface was, of course, in the direction of the incident solar radiation, which was the result of the applied solar tracking. This makes the concentrating solar collectors significantly different from typical collectors which are unmovable and where the incident solar radiation angle is variable. Also, in practice, solar concentrators use only a fraction of total solar radiation, which is the direct radiation (and noted as G). Typical relation between direct and diffuse solar radiation for a sunny day is presented in Figure 6.17. Direct radiation has an intensity a few times larger.

Another important difference is that in most cases it is impossible to insulate well the receiver, especially the part exposed to the source of concentrated solar radiation, due to very high temperatures. Therefore, U_L coefficient will be larger than for "normal" collectors and convective heat losses dominate (radiative heat losses cannot be neglected). Convective heat losses are sensitive to ambient air velocity (hot surface is exposed to ambient air). Therefore, this topic will be described here in greater detail [34].

The U_L coefficient can be calculated taking into account the radiative (described by the h_r coefficient) and convective (given by the h_k coefficient) heat losses according to the formulas:

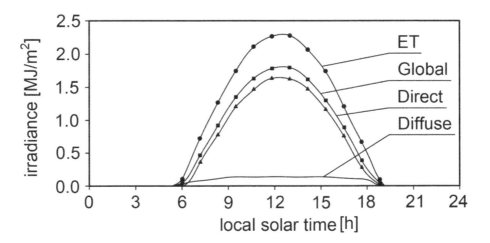

FIGURE 6.17 Exemplary direct and diffuse solar radiation for a sunny day.

$$U_L = h_r + h_k \tag{6.21}$$

$$h_r = \sigma \epsilon_p \left(T_m^2 + T_s^2\right)(T_m + T_s) \tag{6.22}$$

$$h_k = 5.7 + 3.8 \cdot v \tag{6.23}$$

where: σ – Stefan-Boltzmann constant, ϵ_p – the receiver surface emissivity, T_m – average medium temperature inside the receiver, T_s – the so called effective sky temperature, v – wind velocity. Effective sky temperature can be expressed using ambient temperature e.g. as:

$$T_s = 0.0522 T_a^{1.5} \tag{6.24}$$

The average medium temperature T_m was given as $T_m = (T_o + T_{in})/2$ (arithmetic average of the inlet and the outlet temperatures). In the algorithms based on flat low-temperature solar collectors, the outlet medium temperature is calculated in two steps:

- initially the heat power Q transferred to the receiver is calculated assuming that U_L is calculated for non-heated medium parameters,
- the outlet temperature T_o is calculated and subsequently, the average temperature T_m and the U_L coefficient are updated.

However, for concentrated solar radiation, the temperature of surface may be higher (even by several tens of degrees, depending on the medium flow) than the average medium temperature T_m. This leads to an underestimation of the radiative heat losses, but the final error is negligible.

According to the definition of the Incidence Angle Modifier (IAM), this parameter defines the ratio of the collector efficiency at any angle of incidence to that at normal incidence.

It is important to note that the IAM for that case can be taken as 1.0, because there is no glass cover and the maximum angle of reflected solar radiation from the concentrator is practically no more than ~50° (with the geometry of concentrator and receiver). For this angle, even if one glass cover is present, the effect is marginal (typically less than 0.1), thus, can be neglected.

REFERENCES

1. J. W. Ring, Windows, baths, and solar energy in the roman empire, *American Journal of Archaeology*, 100 (4) (1996), 717–724.
2. Passive solar history, http://californiasolarcenter.org/old-pages-with-inbound-links/history-passive/ (last access: 14.04.2022).
3. Ancient solar: How ancient civilizations harnessed the sun's energy, https://cleanchoiceenergy.com/news/Ancient_Solar (last access: 14.04.2022).
4. T. B. Johansson et al., *Renewable Energy. Sources for Fuels and Electricity*, Island Press, Washington, 1993.
5. L. Szabó, The history of using solar energy, *The 7th International Conference on Modern Power Systems (MPS)*, Cluj-Napoca, Romania, 2017.
6. J. Perlin, *Let it Shine. The 6,000-Year Story of Solar Energy*, New World Library, Novato, 2013.
7. Statistics Data of the International Renewable Energy Agency, www.irena.org (last access: 01.05.2022).
8. B. Sørensen, *Renewable Energy. Its Physics, Engineering, Environmental Impacts, Economics & Planning*, Academic Press, Cambridge, 2000.
9. L. Evangelisti, R. De Lieto Vollaro, F. Asdrubali, Latest advances on solar thermal collectors: A comprehensive review, *Renewable and Sustainable Energy Reviews*, 114 (2019), 1–20 (109318)
10. F. Wang et al., Progress in concentrated solar power technology with parabolic trough collector system: A comprehensive review, *Renewable and Sustainable Energy Reviews*, 79 (2017), 1314–1328.
11. F. Q. Wang et al., Transient thermal performance response characteristics of porous–medium receiver heated by multi–dish concentrator, *International Communications in Heat and Mass Transfer*, 75 (2016), 36–41.
12. J. Li, Scaling up concentrating solar thermal technology in China, *Renewable and Sustainable Energy Reviews*, 13 (2009), 2051–2060.
13. E. Przenzak, M. Szubel, M. Filipowicz, The numerical model of the high temperature receiver for concentrated solar radiation, *Energy Conversion and Management*, 125 (2016), 96–106.
14. S. Gorjian et al., A review on recent advancements in performance enhancement techniques for low – temperature solar collectors, *Energy Conversion and Management*, 222 (2020), 113246.
15. S. A. Kalogirou, Solar thermal collectors and applications, *Progress in Energy and Combustion Science*, 30 (2004), 231–295.
16. Y. Tian, C. Y. Zhao, A review of solar collectors and thermal energy storage in solar thermal applications, *Applied Energy*, 104 (2013), 538–553.
17. L. Kumar, M. Hasanuzzaman, N. A. Rahim, Global advancement of solar thermal energy technologies for industrial process heat and its future prospects: A review, *Energy Conversion and Management*, 195 (2019), 885–908.
18. J. A. Duffie, W. A. Beckman, *Solar Engineering of Thermal Processes*, 4th edition, John Wiley & Sons, Inc., Hoboken, NJ, 2013.
19. K. M. Pandey, R. Chaurasiya, A review on analysis and development of solar flat plate collector, *Renewable and Sustainable Energy Reviews*, 67 (2017), 641–650.
20. A. Jamara et al., A review of water heating system for solar energy applications, *International Communications in Heat and Mass Transfer*, 76 (2016), 178–187.
21. A. A. El-Sebaii, S. M. Shalaby, Solar drying of agricultural products: A review, *Renewable and Sustainable Energy Reviews*, 16 (1) (2012), 37–43.
22. L. Ayompe et al., Comparative field performance study of flat plate and heat pipe evacuated tube collectors (ETCs)for domestic water heating systems in a temperate climate. *Energy*, 36 (2011), 3370–3378.
23. M. A. Sabiha et al., Progress and latest developments of evacuated tube solar collectors, *Renewable and Sustainable Energy Reviews*, 51 (2015), 1038–1054.
24. R. S. Gonçalvesa, A. I. Palmero-Marrerob, A. C. Oliveira, Analysis of swimming pool solar heating using the utilizability method, *Energy Reports*, 6 (2020), 717–724.
25. *Biomass in Small-Scale Energy Applications. Theory and Practice*, edited by: M. Szubel, M. Filipowicz, CRC Press Taylor & Francis Group, Boca Raton, FL, 2019.
26. Z. Dong et al., Thermal economic analysis of a double-channel solar air collector coupled with draught fan: Based on energy grade, *Renewable Energy*, 170 (2021), 936–947.
27. A. G. Fernandez et al., Mainstreaming commercial CSP systems: A technology review, *Renewable Energy*, 140 (2019), 152–176.
28. M. I. Hussain, C. Ménézo, J.-T. Kim, Advances in solar thermal harvesting technology based on surface solar absorption collectors: A review. *Solar Energy Materials and Solar Cells*, 187 (2018), 123–139.

29. D. Barlev, R. Vidu, P. Stroeve, Innovation in concentrated solar power, *Solar Energy Materials and Solar Cells*, 95 (10) (2011), 2703–2725.
30. A. Fernández-García et al., Parabolic-trough solar collectors and their applications, *Renewable and Sustainable Energy Reviews*, 14 (7) (2010), 1695–1721.
31. S. Suman, M. K. Khan, M. Pathak, Performance enhancement of solar collectors – a review, *Renewable and Sustainable Energy Reviews*, 49 (2015), 192–210.
32. K. Papis-Fraczek, M. Żołądek, M. Filipowicz, The possibilities of upgrading an existing concentrating solar thermal system — case study, *Energy Reports*, 7 (3) (2021), 28–32.
33. A. M. El-Nashar, *Evacuated Tube Collectors, Renewable Energy Systems and Desalination* – Vol. II – Evacuated Tube Collectors, e-book available at http://www.desware.net/ (last access: 14.04.2022).
34. S. Baljit et al., Mathematical modelling of a dual-fluid concentrating photovoltaicthermal (PV-T) solar collector, *Renewable Energy*, 114 (2017), 1258–1271.

7 Tutorial 1 – Flat-Plate Solar Collector

7.1 EXERCISE SCOPE

The goal of this exercise is to design a model of the simplified flat-plate solar collector, as in Figure 7.1, which consists of an absorber plate (1) and the working fluid (2) receiving and transporting heat below the absorber. This model allows to observe the medium flow characteristics and temperature distribution in the solid and the fluid domain, as well as to calculate the collector thermal power and efficiency.

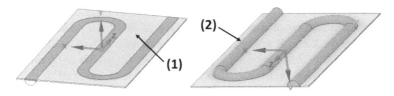

FIGURE 7.1

Try to recall from the theoretical part of this chapter (Section 6.3.1) the construction details of FPC's, as well as the fundamentals of energy conversion and balance (Section 6.4.1). The absorber temperature distribution and its thermal efficiency depend on the number and the arrangement of the working fluid tubes. The considered model allows to study these relations.

Among others, this exercise can teach you how to:

- create complex-shape geometries using the sweep pulling mode in *Ansys SpaceClaim DirectModeler*;
- achieve good quality mesh cells by an appropriate division of spatial geometry and application of the advanced meshing methods in *Ansys Meshing*;
- create new material (working fluid) with defined properties, using functionalities of the *Ansys CFX* solver;
- determine the FPC thermal power using expressions in *Ansys Results*;
- display different types of visualizations in this postprocessing module.

Pay special attention to the impact of spatial geometry division on the computational grid characteristics and the specifics of defining boundary conditions in the *Ansys CFX* solver.

All of the images included in this tutorial use courtesy of ANSYS, Inc.

7.2 PREPROCESSING – GEOMETRY

Find the *Geometry* module in the LHS toolbox, as in Figure 7.2. Then drag (*LMB*) and drop it in the project workspace. The *Geometry* module titled *A*, as in the figure, should appear in the *Project Schematic* window. The geometry module allows to create a spatial model of the computational domain, using two different tools: *Ansys SpaceClaim Direct Modeler (SCDM)* or *Ansys DesignModeler (DM)*. To open the first one, click *2xLMB* on *A2 (1)* cell and wait a moment for the *SCDM* launch.

DOI: 10.1201/9781003202226-9

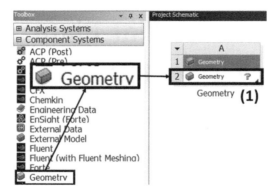

FIGURE 7.2

Like in the case of many other CAD tools, creating spatial objects in *SCDM* consists in making a profile sketch and its pulling. Let's start with creating an absorber plate shape. Select the ***Rectangle (1)*** tool from the ***Create*** section of the top toolbar. One of the three planes available for sketching now should appear on *X* and *Z* axes of the global coordinate system triad. Normal view of the sketching plane, as in Figure 7.3, can be set using ***Plan View (2)*** button or by pressing *V* key from the keyboard. Click the ***Define rectangle from center (3)*** option and start drawing the rectangle from the global coordinate center *(4)*. Dimensions can be introduced from the keyboard during sketching the object (switching between dimensions by the ***TAB*** key, confirming by the ***Enter*** key) or after sketching, using the ***Dimension (5)*** tool. Remember to click ***End Sketch Editing (6)*** when the rectangle sketch is complete.

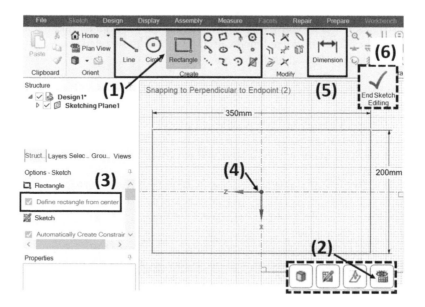

FIGURE 7.3

3D operations are available in the ***Design (1)*** tab on the top of the *SCDM* window. Click ***Pull (2)*** tool thumbnail to extrude the created rectangle sketch. Then click and keep ***LMB*** on the rectangle surface (surface is created automatically when you click the ***End sketch editing*** button and it is visible in the structure window on LHS of the *SCDM* window, as in Figure 7.4). Move the mouse a bit up (do not release ***LMB*** now!) – you'll see how the 3D object is "growing". Now release the mouse

Tutorial 1 – Flat-Plate Solar Collector

button and set *2 mm* height. Confirm it by *Enter*. Now the first 3D object (*Solid*) can be seen in the *Structure (4)* window.

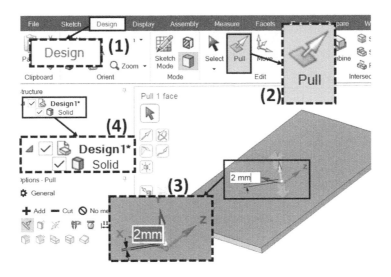

FIGURE 7.4

To create a working medium domain below the absorber plate, another sketch has to be created. First, select the *Construction line (1)* tool from the *Create* section (*Sketch* tab), as in Figure 7.5. Then select an appropriate sketching plane by clicking *1xLMB* on the thin side wall *(2)* of the absorber. A checkered sketching plane as in Figure 7.5 should appear. Triad of the global coordinate system *(3)* might be helpful in determining the appropriate side for sketching the fluid domain profile.

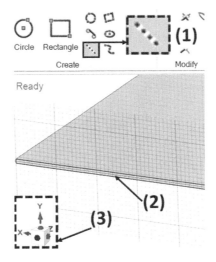

FIGURE 7.5

Create a *25 mm* long construction line *(1)* on the bottom edge of the plate side wall as in Figure 7.6. A construction line is useful to mark the arc midpoint. Select the *Sweep arc* sketching tool (from the top toolbar, *Create* section). First, click *1xLMB* on the construction line end point (arc midpoint), then the arc start point and finally the endpoint, to create the arc according to *(2)*. Create line *(3)* between the arc start point and end point.

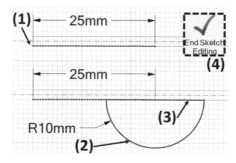

FIGURE 7.6

Have a look at the **Structure** *(1)* window. You can find three different objects (or an object group) there, as in Figure 7.7. **Solid** is the 3D collector plate. **Surface** relates to the just created fluid profile *(2)*. **Curves** *(3)* come from certain sketching operations and can be removed from the project now – select **Curves** by *1xLPM* and press **Delete** on the keyboard (it is not the same as just deactivating **Curves** in the **Structure** window!).

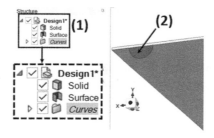

FIGURE 7.7

It is time to create a path as in Figure 7.8, which will be then used as a trajectory to pull the 3D working medium domain. Use the already acquired knowledge to activate a new sketching plane on the bottom surface of the plate (have a look at the coordinate system in the bottom left corner of the figure – sometimes it is convenient to move the triad a bit to ease orientation in space). The previously created construction line (arc midpoint) is still visible, so use it to start sketching the first straight line *(1)*. Construction lines *(2, 3)* might be useful while sketching arcs. Remember to click **End Sketch Editing** *(4)* at the end.

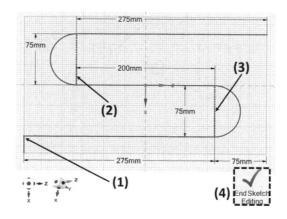

FIGURE 7.8

Tutorial 1 – Flat-Plate Solar Collector

When the sketching is over, the new curve group appears in the *Structure* window, as in Figure 7.9- it is a set of curves creating the pulling trajectory. Select again the *Pull (1)* tool (*Design* tab) and click *1xLMB* on the pulling profile *(2)* – the semicircle. Then click the *Sweep (3)* option and select all of the trajectory lines and arcs *(4)*. Although you can click sections of the trajectory one by one with *LMB+Ctrl*, it is faster to click *2xLMB* on any line or arc – then all the connected trajectory sections will be selected together automatically. Before you pull the fluid domain, click the *No Merge (5)* option – otherwise, as a result of pulling, you will get solid and" fluid bodies merged (you can try and then click *Ctrl+Z* to undo). Then hold (LMB) the yellow arrow (gray arrow) (6) that has appeared on the semicircle and pull it through the whole trajectory.

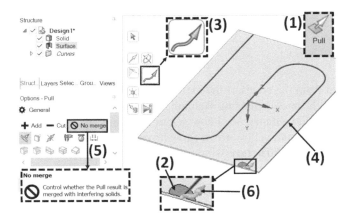

FIGURE 7.9

Have a look at the *Structure (1)* window (Figure 7.10). After pulling for the second time, two solids are available there: the collector plate *(2)* and the working fluid *(2)* underneath *(3)*. As previously, there are also certain curves (coming from sketching) that should be deleted now *(4)*.

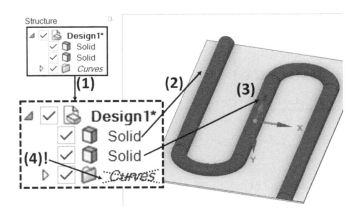

FIGURE 7.10

A brand new set of curves is required now, but it is unnecessary to sketch anything. First, deactivate the first *Solid* in the *Structure* window (click the small square next to the first *Solid* thumbnail). The collector plate should disappear – as in Figure 7.11. Then select all the top edges of the fluid domain (one by one with *LMB+Ctrl* or just by *2xLMB*). Copy your selection by *Ctrl+C → Ctrl+V*. New curves should appear in the *Structure* window, as shown in the figure.

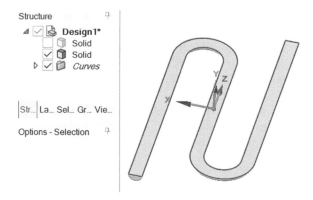

FIGURE 7.11

The curves copied from the edges of the existing spatial object can be used as a profile for the pulling operation, like the ones created by sketching. Both domain bodies (both **Solid** objects in the **Structure** window) should be deactivated now *(1)*, as in Figure 7.12. Click the **Pull** *(2)* tool again. Select all curves, hold the small yellow arrow (gray arrow) and pull curves up by *2 mm (3)*. The new **Surface** should appear in the **Structure** window. Pulling has resulted in the creation of something that can be compared to "a cookie cutter" which has the shape of our working fluid domain.

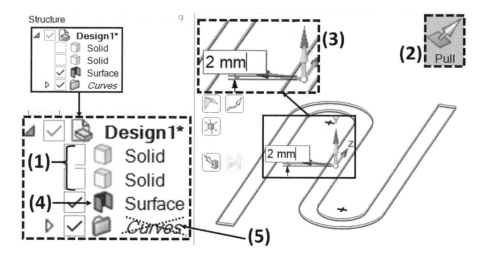

FIGURE 7.12

To get a high-quality mesh and to reduce the cell number, it is recommended to cut the absorber plate into several parts reflecting the shape of the working fluid domain underneath the plate The "Cookie cutter" **Surface** will be applied here. Select the **Combine** *(1)* tool (Figure 7.13) from the top toolbar of the **Design** tab. Use the **Select Target** *(2)* option to select (*1xLMB*) the cutting target – the whole absorber plate *(3)*. Then click **Select Cutter** *(4)* – a saw thumbnail. Click *1xLMB* on the **Surface** *(5)* in the **Structure** window. Now it is recommended to find and click the **Select** button on the top toolbar to deactivate the **Combine** tool (you can also press the **Escape** key on the keyboard twice). Delete the **Surface** *(6)* from the **Structure** window. Note that additional solids have appeared *(7)*. Activate all of them now.

Tutorial 1 – Flat-Plate Solar Collector

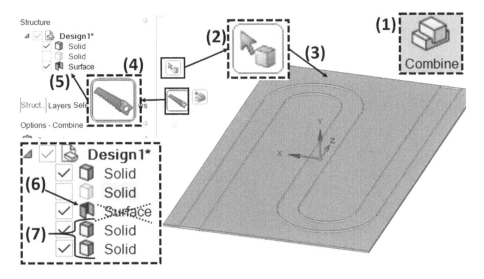

FIGURE 7.13

Figure 7.14 presents now all the parts creating spatial geometry of the considered flat-plate collector. Parts (a) to (c) come from the Combine operation and create the absorber plate. Part (d) is the working fluid.

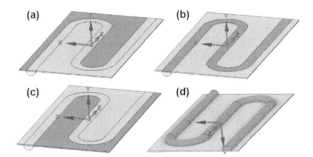

FIGURE 7.14

Two additional divisions will be done now. This time two new planes have to be created, as presented in Figure 7.15 – they will be used as cutting tools. Select *Plane (1)* tool from the *Create* section of the *Design* tab. Move the mouse cursor to any one of the arc edges visible in the place of joining the straight section of the fluid domain with the elbow *(2)*. You'll see that *SCDM* suggests where the new plane can be created. If the plane looks like one of the two in Figure 7.15, just click *1xLMB*. The new plane covers two contacts of the elbow and the straight fluid sections *(3)*. Now create the second plane analogously. Two planes should appear in the *Structure (4)* window.

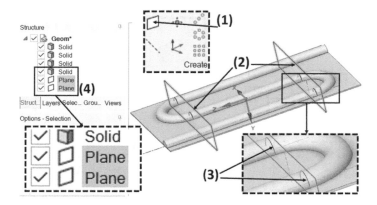

FIGURE 7.15

Select the ***Split Body*** tool from the ***Intersect*** section of the ***Design*** tab (next to the ***Combine*** tool). Click ***Select Target (1)*** option, as in Figure 7.16 and select all the parts of the geometry *(2)* ***with the Ctrl key***. The cursor thumbnail switches automatically to the saw – the ***Select Cutter (3)*** mode is active (if not – click the saw thumbnail). Now click *1xLMB* on one of the two created planes *(4)*. The first cut is done. Click the ***Select*** tool from the ***Edit*** section. Then click again the ***Split Body*** tool, select an appropriate target and cut off the second part using the second plane. As a result of applying the ***Split Body*** tool, two external individual parts should be created *(5)*. Of course, the absorber plate (including the part above the duct) is additionally divided (Figure 7.14).

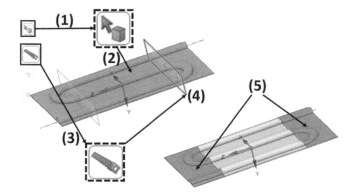

FIGURE 7.16

This project assumes a consistent mesh for the whole geometry (solid and fluid domains). Thus, it requires application of shared topology. Click on the ***Geom* (1)*** component in the ***Structure*** window, as in Figure 7.17. Then have a look at the ***Properties*** table at the bottom. Find the ***Share Topology*** setting in the ***Analysis*** tab and change the option to ***Share (2)***.

Tutorial 1 – Flat-Plate Solar Collector

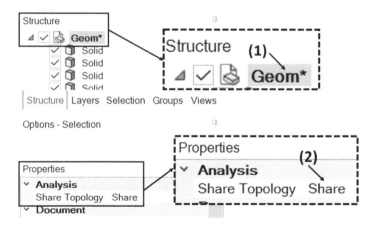

FIGURE 7.17

The basic solar thermal collector considered in frames of the exercise is based on the working fluid duct of a meander shape. You can use the already acquired skills to design another project as in Figure 7.18: the absorber plate *(1)* with the harp shape duct *(2)*.

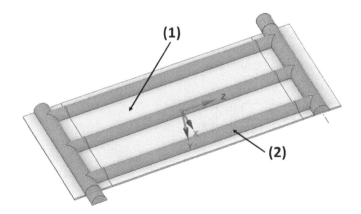

FIGURE 7.18

To make it easier to compare the selected results of the simulations, you can apply certain analogous dimensions – an example is presented in Figure 7.19. As in case of the meander duct, the harp duct includes three straight sections, deployed at the same distance as in the first design. The absorber plate dimensions are the same as well. Furthermore, the same shape and dimensions of the working fluid domain cross section are applied.

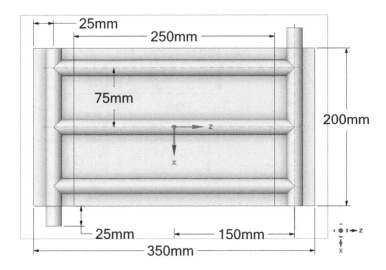

FIGURE 7.19

Remember to cut the geometry into several parts to achieve better meshing conditions. Figure 7.20 presents an exemplary set of parts (alternately brighter and darker to make it easier to distinguish subsequent components). You can use the **Combine (1)** tool if you apply any cutting surface (do you remember cookie cutter?) and/or create **Plane (2)** (or planes) to use it as the cutting plane with the **Split Body (3)** tool. Close the SCDM module and save the whole project from the Ansys Workbench main window (File -> Save).

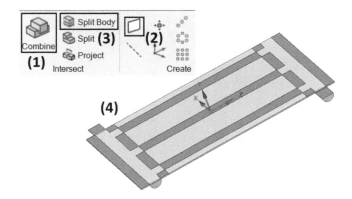

FIGURE 7.20

7.3 PREPROCESSING – MESHING

Find the **Mesh (1)** module in the Component Systems tab of the LHS **Toolbox**, as in Figure 7.21. Drag and drop it in the **Project Schematic** window *(2)*. Hold **LMB** on *A2* cell and drag the connection to *B2* cell to get the link *(3)*. Then click *2xLMB* on *B3* (*Mesh*) cell *(4)* to launch the *Ansys Meshing* module.

Tutorial 1 – Flat-Plate Solar Collector

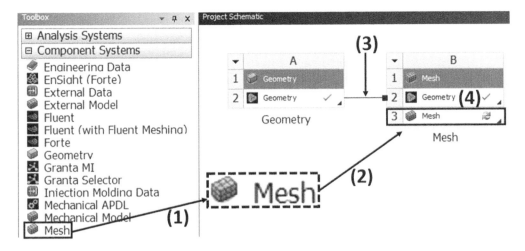

FIGURE 7.21

The project tree *(1)* which describes all the mesh settings is located on the LHS in the **Outline** window, as in Figure 7.22. The **Geometry** tab includes the **Geom** *(2)* tab, where all the components of the spatial geometry are listed. Different parts have different colors *(3)* and can be edited individually if required. Click **Mesh** on the project tree to see the **Details of "Mesh"** *(4)* table allowing to set global mesh settings.

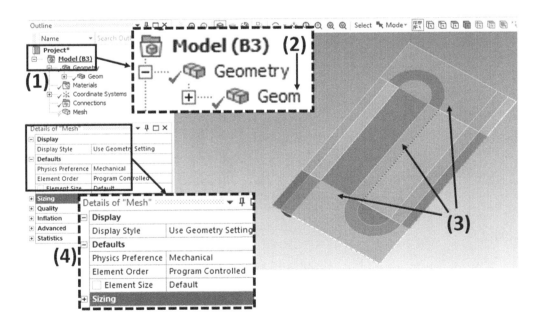

FIGURE 7.22

Global mesh settings apply to the whole model. In this project, **Physics Preference** *(1)*, **Solver Preference** *(2)* and **Element Size** *(3)* have to be set according to Figure 7.23. **Physics Preference** allows to automatically apply preliminary settings desirable from the CFD simulation point of view (these settings differ from the ones that are required for the modeling of structural or electromagnetic phenomena). **Solver Preference** adjusts mesh features to the requirements of a specific solver. The **Element size** setting limits the maximum dimension of the mesh cell.

FIGURE 7.23

Unlike global settings, the local ones apply only in selected areas of the computational domain. They influence mesh generation algorithms and mesh features in these regions. Click *1xLMB* on the *Mesh (1)* tab in the project tree (Figure 7.24). Go to the Mesh *(2)* tab (top toolbar). Find selection mode options – four cubes *(3)* and click the last one – Body *(4)*. It allows to select volumes. Select the fluid parts (use *Ctrl+LMB*) according to *(5)* and click *Method (6)*. Change method to *Sweep (7)* (drop-down list). Click *1xRMB* on *Mesh (1)* → *Generate Mesh*.

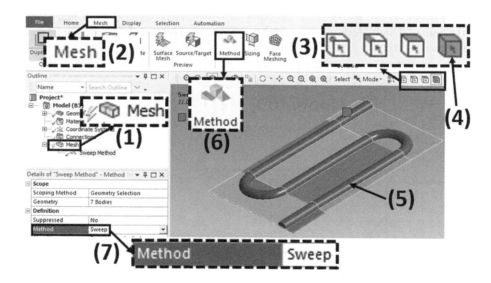

FIGURE 7.24

Figure 7.25 presents selected details of the created mesh. Applied **Element Size** allowed to obtain two mesh cells across the absorber plate *(1)*. It is unacceptable to model heat transport phenomena in thin domains with only one cell across it. Due to shared topology, all mesh nodes on the boundary between solid and fluid bodies are consistent *(2)*. Due to specific division of the absorber plate, most cells are hexahedral there. However, due to the curves coming from the fluid duct shape, certain cells are deformed *(3)*. Due to the use of the *Sweep* local meshing method, the mesh in the fluid domain is structural.

Tutorial 1 – Flat-Plate Solar Collector

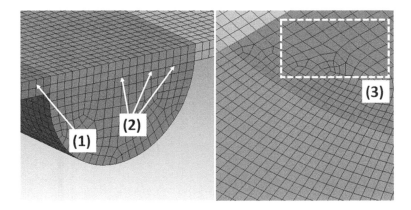

FIGURE 7.25

To analyze the selected mesh quality indicator click *1xLMB* on the *Mesh (1)* tab (project tree – see Figure 7.26) and in the *Details of Mesh* table go to *Mesh Metric (2)*. Select *Orthogonal Quality* from the drop-down list (where *None* is the default). Note that the orthogonal quality should be higher than 0.05. Information concerning the number of nodes and elements in the generated mesh is available in the *Statistics (3)* tab.

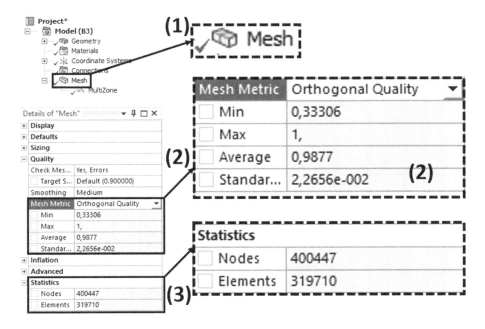

FIGURE 7.26

Additionally, when any quality indicator is selected, a *Metric Graph* appears (Figure 7.27). This is the graph presenting the histogram of the mesh quality distribution, taking into account different cell types. In this mesh *Hexahedrons Hex8 (1)* and *Wedges Wed6 (2)* are present. The highest bar *(3)* corresponds with high-quality hexahedrons. Due to a low percentage of wedges, the vertical axis range has to be significantly reduced using **Controls (4)** to see the histogram bars corresponding with these cells. The location of cells linked with a certain bar in the grid can be checked by clicking the bar.

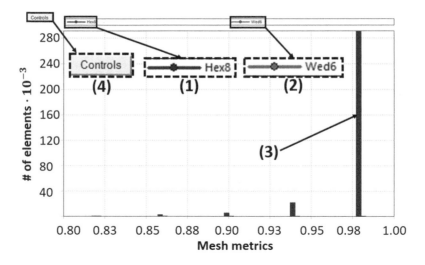

FIGURE 7.27

The Named Selections option allows to define model physics fast and efficiently. It is recommended to apply this functionality. To define a new *Selection Name* click *Face (1)* selection mode. Select the working fluid inlet face *(2)*. Pay attention to the triad *(3)* position – make sure that an appropriate face has been selected (not the outlet one). Press *N* key and write *inlet* in the *Selection Name* dialog box *(4)*, as in Figure 7.28. Click *OK*. The new *Selection Name* is now available in the *Named Selections (5)* tab of the project tree.

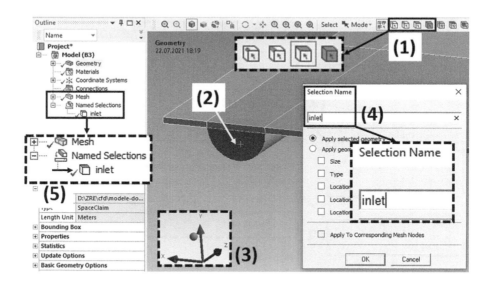

FIGURE 7.28

Define *Selection Names* for the remaining faces (triad *(1)* might be useful) analogously – according to Figure 7.29. First, *outlet (2)* and *wall hot (3)* for the whole top absorber plate face (use *Ctrl* for multiple selections). Then turn the geometry to see its bottom side *(4)*. Define *Selection Names* INDIVIDUALLY for solid and fluid faces: *wall cold solid (5)* (bottom faces and thin side faces) and *wall cold fluid (6)*. Do not define any *Selection Names* for the contact face between the solid and the fluid domain!

Tutorial 1 – Flat-Plate Solar Collector

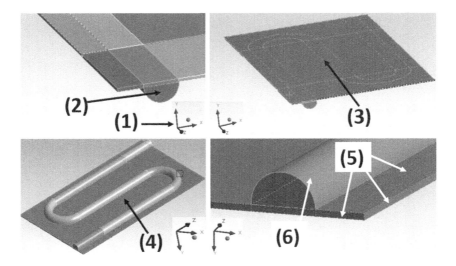

FIGURE 7.29

The required set of *Selection Names (1)* is presented in Figure 7.30. All external (visible without hiding parts) surfaces of the geometry should be covered by Named Selections. Legend with locations (required any selection from the project tree!) is also available in the *Ansys Meshing* workspace *(2)*.

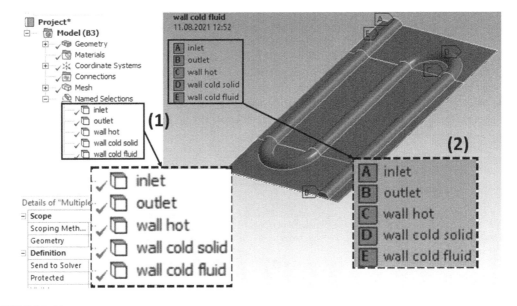

FIGURE 7.30

The Named Selections tool can be applied to define faces, as well as edges and whole volumes. Defining *Selection Names* for volumes helps to define individual domains during setting up the solver preprocessor. Two domains exist in this project: fluid (working medium) and solid (metal absorber plate). Change the selection mode to *Body (1)* and select the whole fluid domain *(2)*, according to Figure 7.31. Insert *Selection Name*: *fluid*. The new object should appear in the *Named Selections* section of the project tree. Now, analogously define *Selection Name* for the absorber plate *(3)*: *solid*.

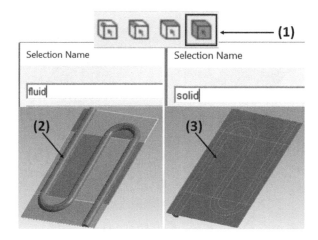

FIGURE 7.31

Now close the Ansys Meshing module and save the whole project from the Ansys Workbench main window. Time to set up the solver preprocessor.

7.4 PREPROCESSING – SOLVER SETTINGS

In the same way as in the case of *Geometry* and *Mesh* modules, find *CFX (1)* module (Figure 7.32) in the *Component Systems* toolbox and add it to the *Project Schematic* window *(2)*. Create a connection between *B3* (*Mesh*) and *C2* (*Setup*) cells *(3)*. When the connection is created, lighting thumbnail appears in *B3* cell *(3)*. It means that the mesh data have to be exported to the *CFX* solver. Click *1xRMB* on *C3* cell → *Update (5)*. Wait until the lighting thumbnail changes to a green check mark (as in *A2* and *B2* cell). Then launch *Ansys CFX* by clicking *2xLMB* on *C2* cell. Before you start the model setup, make sure that the *Automatic Default Interfaces* option is activated. Click *2xLMB* on *General* (in the *Case Option* tab) at the top of the project tree and mark all checkboxes below the *Auto Generation* inscription. Then *Apply* the settings and close the window.

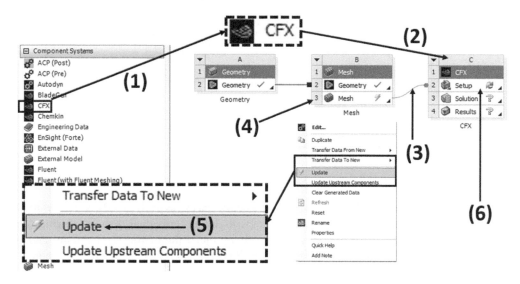

FIGURE 7.32

Tutorial 1 – Flat-Plate Solar Collector

To define the material of the working fluid with temperature-dependent density, an expression describing this function has to be added. Find in the project tree (Figure 7.33) and open the *Expressions, Functions and Variables* tab and open it. Click *1xRMB* on *Expressions (1)*, and select *Insert → Expression*. Enter *rho (2)* name for the new expression.

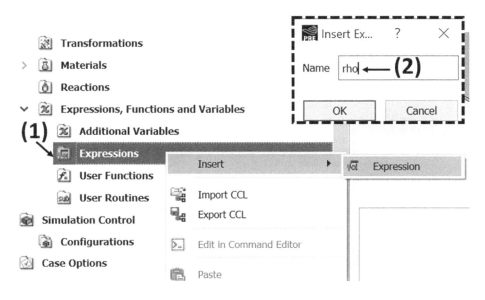

FIGURE 7.33

You can enter the *rho* expression *(1)* in the *Details of rho* window (Figure 7.34):

$$-0.0016[kg*m^{-3}*K^{-2}] *T^2 + 0.3326[kg*m^{-3}*K^{-1}] *T + 1070.8[kg*m^{-3}]$$

Click *Apply (2)*. The Defined expression should appear above, in the *Expressions* window *(3)*. Now you can close the *Expressions* tab (*x*).

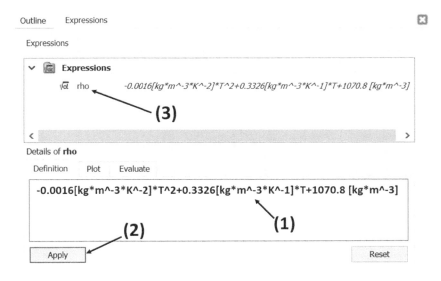

FIGURE 7.34

Apart from other physical properties, the defined density function will be used now to set up a new material of the working fluid. Find in the project tree the *Materials (1)* tab and open it. Although certain materials from the **CFX** library are already available in the *Materials* tab *(2)*, as in Figure 7.35 a brand new one has to be created. Click *1xRMB* on the *Materials* tab → *Insert* → *Material*. Enter *MyFluid (3)* name into the *Insert Material* dialog box. Confirm by *OK*. Further hints are below.

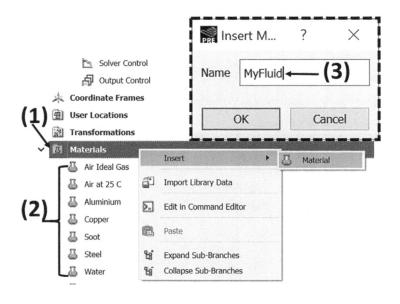

FIGURE 7.35

Material: MyFluid window should open automatically. Pass to the *Material Properties* tab. Enter *Molar Mass*: **40.5 [kg kmol^–1]** and *Density*: **rho**, as in Figure 7.36. It is possible to insert expression name to the *Density* cell when you press the square root of alpha mark on the RHS of the cell. Note that *rho* is the name of the previously defined expression describing the density function. Instead of writing the *rho* name directly, you can also click *1xRMB* in the *Density* cell, then select *Expressions* from the drop-down list and the *rho* expression (according to the Figure 7.36).

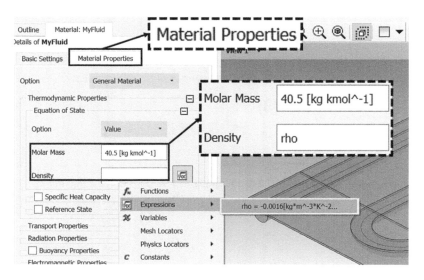

FIGURE 7.36

Tutorial 1 – Flat-Plate Solar Collector

Figure 7.37 presents further required settings of the material properties. Activate and enter **Specific Heat Capacity (1)**: **3700 [J kg^–1 K^–1]**. Use+to open the **Transport Properties** tab to enter **Dynamic Viscosity (2)**: **2.1673E–03 [Pa s]** and **Thermal Conductivity (3)**: **0.48 [W m^–1 K^–1]**. It remains to activate **Buoyancy Properties** and enter the value of **Thermal Expansivity (4)**: **3.4112E–03 [K^–1]**. Confirm the setup by **Apply (5)**.

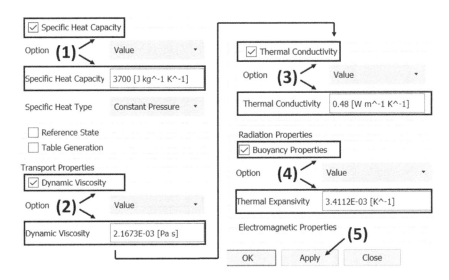

FIGURE 7.37

Before you add any domain (fluid/solid), all physical elements of the model are included in the **Default Domain (1)** tab of the project tree. Find and select the **Domain (2)** option (top toolbar), as in Figure 7.38. Enter *fluid (3)* name. Note that it is the same name as the one defined for the fluid domain in **Ansys Meshing** (defining **Named Selections**). Confirm the name by *OK*. A new setting window *Details of fluid in Flow Analysis 1* should appear automatically.

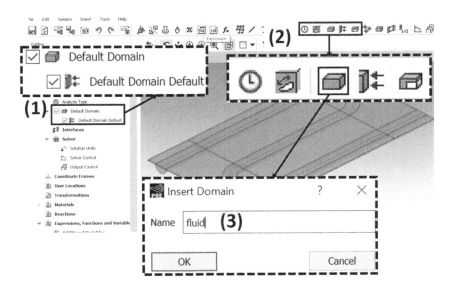

FIGURE 7.38

If the given domain name is consistent with the one defined by **Selection Name** (in **Ansys Meshing**), **Ansys CFX** recognizes automatically the correct **Location (1)** of the domain (geometry representing working fluid), as in Figure 7.39. Choose an appropriate domain material from the **Material** drop-down list – **MyFluid (2)**. Set the **Reference Pressure (3) 1 [atm]** for the domain. 45° absorber slope can be taken into account by defining appropriate gravity vector components. Select **Buoyant** from the **Option** drop-down list in the **Buoyancy Model (4)** frame. Then subsequent vector components can be defined: *0 [m s^--2]* for *X* direction, *−6.94 [m s^−2]* for *Y* direction and *−6.94 [m s^−2]* for *Z* direction. It is recommended to set **Buoyancy Reference Density** as the lowest existing in the modeled system – in this case, it is around *1037 [kg m^−3]*.

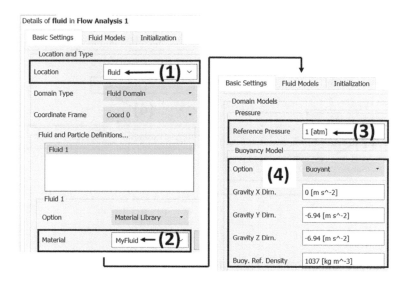

FIGURE 7.39

Go to the **Fluid Models (1)** tab and activate the energy equation for low-Mach-number flows by selecting **Thermal Energy (2)** from the **Option** drop-down list in the **Heat Transfer** frame, according to Figure 7.40. Based on the Reynolds number calculated for this case, the **None (Laminar) Option** should be selected in the **Turbulence (3)** frame. Confirm all settings by **Apply (4)** and close the **Details of fluid in the Flow Analysis 1** window.

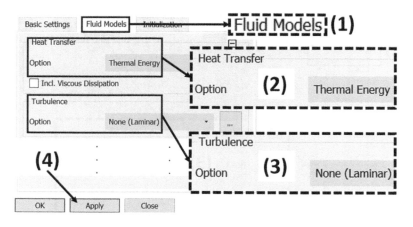

FIGURE 7.40

Tutorial 1 – Flat-Plate Solar Collector

Analogously to the previous example (fluid domain) add and define the absorber plate domain (solid body), according to Figure 7.41. Use the *solid (1)* name (as in *Named Selections*) to set the *Location* automatically. Obviously, this time *Domain Type* is *Solid Domain (2)*. The appropriate material is *Copper (2)*. Then confirm settings by OK and Apply and close the domain window.

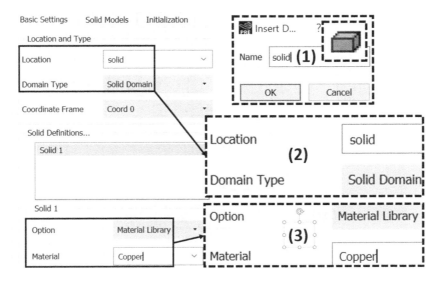

FIGURE 7.41

After creating the *fluid* and *solid* domain, have a look at the project tree. The new object has been added automatically below the *Interfaces* tab: *Default Fluid Solid Interface (1)*, as in Figure 7.42. It is a specific kind of boundary condition allowing to transfer data between two different domains (like our *fluid* and *solid*). Both sides of the interface (surfaces creating it) are also visible below *fluid* and *solid* tabs *(2) respectively*. Temporary errors coming from the manual change of the domain type from fluid to solid (during absorber domain setting up) may appear in the console *(3)*. If so, click *2xLMB* on any information about the error and confirm any settings displayed on LHS (whatever they are) by *Apply*. It will fix the issue.

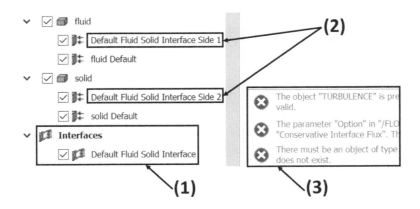

FIGURE 7.42

When any domains already exist in the model, it is possible to define boundary conditions (BC). Add first BC to the *fluid* domain. Find *fluid (1)* domain on the project tree and click *1xRMB* on it. Select *Insert (2)* from the drop-down list and *Boundary (3)*, according to Figure 7.43a. An alternative method to add a new BC is to click *Boundary (4)* from the top toolbar and select the domain (*in fluid (5)* in this case) in which the BC has to be created (Figure 7.43b).

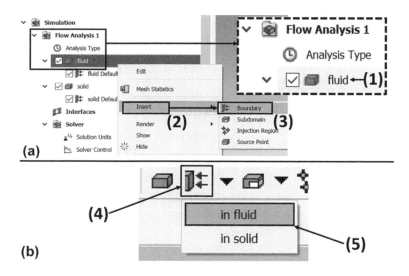

FIGURE 7.43

Enter the *inlet (1)* name for this BC – according to *Named Selections* defined in *Ansys Meshing*. Thanks to this, *Ansys CFX* correctly recognizes *Boundary Type* as *inlet* BC *(2)* and links it with the appropriate *Location (3)* (fluid inlet), as in Figure 7.44a. Go to the *Boundary Details (4)* tab. Set *Option* in the *Mass and Momentum (5)* frame to *Normal Speed*, according to Figure 7.44b. The normal speed magnitude is *0.025 [m s^–1]*. Set *Static Temperature (6)* below *283.15 [K]*. Click *Apply (7)* and close the window (*OK* or *x*).

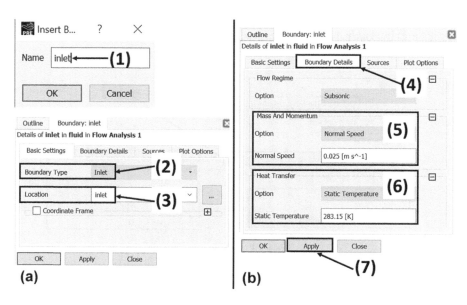

FIGURE 7.44

Tutorial 1 – Flat-Plate Solar Collector

Analogously, add outlet BC. This time **Boundary Type** should be set to **Opening (1)**, while the location is of course **outlet (2)**, according to Figure 7.45a. **Opening** BC allows for reversed flow through the boundary. This makes the model more flexible. In the **Boundary Details (3)** tab select **Opening Pres. and Dirn** from the **Option** tab in the **Mass and Momentum** frame and set **O [Pa] (4)** value. Due to the possibility of the reversed flow, the backflow temperature has to be determined. Let's imagine that just after the *outlet* from the domain the temperature is almost the same or even exactly the same as on the *outlet* face. Then the assumption that the backflow temperature is equal to the average one on the *outlet* is a good approximation. To express this situation, an appropriate expression has to be written in the **Opening Temperature (5)** cell: **areaAve(T)@outlet** (Figure 7.45b). Remember that the small alpha square root mark *(6)* has to be clicked to make it possible to write the expression in the cell. Remember to click **Apply** and **OK** at the end. In more simplified case, the backflow temperature can be expressed as a constant value instead of an average. Have a look at the model in the workspace – *inlet* (inlet type) and *outlet* (opening type) BCs are shown, respectively, by one- *(7)* and two-way *(8)* arrows.

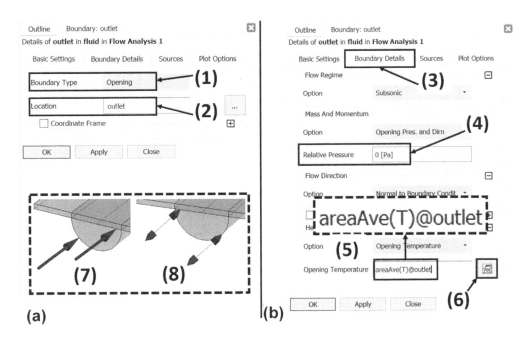

FIGURE 7.45

Add the heat source in the system – on the absorber plate top face. Add **wall hot (1)** BC in the solid domain. Go to the **Boundary Details (2)** tab. Select **Heat Flux** from the **Option** drop-down list in the **Heat Transfer (3)** tab, as in Figure 7.46a. Set *1000 [W m^–2]* value. Analogously, (in the solid domain) add **wall cold solid (4)** BC. In the Boundary Details (5) tab, change **Option** in the **Heat Transfer (6)** frame to **Heat Transfer Coefficient**. Set **Heat Transfer Coeff.** to *15 [W m^–2 K^–1]* and **Outside Temperature** to *283.15 [K]*.

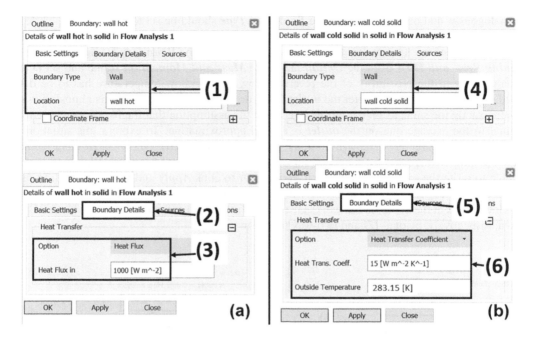

FIGURE 7.46

One more BC has to be added to the fluid domain – **wall cold fluid**. Do it as in the case of **wall cold solid** (heat loss will be the same in this project for the bottom face of the absorber and duct walls). Including the interface sides, there should be four BCs in the *fluid* domain *(1)* and three BCs in the *solid* domain *(2)*, according to Figure 7.47.

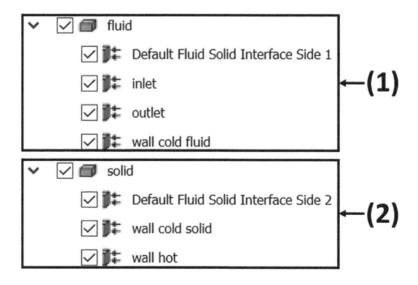

FIGURE 7.47

Tutorial 1 – Flat-Plate Solar Collector

Before the simulation launch, always at least a couple of basic numerical parameters have to be reviewed and, optionally changed if required. Click *2xLMB* on the **Solver Control (1)** (project tree, Figure 7.48). Find **Max. Iterations (2)** setting and change it to *1500*- it will be enough to reach the convergence of all the governing equations. Check **Residual Target (4)**. Although, in the case of the engineer computations even *1.E–3* is sometimes acceptable, this time it is recommended to go with *1.E–5*. Close the **Solver Control** tab (*Apply* and *OK* or *Apply* and *Close*).

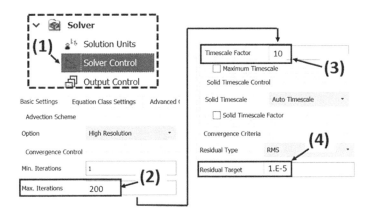

FIGURE 7.48

Creating at least one monitor point to control the selected computation results during simulation is recommended. Let's add a monitor for the fluid *outlet* temperature. Click *2xLMB* on the **Output Control (1)** tab (project tree, Figure 7.49a). The **Details of Output Control in Flow Analysis 1** window appears automatically. Go to the **Monitor (2)** tab and activate **Monitor Objects (3)**. Click **Add new item (4)**, next to the **Monitor Points and the Expressions** window. Enter the name of the new monitor, for example, *MyTempOut (5)* and click *OK*. First, the monitor point should appear in the **Monitor Points and Expressions** window *(6)*, as in Figure 7.49b. Select **Expression** from the **Option** drop-down list *(7)*. Then write the expression calculating the area of averaged temperature on the fluid *outlet* face *(8)*: *areaAve(T)@outlet*. Do not forget to click the alpha root square mark first *(9)*. Confirm settings by *Apply* and close the window.

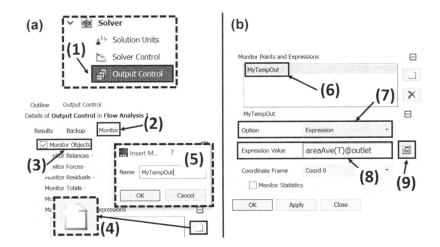

FIGURE 7.49

Close *Ansys CFX* and save the whole project in the *Ansys Workbench* main window. Now click *2xLMB* on *C4 (1)* cell (*Solution*) to launch *Solver Manager* (Figure 7.50). The *Ansys Workbench* student version allows to run parallel computation processes. However, this license limits the available number of processes to four. In the *Define Run* window find *Run Mode (2)* drop-down list and change settings to *Intel MPI Local Parallel*. Then use *+(3)* to increase *Partitions* to *4*. The model will be physically divided into four parts that will be considered in parallel during the simulation. Click *Start Run (4)* to launch the simulation.

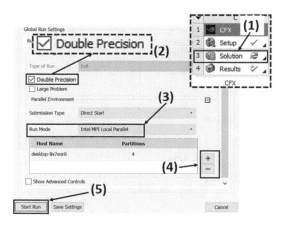

FIGURE 7.50

The simulation should start automatically. Have a look at the tabs available at the top of the *Solver Manager* window *(1)*, as in Figure 7.51. They concern different classes of governing equations. Roots mean square (RMS) for residual (imbalance) representing each governing equation in subsequent iterations of the solution process is represented by residual curves *(2)*.

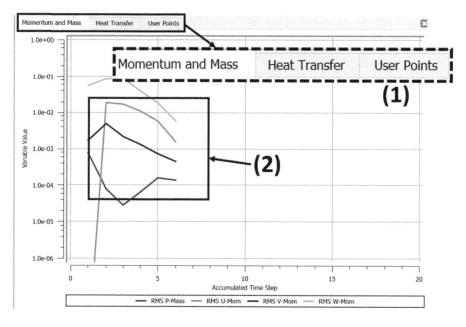

FIGURE 7.51

Tutorial 1 – Flat-Plate Solar Collector

According to the solution, each residual curve has to reach at most the level of the residual target *(1)* that has been set in the solver preprocessor (remember **Residual Target** setting in **Solver Control** – Figure 7.48), as in Figure 7.52. The outlet fluid temperature plotted in the **User Points** tab *(2)* changes in subsequent iterations. The stabilized value at the end of the computations confirms that the model has been solved. Of course, minor changes in the temperature would be still observed in case of further solving (for example with a lower residual target), but this accuracy is fine for this example's purposes.

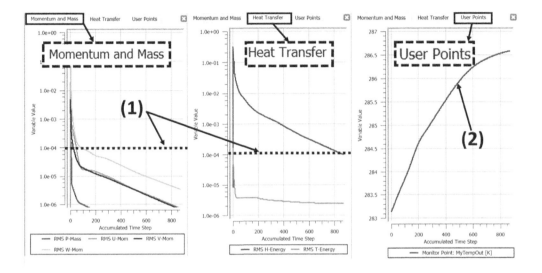

FIGURE 7.52

The progress of the model solution is reported in the **Out File** tab, as in Figure 7.53. The first column includes the names of the solved governing equations. The most important information given in the console during solution is timescales *(2)*, number of current outer loop iterations *(3)*, computation time *(4)*, max. residuals *(5)* and quality of the linear solution (**OK** means that everything is fine). When the solution is complete, **CFX Solver Manager** displays the information: **Solver Run Finished Normally**. Then click **OK**, close the **CFX Solver Manager** window and save the whole project again.

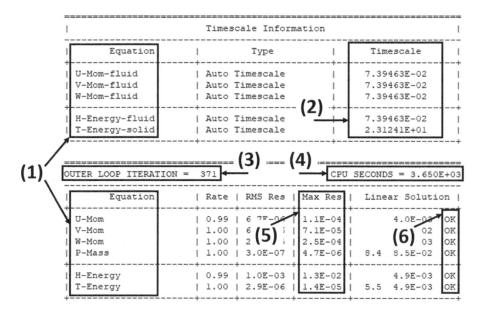

FIGURE 7.53

7.5 POSTPROCESSING

In case of the projects involving the *Ansys CFX* solver, it is required to use the *CFD-Post* module to postprocess simulation results. Click *2xLMB* on *C4 (1)* cell to launch *Ansys CFD-Post*. Each object in the model can be visually activated and deactivated by marking it on the project tree – BCs are listed below *fluid* and *solid* tabs *(2)*, as shown in Figure 7.54. Basic result visualization tools are available on the top toolbar *(3)*. Settings of the legend and wireframes can be modified in *Default Legend View 1* and *Wireframe* tabs *(4)* of the project tree, respectively.

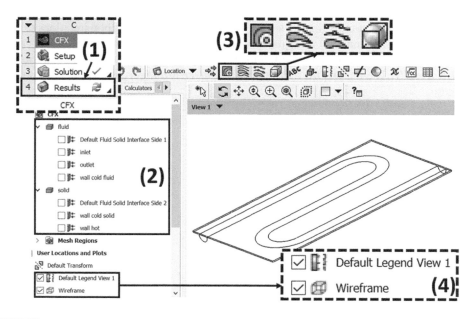

FIGURE 7.54

Tutorial 1 – Flat-Plate Solar Collector

Create the first contour. Click the ***Contour (1)*** tool from the top toolbar and confirm the default name *(2)* by ***OK*** (or enter any name). Go to the ***Details of Contour 1*** window below and set the ***Location*** of the contour: ***wall hot (3)***. It means that the selected variable will be visualized on this face. Set ***Variable*** to ***Temperature*** and ***Range*** to ***Local (4)***. The ***Local*** range allows to adjust the color range to variable changes only in the selected location (in this case ***wall hot***). Increase ***# of Contours*** to ***50 (5)*** allows to improve contour resolution. Click ***Apply*** to display the result as in Figure 7.55.

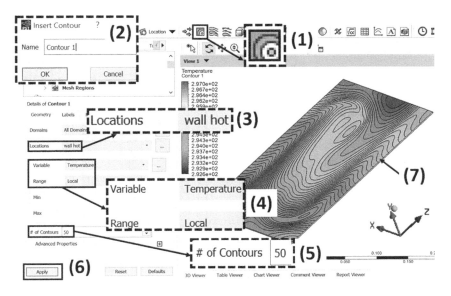

FIGURE 7.55

Volume Rendering (1) gives 3D visualization of the selected variable in the whole domain or its part. Select this tool (Figure 7.56) and in ***Details of Volume Rendering 1*** window set ***Domains*** to ***fluid***. Then again set ***Variable*** to ***Temperature*** and ***Range*** to ***Local (3)***. Increase ***Resolution*** to ***30*** and click ***Apply***.

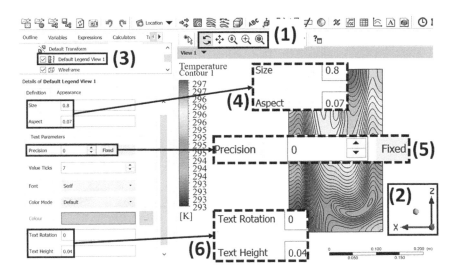

FIGURE 7.56

Although the mouse scroll button or navigation tools *(1)* can be used to turn the geometry, sometimes it's faster to click the triad axis *(2)*. This way an object can be automatically set to the position normal to the certain axis (Figure 7.57). If you want to resize the legend or do not want its scientific mode, you can edit it in **Default Legend View 1** window – click **2xLMB** on tab *(3)* of the project tree. You can change the legend bar height and/or aspect *(4)* and change the value display mode from **Scientific** to **Fixed** *(5)*. **Precision** setting allows to increase or decrease the number of decimal places. You can also edit the text parameters *(6)*. New settings have to be confirmed by **Apply** button.

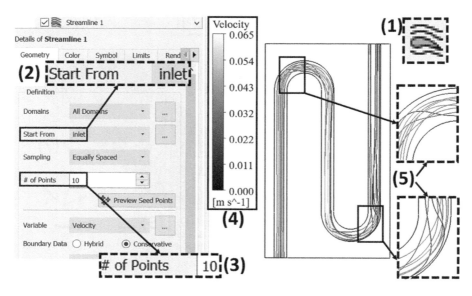

FIGURE 7.57

Now add the **Streamline (1) with the Ctrl key** visualization reflecting the fluid velocity and trajectories of its selected elements. Set **Start From**: *inlet (2)*. Reduce **# of Points** to **10 (3)**, to make the visualization more transparent. Click **Apply** and adjust the legend *(4)*. **Streamline** tool allows to analyze specific fluid behavior, for example on the arched duct sections *(5)*, as in Figure 7.58.

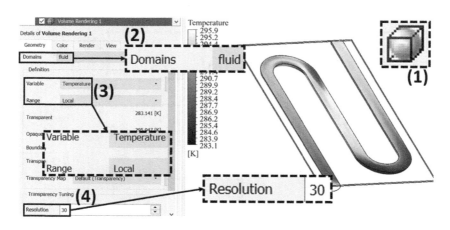

FIGURE 7.58

Tutorial 1 – Flat-Plate Solar Collector

Applied tools allow to display different data: **Contour 1**- temperature distribution on the absorber plate, **Volume Rendering 1**- temperature of the working fluid, **Streamline 1**- fluid velocity and flow characteristics. Different visualizations can be displayed together, to make it easier to study certain relations and system behavior. The selected objects have to be activated below the *User Locations and Plots* tab of the project tree *(1)* –for example, *Contour 1* together with *Streamline 1*, as in Figure 7.59.

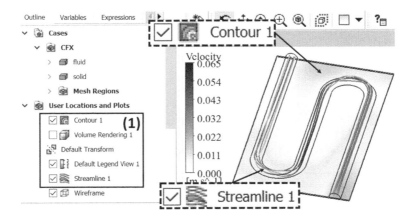

FIGURE 7.59

ANSYS CFD-Post allows to obtain a wide range of numerical data. Information concerning flow characteristics can be obtained above all by expressions, already known from the solver preprocessor. Let's check the outlet temperature that was tracked by the monitor point during the simulation. Find the **Expressions** *(1)* tab just above the project tree window, as in Figure 7.60. Click *1xRMB* on the **Expressions** heading of the existing expression list. Click *New (2)* and enter *MyTempOut (3)* name. Then confirm by *OK*. Go to the *Details of MyTempOut (4)* window below the project tree. The expression can be written in the blank white space below. Of course, you can just write an appropriate expression, however, in case of a more complicated syntax, it is better to use a more automated procedure. This will be presented now. Click *1xRMB* in the white blank space à *Functions* → *CFD-Post* → *areaAve (5)*. *areaAve()@* appears in the expression window. Try to recall the syntax of the same expression in the solver preprocessor. Two items of information have to be supplied here: variable – in brackets and location after @ mark.

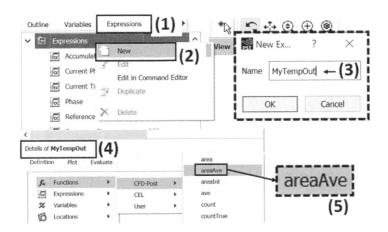

FIGURE 7.60

Click *1xRMB* between the brackets: *Variables (1)* → *Temperature (2)*. The Expression is extended now to *areaAve(Temperature)@*, as in Figure 7.61a. As you can see, variables can be selected from the drop-down list. In case of temperature, both the above mentioned syntax and *areaAve(T)@* work in the same way. Analogously, an appropriate location after @ mark can be added. Click *1xRMB*, this time after the @ mark: *Locations (3)* → *Outlet (4)*. The complete expression (after adding the location) should look as in Figure 7.61b. Click *Apply* to see the computed result.

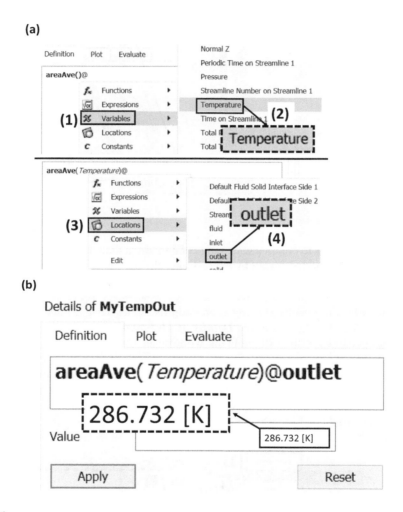

FIGURE 7.61

Each added expression appears on the expression list. Now use the above presented procedure to define further expressions: fluid specific heat *MyCp (1)*, fluid mass flow rate *MyMFrate (2)*, inlet temperature *MyTempIn (3)* (alternatively, a known, fixed value can be set here – as in case of *MyCp*). The syntax of each expression is given in Figure 7.62. Note that the expression for the outlet fluid temperature *MyTempOut (4)* has been already defined (Figure 7.61a and b). When the expression includes a fixed value, as in the case of *MyCp*, an appropriate unit in square brackets has to be taken into account. Note that there are no fractions here – denumerators are expressed by using appropriate negative power, for example *[kg^–1]* or *[K^–1]*. All of the added expressions can be used to calculate the thermal power of the considered flat-plate solar collector.

Tutorial 1 – Flat-Plate Solar Collector

| Outline | Variables | Expressions | Calculator |

- MyCp — *(1)* 3700[J kg^-1 K^-1]
- MyMFRate — *(2)* massFlow()@inlet
- MyTempIn — *(3)* areaAve(Temperature)@inlet
- MyTempOut — *(4)* areaAve(Temperature)@outlet

FIGURE 7.62

Of course, to calculate thermal power one long expression can be developed instead of a set of shorter ones. However, in such a situation a mistake is more likely. Furthermore, once developed "partial expressions" can be used in many other, more complicated calculations. Thus, using the above presented approach speeds up data postprocessing. Insert new *MyPower (1)* expression. Thermal power is a product of the mass flow rate *MyMFRate*, specific heat *MyCp* and fluid temperature difference *MyTempOut–MyTempIn*, according to *(2)* in Figure 7.63. Click *1xRMB* in the white space of the *Details of MyPower* window. Then select *Expressions (3)* – you can add the subsequent required expressions from the drop-down list *(4)* and just put multiplication signs between them. At the end click *Apply* to see the result below.

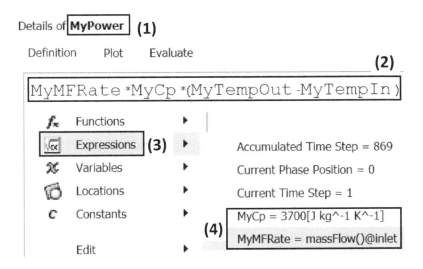

FIGURE 7.63

As shown above, expressions allow to calculate a wide range of numerical results. The one really important in the case of any energy device is total efficiency. Regardless of technology, in general, energy efficiency comes from the input energy to output energy ratio. In the case of the thermal solar collector, it is the working fluid thermal power to the total power of the supplied heat flux (from the absorber plate) ratio. Thus, to calculate the efficiency of the considered system, the total power of the above mentioned heat flux has to be found first. Add a new *MyInHeatFlux (1)* expression. Figure 7.64 presents two alternative expressions that can be used here. Expression *(2)* calculates the power as the product of the absorber plate area and unit heat flux (which is known and defined as the *wall hot* BC). The second approach *(3)* calculates the power on the *wall hot* BC directly, as the area integral. You can check whether both options give the same result and finally choose one of them.

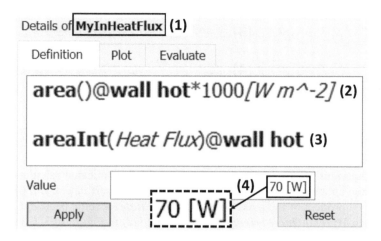

FIGURE 7.64

Now you can make any interesting calculations. For example, you can check the quantity of the mass imbalance in the system expressed as a difference between the inlet and the outlet mass flow rate. Furthermore, you can confirm the correctness of the laminar flow approach by calculating Reynolds number (hydraulic diameter is known, and other required parameters can be found using expressions). You can do it for inlet and outlet conditions, as well as for average and maximum parameter values to compare the result and make conclusions.

8 Tutorial 2 – Evacuated Tube Solar Collector

8.1 EXERCISE SCOPE

The goal of this tutorial is to present the idea of the evacuated tube collector operation, with respect to preventing heat losses by the reduction of thermal conduction and convection inside the collector tubes. As shown in Figure 8.1, the model developed in the course of this exercise allows to compare temperature distribution on the heat pipe *(1)* and the occurrence of possible fluid flow phenomena *(2)* in the case of the air-filled tube and the evacuated tube.

As you could find out from the theoretical background (see Section 6.3.2), in the simplest terms, the vacuum space (actually strong underpressure) is created between the external glass and heat pipes in each glass tube of the collector, to avoid heat losses. This model helps to assess the effectiveness of such approach.

Apart from many other interesting software functionalities, this exercise presents the methodology of:

- creating 3D objects from a sketch by the revolve pulling mode and dividing objects using cutting planes in **SpaceClaim DirectModeler**,
- generating polyhedral mesh and evaluating its quality in the **Fluent Meshing** module,
- applying the **Discrete Ordinates** radiation model with the additional **Solar Load** sub-model, allowing to take into account the location and position of the collector and the day of the year.

Pay special attention to the fact that, as shown in this tutorial, it is possible to activate and deactivate basic transport phenomena, which allows to simulate strong under pressure (void) in the simplified vacuum-tube collector model.

All of the images included in this tutorial use courtesy of ANSYS, Inc.

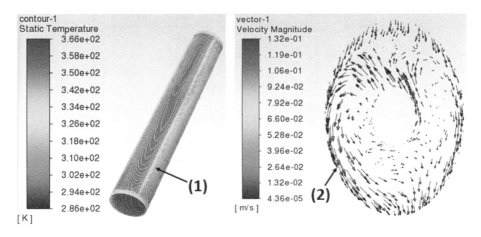

FIGURE 8.1 Temperature distribution on the heat pipe surface (1) and air circulation (2) obtained from the simulation considered in the exercise.

DOI: 10.1201/9781003202226-10

8.2 PREPROCESSING – GEOMETRY

Find the *Toolbox* window (LHS) (Figure 8.2). Then drop down the *Analysis System* list, hold *1xLMB* on *Fluent Flow (Fluent with Fluent Meshing) (1)*, drag and drop it in the *Project Schematic* window (in displayed green (light gray) rectangle). The *Fluent Flow…* module, consisting of a set of six cells (*Geometry*, *Mesh*,…, *Results*) should appear. Click *2xLMB* on *A2 Geometry (2)* cell to launch *Ansys SpaceClaim DirectModeler*.

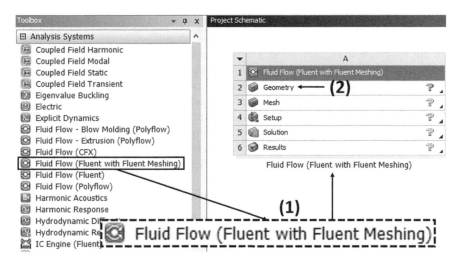

FIGURE 8.2

Find and click (*1xLMB*) the Sketch *(1)* tab from the top bar. Just below, you can find several sketching tools – as in Figure 8.3. Click the *Line (2)* tool. You have to select the sketching plane before creating the first sketch. Go to the functions available at the bottom of the workspace and click *Select New Sketch Plane (3)*. To select a certain plane, move the mouse cursor on the coordinate system axis normal to this plane (the plane should appear) and click *1xLMB* to confirm. In this example you should select the *X-Y (4)* plane, so click the *Z* axis. Then press the *V* key on the keyboard to set the plane view.

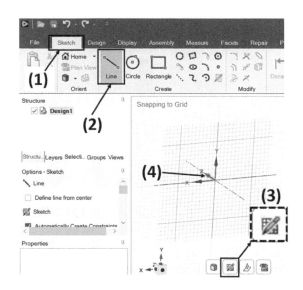

FIGURE 8.3

Tutorial 2 – Evacuated Tube Solar Collector

The coordinate system triad is set now according to *(1)* in Figure 8.4. Use the **Line** *(2)* tool to draw four sections creating a rectangle *(3)*. Use the **Dimension** *(4)* tool to set the appropriate section lengths (***200x35 mm***) and the angle between the bottom, longer section and the ***X-axis*** (***45°***). Then select the **Construction line** *(5)* tool and create a construction line *(6)* parallel to the longer rectangle section (***15 mm*** distance).

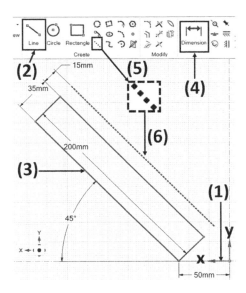

FIGURE 8.4

Click **End Sketch Editing** *(1)*, to activate the 3D operation mode. Select the **Pull** *(2)* tool (Figure 8.5) from the top toolbar. Then click the surface *(3)* that has been automatically created from the rectangle sketch. Select the **Revolve** *(4)* function and click the rotation axis *(5)*. Then hold ***LMB*** on the small, yellow arrow that has just appeared and drag it a bit. Set ***360°*** in the dimension window and press **Enter** on the keyboard to create a 3D object by rotation. Go to the **Structure** window (LHS), click ***1xRMB*** on the **Curves** and **Delete** them. Only **Solid** should remain there.

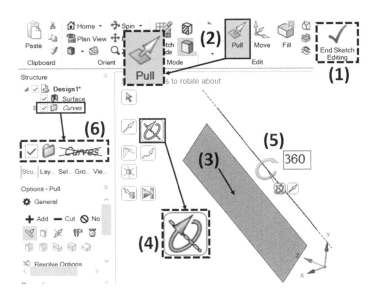

FIGURE 8.5

Although the polyhedral mesh generated in Fluent Meshing works well for cylinders, let's practice dividing the geometry into smaller parts. This step may be helpful when the consistent mesh has to be generated in the ANSYS meshing module. Create two planes that next will be used to divide the already created geometry. Usually dividing object helps to create a good quality mesh with a relatively low number of control cells. Find and click the **Plane (1)** tool on the top toolbar. Then move the mouse cursor to the rotation axis of the cylinder *(2)*. It can be invisible before you point it. The plane crossing the cylinder, as in the top part of Figure 8.6, should appear. If so, click on the axis *1xLMB* to confirm the plane. Now click the **Plane (1)** tool again and move the mouse cursor to the *X-axis (3)* and click it, to create the second plane (perpendicular to the first one), as in the bottom part of Figure 8.6. So the second plane is built on the *Y* and *Z* axes.

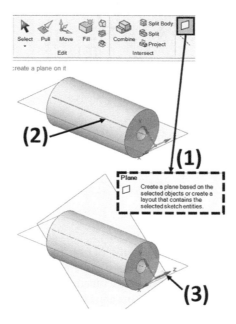

FIGURE 8.6

Use the **Split Body (1)** tool (Figure 8.7) to divide the cylinder. Click the **Select Target (2)** function top left corner of the workspace) and select the cylinder (*1xLMB*) *(3)*. Now switch the function to **Select Cutter (4)**. It might switch automatically – if the saw thumbnail appears next to the mouse cursor, the **Select Cutter** function is already active. Now select (1xLMB) the cutting plane (5) – one of the two previously created. Have a look at the Structure window – two Solid objects should be available there after the splitting operation. If so, continue dividing – click Select Target (6) again and select with the Ctrl key two cylinder halves. Switch to the Select Cutter (8) function and click the second plane (9). Four Solid objects should be available in the Structure window now. Delete planes (10) from there.

Tutorial 2 – Evacuated Tube Solar Collector

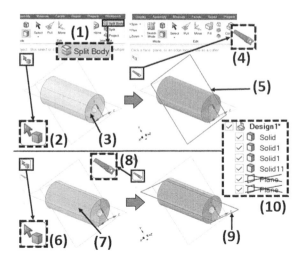

FIGURE 8.7

Creating a consistent mesh is possible for a set of objects if they share topology. Create shared topology for four quarters of the cylinder now. Go to the **Workbench** *(1)* tab of the top bar, according to Figure 8.8. Click **Share** *(2)*. Contact surfaces will be found and highlighted *(3)*. Click **Complete** *(4)* to confirm shared topology.

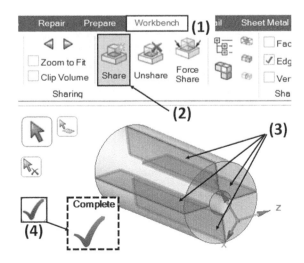

FIGURE 8.8

Creating **Named Selections** helps to define local meshing functions and boundary conditions during work with the solver preprocessor. Go to the **Groups** *(1)* tab (below the **Structure** window). Then click the whole internal wall *(2)* of the cylinder, as in Figure 8.9. Click **Create NS** *(3)*. New **Selection Name** appears on the list below with a default name (**Group1**). Change it – for example to **wall-int**. Now repeat the procedure for all of the external cylinder walls *(4)* – select them, click **Create NS**, and change the name to **wall-ext**. Two **Selection Names** *(5)*: **wall-int** and **wall-ext** are available on the list. Now select the volume objects (cylinder quarters) – click **3xLMB** on any quarter, hold **Ctrl** key and select the other three in the same way *(6)*. Create the third **Selection Name** – **fluid**. Now three **Selection Names** should be available *(7)*. If so, **Ansys SpaceClaim DirectModeler** can be closed.

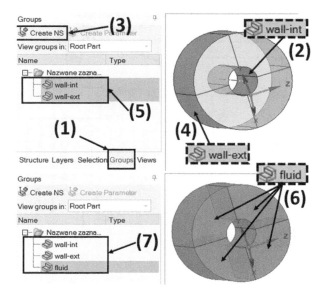

FIGURE 8.9

8.3 PREPROCESSING – MESHING

The whole project can be saved now using the disc thumbnail *(1)*, according to Figure 8.10 (or: *File → Save as* in the *Workbench* project window). Next, click *2xLMB* on *A3 Mesh (2)* cell to launch *Fluent Meshing*.

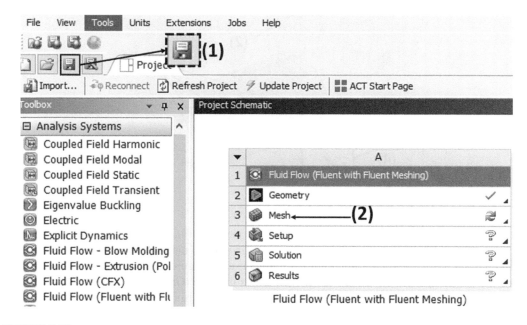

FIGURE 8.10

Tutorial 2 – Evacuated Tube Solar Collector

In the *Fluent Launcher* window (Figure 8.11), set the number of *Parallel Meshing Processes (1)* to *4*- this is the limit for your *Ansys Workbench* student license. Set analogously *Parallel Solver Processes (2)*. Parallel meshing/solving means division of a computational domain into a certain number of partitions that are processed simultaneously. Click *Start* now to launch *Fluent Meshing*.

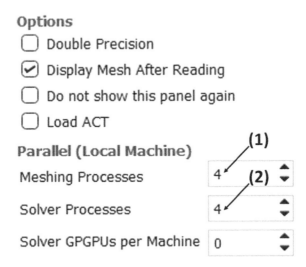

FIGURE 8.11

First, the designed geometry has to be imported from *Ansys SpaceClaim DirectModeler*. Click *1xLPM* on the *Import Geometry (1)* tab on the *Workflow* list, as in Figure 8.12. When the project is prepared in the *Analysis Systems* mode (see Figure 8.2) the file path is usually correct. Thus, there is no need to browse any folders for the geometry file. Just click *Import Geometry (2)* and wait until the geometry importing is done (information in the console).

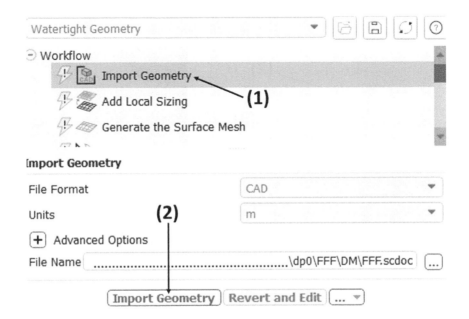

FIGURE 8.12

After importing the model, it appears in the workspace window *(1)*, as in Figure 8.13. The next tab below the **Import Geometry** tab is titled Add Local Sizing (2) – click it and press **Update** at the bottom of the setting window (right bottom corner) – no local sizing is required in case of such simple geometry. Then go to the **Generate the Surface Mesh** tab of the **Workflow** list. Set **Maximum Size** to *0.003 m (3)*. Random mesh cells with a determined size are visualized in the workspace, on the model surface *(4)*. Set **Size Function** *(5)* to **Curvature** – it can be said that mesh cells will be finer on any curvatures. Then click **Generate the Surface Mesh** *(6)* to proceed.

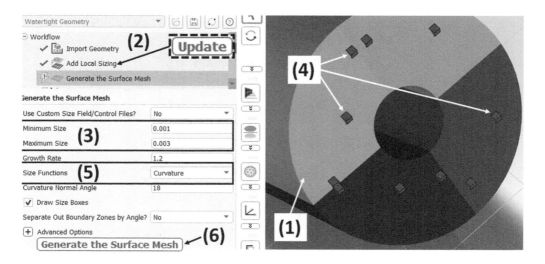

FIGURE 8.13

When the surface mesh is generated, it can be seen at the model surface *(1)* (Figure 8.14) – note that for now the cells are triangular. Later they will be used to create polyhedral, volume cells. Go to the **Describe Geometry** *(2)* tab. Switch **Geometry Type** to **The geometry consist of only fluid regions…** *(3)*. Then in **Change all fluid-fluid boundary types…**, select **Yes** *(4)*. There is no need to apply Share Topology *(5)* – it has already been done. Click **Describe Geometry** *(6)* to continue.

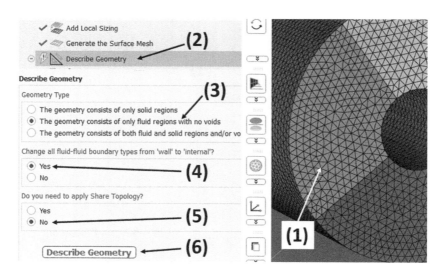

FIGURE 8.14

Tutorial 2 – Evacuated Tube Solar Collector   119

In the ***Update Boundaries (1)*** tab, you can control the boundaries – based on ***Boundary Names (2)*** (in accordance with the defined ***Selection Names*** – see Figure 8.9), in both cases ***Boundary Type*** has been automatically set to a wall. Click ***Update Boundaries (3)*** to proceed (Figure 8.15).

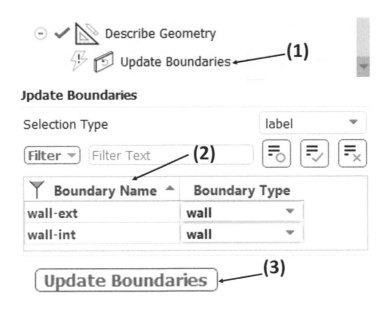

FIGURE 8.15

The ***Update Regions (1)*** tab includes a table (Figure 8.16) with domain regions *(2)* and their types *(3)* (fluid/solid). In this model, there are four regions – four cylinder quarters. All of them should be set to *fluid* in the ***Region Type*** column. Check and click ***Update Regions (4)*** to continue.

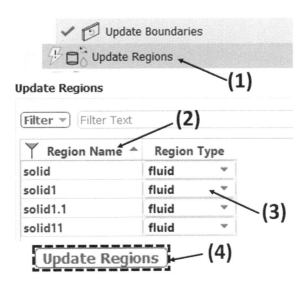

FIGURE 8.16

The Add Boundary Layer (1) tab allows to create a very fine mesh in the near wall regions. Let's add a boundary layer on the cylinder walls. Select *smooth-transition (2)* from the *Offset Method Type* drop-down list, as in Figure 8.17. Set *Number of Layers* to *5 (3)*. *Transition Ratio* and *Growth Rate (4)* can be set as default – in general, these settings control the rate of size changes of the subsequent sublayers in the boundary layer. Set *Grow on* to *only-walls (5)* and click *Add Boundary Layers (6)* to proceed.

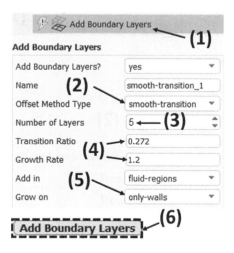

FIGURE 8.17

Smooth-transition_1 boundary layer is created and visible, as in Figure 8.18. If any model requires more boundary layers, they can be added analogously. You can go now to the *Generate the Volume Mesh* tab. Set *Fill With* to *polyhedra (2)*. Additionally, you can change *Max Cell Length* to *0.004 m* – note that it is greater than the face cell size (see Figure 8.13). Click *Generate the Volume Mesh (3)* and wait a moment, until the volume mesh appears in the workspace window (Figure 8.18). Surface mesh cells *(4)* have been converted to polyhedral ones. The boundary layer mesh *(5)* differs from the one distant from the walls.

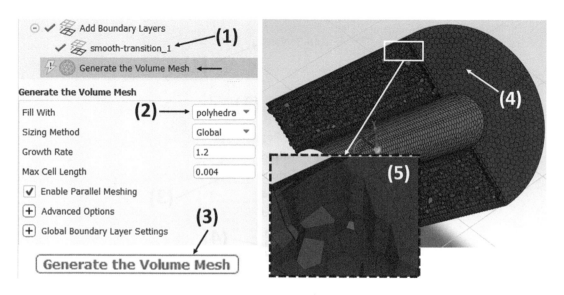

FIGURE 8.18

Tutorial 2 – Evacuated Tube Solar Collector

Have a look at the top toolbar, just above the workspace window. If the *Insert Clipping Plane (1)* option (Figure 8.19) is active, you can analyze the volume mesh inside the model. Location of the clipping plane can be set using the slide bar *(2)*. Plane orientation (axis normal to plane: *X*, *Y* or *Z*) can be set below *(3)*. Go to the console below the workspace window now. After generating the mesh, its parameters are displayed there. The most important information covers the number of cells with the poorest quality *(4)*, minimum quality indicator value *(5)* and total cell count *(6)*.

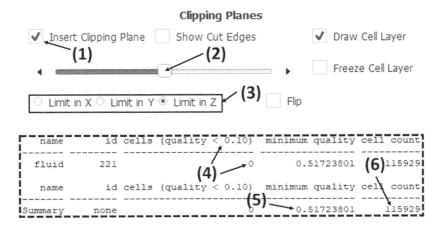

FIGURE 8.19

Accurate control of grid cell quality distribution can be run using the *Display (1)* drop-down list (Figure 8.20) → *Plot (2)* → *Cell Distribution… (3)*. *Options (4)* allows to analyze of cell quality or size. The quality indicator can be set using the *Quality Measure… (5)* options. If the assembly model is considered, the appropriate geometry part can be selected in the *Cell Zones* window. Click *1xLMB* on *fluid (6)* and *Plot (7)*, to display results. You can test different options of the *Cell Distribution…* window and then go to the solver, using the *Switch to Solution (8)* button in the top left corner of the *Fluent Meshing* GUI.

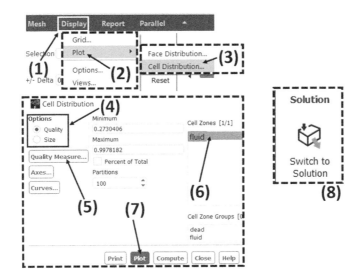

FIGURE 8.20

8.4 PREPROCESSING – SOLVER SETTINGS

After the solver launch, have a look at the ***Outline View*** window (LHS) with a project tree. The project tree includes tabs for subsequent setting categories. Usually, projects do not require passing through all of the available tabs. The ***General (1)*** tab is active as default. The ***Check (2)*** tool displays geometry details in the console. ***Report Quality (3)*** does the same in the mesh quality range. ***Display… (4)*** allows to visualize selected elements (faces, interiors) of the computational domain – it is useful when you want to check if all the required selection names have been defined correctly. Activate Gravity now (bottom of the ***Task Page*** window) and set ***Gravitational Acceleration***: *Y* $[m/s^2] = -9.81$ *(5)* (natural convection) (Figure 8.21).

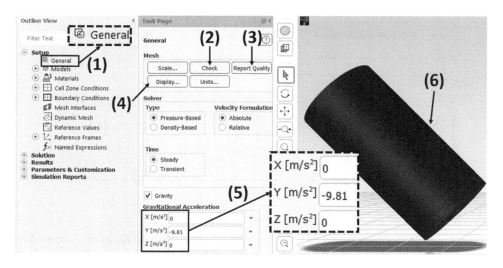

FIGURE 8.21

Click ***2xLMB*** on the ***Models (1)*** tab. It allows to activate and deactivate models and transport equations. Go to the ***Models*** window (LHS, ***Task Page***, as in Figure 8.22) and click ***2xLMB*** on ***Energy (2)***. In the ***Energy*** window activate ***Energy Equation (3)*** – this model is not isothermal. Click ***OK*** and go back to the ***Models*** list. This time click ***2xLMB*** on ***Viscous – SST k-omega (4)*** (this turbulence model is active as default). In the ***Viscous Model*** window change ***Model*** to ***Laminar (5)***. Then click ***OK***.

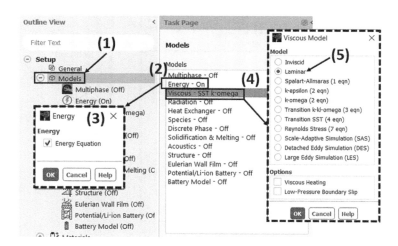

FIGURE 8.22

Tutorial 2 – Evacuated Tube Solar Collector

Stay in the *Models (1)* tab of the project tree. Click *2xLMB* on the *Radiation – Off (2)* tab to define radiation model settings. In the *Radiation Model* window (Figure 8.23) select the *Discrete Ordinates (DO) (3)* radiation model. Settings of the *Angular Discretization* section determine the number of rays emitted from a certain mesh location. Default settings are fine in this case. Have a look at the *Iteration Parameters* section *(5)*. The number of *energy Iterations per Radiation* equation iteration can be set there. *10* is enough, taking into account the stability of the solution. A denser radiation iteration can reduce fluctuations in other transport equations in the case of more complicated projects. Activate the *Solar Irradiation (7)* model and click *Solar Calculator (8)*.

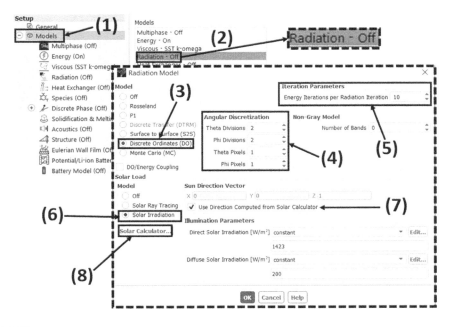

FIGURE 8.23

In the *Solar Calculator* window (Figure 8.24) set the location of the considered tube collector – go to the *Global Position (1)* section and set *Longitude [deg]: 50.066*, *Latitude [deg]: 19.917* and *Timezone (+-GMT): +2*. These are the values of the location of the *Faculty of Energy and Fuels* building at the *AGH University of Science and Technology (Krakow, Poland)*. Of course, you can set any other coordinates if you want to. Then in the *North (2)* section set *X* to *1* (it points the north direction) and in the *East* section set *Z* to *1* (as above). *Date and Time (4)* settings can remain as default. *Solar Irradiation Method (5)* allows to take into account *Fair Weather Conditions* and the cloud impact (*Sunshine Factor* equal *1*–100% clear sky). *Apply (6)* to proceed.

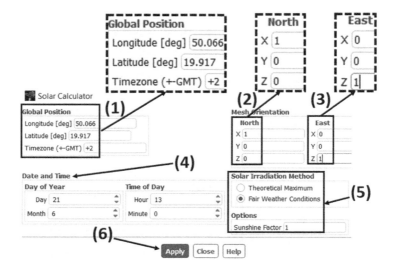

FIGURE 8.24

Let's define a tube glass material. First, click *2xLMB* on the *Materials (1)* tab in the *Outline Tree* window. In the *Materials (Task Page)* window you can see available by default: *air (2)* (default fluid), *aluminum (3)* (default solid). Click *2xLPM* on the latter one. The material libraries can be found in *Create/Edit Material*, under the *Fluent Database (4)* option. Note that, the brand new materials can be defined manually too. Set the *Name (5)*: *glass*. Remove *Chemical Formula (6)* – just leave this cell empty. Now go to the *Properties* section below. Set the new material properties (all *constant*): *Density [kg/m³]*: 2200, *Cp (Specific Heat) [J/(kg K)]*: 800, *Thermal Conductivity [W/(m K)]*: 1.3, *Absorption Coefficient [m⁻¹]*: 0.05, *Scattering Coefficient [m⁻¹]*: 0.02, *Scattering Phase Function*: isotropic, *Refractive Coefficient*: 1.5 (Figure 8.25).

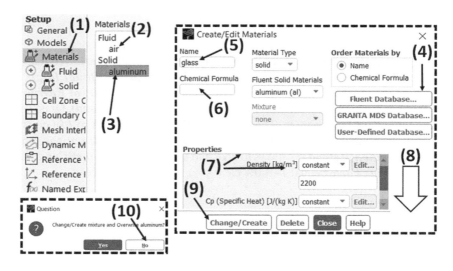

FIGURE 8.25

The created/added materials that already exist in the *Material* window can be edited. Click *air (1)* below the *Fluid* category (Figure 8.26). In the *Create/Edit Materials* window, go to the *Properties* section and find the *Density [kg/m³] (2)* drop-down list. Select *incompressible-ideal-gas (3)* (allowing to consider natural convection). Click *Change/Create*, to confirm settings.

Tutorial 2 – Evacuated Tube Solar Collector

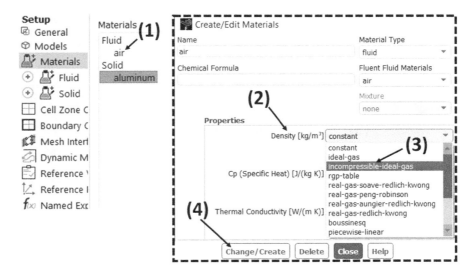

FIGURE 8.26

After creating and editing materials, the **Material** (*Task Page*) window includes one new tab – *glass* (**1**) and two default tabs: ***aluminum*** and modified ***air*** (Figure 8.27). Click ***Cell Zone Conditions*** (**2**) in the ***Outline View*** window. The ***Fluid*** tab below means that only fluid zones are present in the model. You can open the ***Fluid*** tab (+), to check how many zones are there. Due to applying the volume ***Selection Name*** for the whole cylinder (see Figure 8.9), there is only one *fluid* (**3**) zone. If your cell zone, for some reason, is solid, go to the ***Cell Zone Conditions*** (*Task Page*) window, click the zone (**4**) *1xLMB* and change its ***Type*** (**5**) to *fluid*.

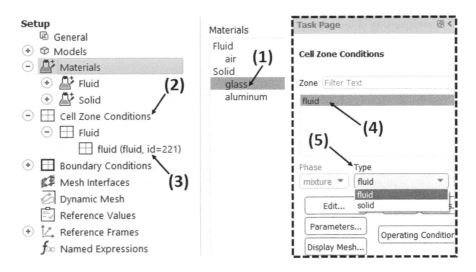

FIGURE 8.27

Time to define boundary conditions (BCs). Click ***2xLMB*** on the ***Boundary Conditions*** tab in the ***Outline View*** window, according to Figure 8.28. In the ***Boundary Conditions*** (*Task Page*) window, click ***1xLMB*** on the first tab: *fluid-wall* – these are the top and bottom walls of the cylinder that have no ***Selection Names***. Their ***Type*** should be *wall* (**3**). ***Interior-fluid*** and ***internal-zones*** (**4**) represent different elements of the domain filling (zones can be displayed using the ***Display*** tool – see Figure 8.21). Click ***wall-int*** (**5**) BC now and ***Edit...*** (**6**). This BC concerns the heat transfer from

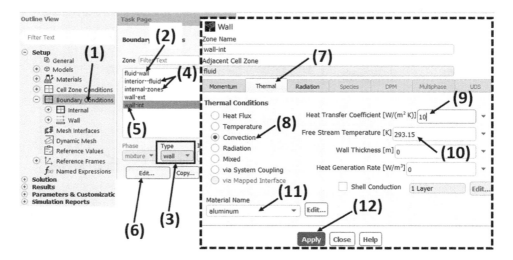

FIGURE 8.28

the absorber pipe to the working fluid. In the **Wall** BC window, go to the **Thermal** *(7)* tab and set this BC as **Convection** *(8)* in the **Thermal Conditions** section. Set **Heat Transfer Coefficient [W/m² K]** to **10** *(9)*. **Free Stream Temperature [K]** *(10)* (working medium) is **293.15**. The material is **aluminum** *(11)*. **Apply** *(12)* settings and close the window to proceed.

Click **1xLMB wall-ext** *(1)* (Figure 8.29) BC and **Edit...** *(2)*. In the **Wall** window, go to the **Thermal** *(3)* tab. This time set **Thermal Conditions** to **Mixed** *(4)* – both convection and radiation will be considered. Set heat transfer parameters *(5)*: **Heat Transfer Coefficient [W/m² K]** to **5**, **Free Stream Temperature [K]** (ambient conditions) to **283**, **External Emissivity** to **0.9** and **External Radiation Temperature [K]** to **283**. Change **Material Name** to **glass** *(6)*. Actually, the real significance of the latter setting comes up when non-zero **Wall Thickness [m]** is set. However, in this case, the material is changed in order to practice this activity. **Apply** settings but do not close the window.

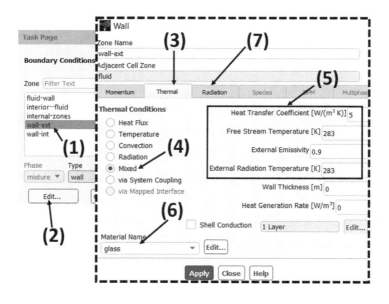

FIGURE 8.29

Tutorial 2 – Evacuated Tube Solar Collector

Go to the ***Radiation (1)*** tab (Figure 8.30) and change ***BC Type*** to ***semi-transparent (2)***. Radiation will be transmitted through the "virtual" glass tube wall. ***Beam Width*** settings can remain as default. In order to use conditions for the already defined location, in ***Solar BC Options*** activate both ***Use Beam Direction from... (3)*** and ***Use Irradiation from... (4)***. ***Apply*** settings and ***Close (5)*** the window.

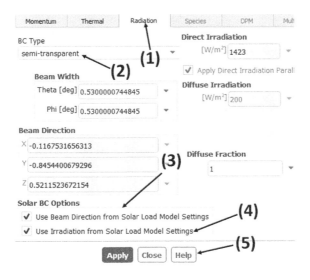

FIGURE 8.30

Go to the ***Methods (1)*** tab (***2xLMB***) of the project tree (as in Figure 8.31) and leave the default ***Pressure Velocity Coupling Scheme*** as ***Coupled (2)*** (recommended for natural convection models). Change the ***Spatial Discretization*** schemes: ***Pressure*** to ***Body Force Weighted*** (natural convection) ***(3)***, ***Momentum (4)*** and ***Energy (5)*** to ***First Order Upwind***, to improve the convergence. Leave the remaining options as default.

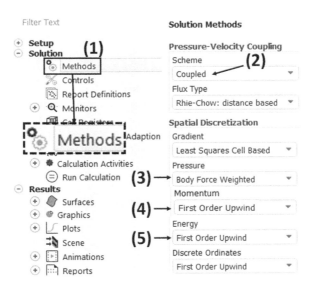

FIGURE 8.31

Now go to the *Initialization (1)* tab and set *Standard Initialization (2)*, as in Figure 8.32.

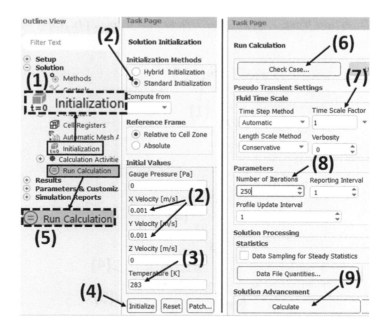

FIGURE 8.32

Set velocity to *0.001 m/s* for *X* and *Y* components *(2)*, set *Temperature (3)* to *283 K* and click *Initialize (4)*. Then go to the *Run Calculation (5)* tab. *Check Case* displays recommendations for the model (you can check it optionally), like checking the mesh or changing discretization schemes to higher, for example, the second order upwind (the model in this exercise is simplified, so such kind of information can appear when you click *Check Case*). *Time Scale Factor (7)* (TSF) is a parameter available in the *Coupled Pressure Velocity Coupling* mode. Reduction of TSF helps to stabilize the solution in the case of more complicated problems, but it can extend the solution time. Set the *Number of Iterations* to *250 (8)* and click *Calculate (9)*.

During the solution, you can track the curves *(1)* corresponding with *Residuals* in the governing equations that are being solved (Figure 8.33). Note that the curve that corresponds to the *do-intensity (3)* equation has a different shape compared to others – it is solved after every ten iterations of the remaining equations, so it has a stair-like shape. The values of the residues in the subsequent equations and the current iteration number can be read in the *console (4)*.

Tutorial 2 – Evacuated Tube Solar Collector

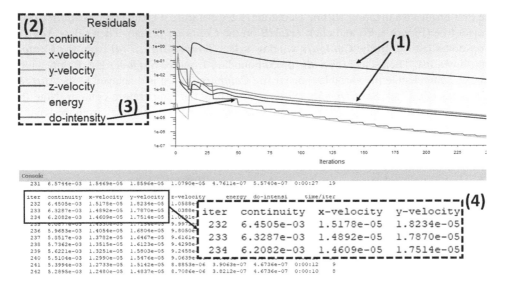

FIGURE 8.33

8.5 POSTPROCESSING

After the solution process, open the (+) *Results (1)* tab in the *Outline View* window. Then open the (+) *Surfaces (2)* tab, to create a new surface that will be used to display visualizations a bit later. Click *1xRMB* on the *Surfaces (2)* tab → *New* → *iso-Surface… (3)*. In the *Iso-Surface* window (Figure 8.34), select *Mesh… (4)* from the *Surface of Constant* drop-down list. Then select *X-coordinate (5)* from the drop-down list just below. Click the *Compute (6)* button to see the available spatial range of the *X*-coordinate. It is displayed in the *Min [m]* and *Max [m]* cells *(7)*. The surface that is being created should be somewhere in the middle of the range – set *Iso-Values [m]* to *0.1 (8)*. The slide bar *(9)* can be used to set this location too. Click *Save (10)* to save the surface. New iso-surface is visible in the project tree below the surfaces tab.

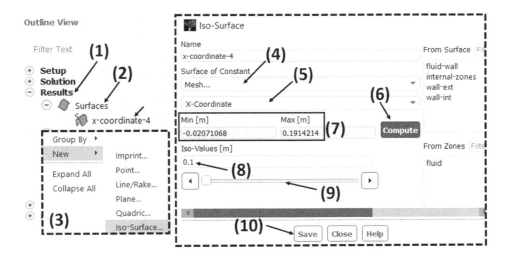

FIGURE 8.34

The first result visualization will be a contour of temperature. Open (+) the ***Graphics (1)*** tab of the project tree (Figure 8.35) and click *1xLMB* on the ***Contours (2)*** tab. Then select ***New…*** from the drop-down list ***(3)***. In the ***Contours*** window select ***Temperature… (4)*** from the ***Contours*** of the drop-down list. Then select the **wall-int *(5)*** boundary from the ***Surfaces*** list. Before saving the contour, you can change the variable range using ***Compute (6)*** – it is displayed in the ***Min [K]*** and ***Max [K]*** cells ***(7)***. Click ***Save/Display (8)*** now, to display the temperature contour. Note that the temperature distribution on the absorber pipe is heterogeneous ***(9)*** – it comes from the solar irradiance direction (top pipe side is exposed).

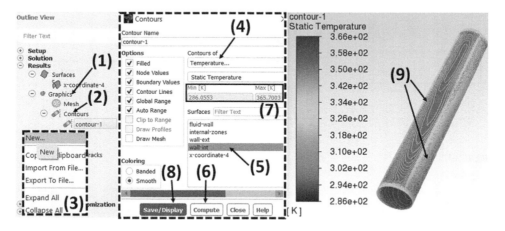

FIGURE 8.35

Saved ***Contour (1)*** is available now below the ***Contours*** tab, as in Figure 8.36, so you can turn back to it whenever required. Time to use the already created surface (see Figure 8.34), to visualize the velocity vector field. Of course, you can use this (and any other) surface to display any contours too. Click *2xLMB* on ***Vectors (2)*** and in the ***Vectors*** window, make sure that the selected variable is ***Velocity Magnitude (3)***. It is recommended to change the ***Style*** to the ***arrow***. Furthermore, you can set ***Scale*** to ***1.5*** (longer arrows) and the ***Skip 4*** vectors ***(4)*** (less vectors – clearer visualization). Select the ***X-coordinate… (5)*** surface from the ***Surfaces*** list and click ***Save/Display (6)***. Let's try to analyze the air circulation – the air heats up close to the absorber pipe ***(7)*** and flows up until it reaches the external glass pipe. Then heat removal through the glass takes place and colder air runs down the walls ***(8)***. Thus, although the air velocities are quite low ***(9)***, its presence inside the collector pipe causes heat losses.

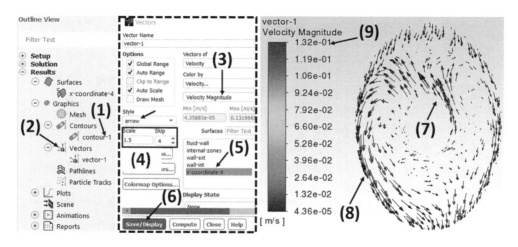

FIGURE 8.36

Tutorial 2 – Evacuated Tube Solar Collector 131

A comparison of surface temperatures can be carried out using reports. Open the (+) *Reports (1)* tab of the project tree (Figure 8.37) and click *2xLMB* on *Surface Integrals (2)*, to open its window. Change *Report Type* to *Area Weighted Average*. *Field Variable* should be *Temperature… (4)* (*Static Temperature* below). Select both the *wall-ext* and *wall-int (5)* boundaries from the *Surfaces* list and click *Compute (6)*. The report is displayed in *Console (7)*. Of course, due to heat losses to the environment, the average *wall-ext* surface temperature is much lower. This confirms the previous conclusions about the impact of the air circulation on the temperature distribution in the domain.

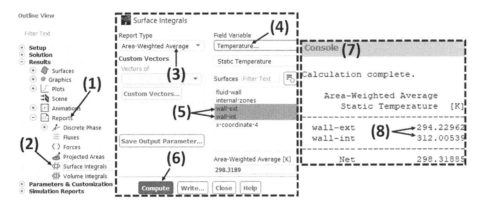

FIGURE 8.37

Let's compare the above considered case with the vacuum-tube collector. The void inside the tube is treated in this example as a new material. To create it, go back to the *Materials (1)* tab (*2xLMB*) of the project tree (Figure 8.38). You'll create it analogously to the glass material (see Figure 8.25), but this time it is fluid. Click *1xLMB* on *air (2)* (*Materials* list) → *Create/Edit… (3)*. Change *Name* to *void (4)* and go to the *Properties (5)* section. From the CFD fundamentals point of view, fluid material can only pretend real vacuum or very thinned air. So, actually, to simulate this case, the selected air properties will be changed: *Density [kg/m] (constant)* to *0.0001*, *Cp (Specific Heat) [J/(kg K)]* to *1e06* and *Thermal Conductivity [W/(m K)]* to *1e-06*. Some properties can remain unchanged because fluid flow transport equations will be deactivated before the simulation. Click *Change/Create (6)* to confirm settings. Do not substitute the air material, but create an additional one – answer *No (7)* to the displayed question. *Void* material is now available on the *Materials* list *(8)*.

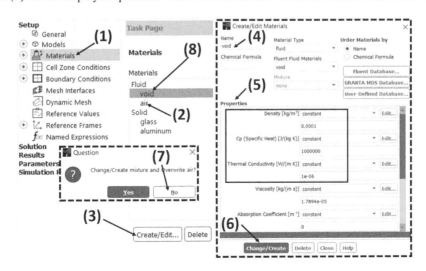

FIGURE 8.38

Click **2xLMB** on the **Cell Zone Conditions (1)** tab to replace its default material with the just created new one. Click **1xLMB** on *fluid (2)* and **Edit… (3)**, according to Figure 8.39. In the **Fluid** window, select *void* from the **Material Name (4)** drop-down list and Apply (5).

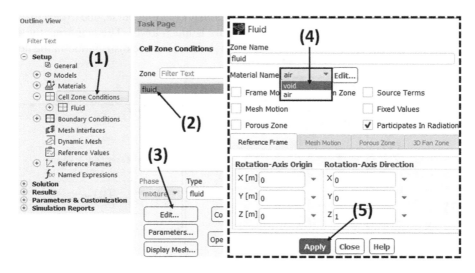

FIGURE 8.39

Click **2xLMB** on the **Controls (1)** tab in the **Outline View** window (Figure 8.40). Then click **Equations… (2)**, to deactivate the **Flow (3)** equations (just click *1xLMB* on **Flow**). Thanks to this, natural convection inside the tube will not be possible. Click **OK** in the **Equations** window to proceed.

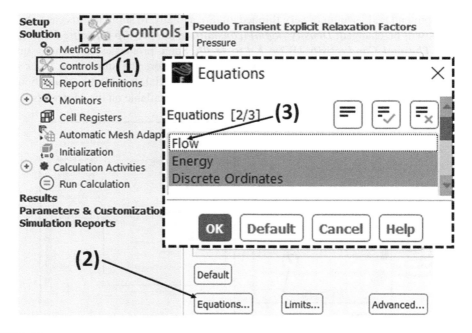

FIGURE 8.40

Tutorial 2 – Evacuated Tube Solar Collector

Go to the ***Initialization (1)*** tab and insert velocity of ***0 [m/s]*** in each direction (***X, Y, Z***) ***(2)***, as in Figure 8.41. Then click ***Initialize (3)***, to update initial conditions. Go to the ***Run Calculation (4)*** tab and run the solution (***Calculate*** button) – same as the first time (see Figure 8.32).

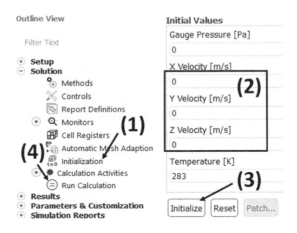

FIGURE 8.41

Note that during the solution process, only two equations are solved this time (Figure 8.42) – the ***energy*** equation and the discrete ordinates radiation model (***do-intensity***) equation. As previously, the curve corresponding with the latter has a stair-like shape.

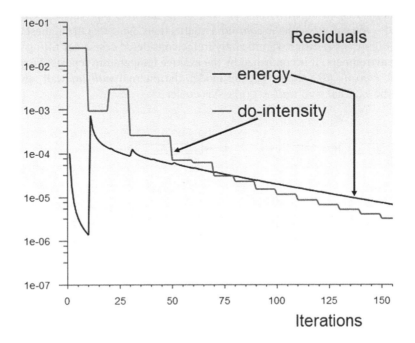

FIGURE 8.42

After the solution, try to create the same visualizations and reports as after the first simulation, to compare the results (Figure 8.43). You can use the already existing temperature *contour-1 (1)*. Actually, in this case, velocity *vector-1 (2)* has no practical sense due to no solution of continuity and momentum equations (flow equations). Check temperature at the *wall-int* and *wall-ext* surfaces, using *Surface Integrals (3)* (see Figure 8.37).

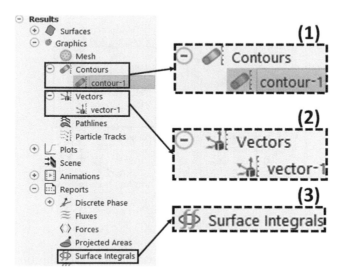

FIGURE 8.43

Regardless of the tube filling (*air/void*), one side of the absorber pipe surface is always hotter *(1)* due to solar load – you can see it on the *contour* visualizations. Note that the highest temperature on the legend *(2)* (Figure 8.44) differs significantly in the considered cases. Air filling intensifies heat transfer to the environment. It is confirmed by the average temperature reports (*Surface Integrals*) displayed in the *console (3)* – in the second model, the internal *wall-int* wall (absorber pipe) is warmer, while the external wall *wall-ext* (glass) is cooler.

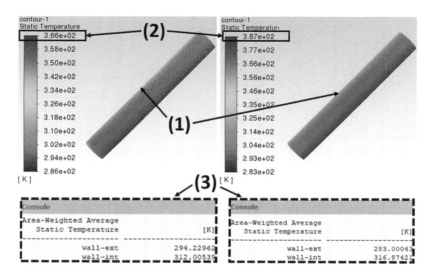

FIGURE 8.44

9 Tutorial 3 – Heat Receiver for a Solar Concentrating System

9.1 EXERCISE SCOPE

This tutorial presents the procedure of simulating the operation of the heat receiver dedicated to concentrating systems with a parabolic dish (see classification in Section 6.3.4). Due to the presence of an optical element *(1)* in the real installation, the irradiance distribution on the receiver *(2)* is uneven and depends on the receiver location in the concentrating system (Figure 9.1). In the course of the simulation, it is possible to analyze the dependence between the distribution of irradiance and the thermal operation of the heat receiver.

Try to recall from the theoretical background (Section 6.4.3) the factors which influence the operation of concentrating systems and how to calculate their basic work parameters.

Among others, this exercise can teach you how to:

- use the sweep tool to create the internal channel,
- project the sketch into geometry in ***SpaceClaim DirectModeler***,
- refine your mesh according to its appearance and general meshing recommendations, using different functionalities of ***Ansys Meshing***,
- create new material (define its properties) in ***Ansys Fluent***,
- calculate the power of a heat receiver using ***Expressions*** in ***Ansys Results***.

Note how the geometry is prepared to imitate the irradiance distribution over the absorbing surface.
All of the images included in this tutorial use courtesy of Ansys, Inc.

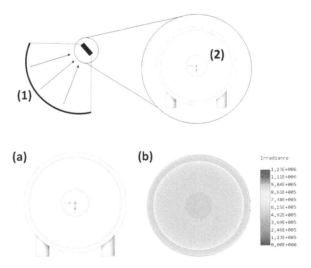

FIGURE 9.1 (a and b) Location of the analyzed heat receiver in a parabolic dish concentrating system.

DOI: 10.1201/9781003202226-11

9.2 PREPROCESSING – GEOMETRY

Drag the *Geometry* module *(1)* to the Ansys Workbench *Project Schematic* (Figure 9.2). To open the geometry editor, double click the left mouse button (LMB) on the A2 cell *(2)*.

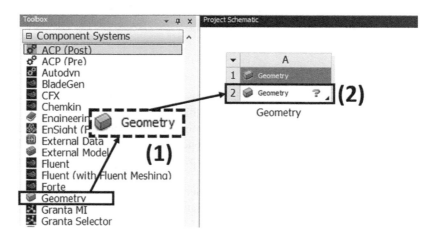

FIGURE 9.2

The *Sketch* tab will open (Figure 9.3). Firstly, click LMB on the *Plan View (1)* to view the XZ grid head-on. Then, draw a single circle using the *Circle* tool from the *Create* ribbon *(2)*. Start drawing from the coordinate system origin *(3)* and set the diameter to 150 mm. You can check the sketch dimensions with the *Dimension* tool *(4)*. Finish your work by choosing the *End Sketch Editing* icon *(5)* so you will be automatically switched to the *Design* tab.

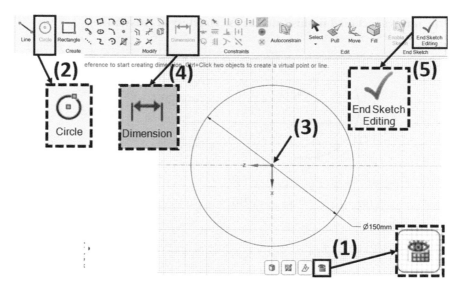

FIGURE 9.3

In the *Design* tab click on the *Pull* tool *(1)* from the *Edit* ribbon, which will allow you to create a cylinder from the previously prepared sketch. Firstly, select the green (light gray) circular face (it should change its color to orange) and from the popup window choose the *Pull Both Sides* option *(2)*, as in Figure 9.4. Then set the cylinder thickness to 30 mm. After that, come back to the *Sketch* mode *(3)*.

Tutorial 3 – Heat Receiver for a Solar Concentrating System

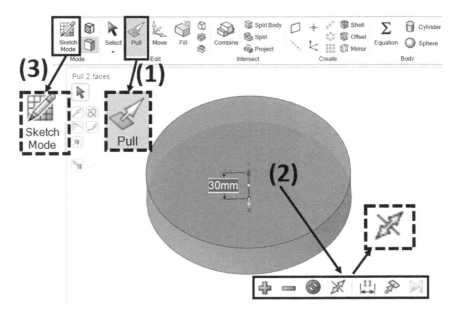

FIGURE 9.4

The main body of the receiver is ready, so it is time to prepare the inlet and outlet channels (Figure 9.5). Start by choosing the ***Select New Sketching Plane (1)***. It should be oriented in the YZ plane, so click on the X-axis. Orient the sketch view perpendicularly using the triade *(2)* visible in the bottom-left corner of the workspace. Draw a ***Circle*** with a 20 mm diameter *(3)* – place its center on the Z-axis. Set the distance between the circle center and the Y-axis to 48 mm using the ***Dimension*** tool *(4)*. To prepare the second circle use the ***Mirror*** tool *(5)* from the ***Modify*** ribbon. Set the Y-axis as the mirror plane and then the newly drawn circle as the body to be mirrored. Finish your sketch *(6)*.

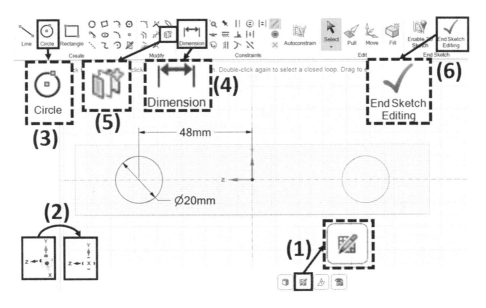

FIGURE 9.5

Select the **Surface** object from the **Structure** of your project *(1)* to see the prepared sketch (Figure 9.6). Using the **Pull** tool *(2)* extrude the two circles to cylinders of 80 mm length. During this operation, it is important to click on the white background of the workspace, not on the geometry. Finally, go to the **Sketch mode** *(3)*.

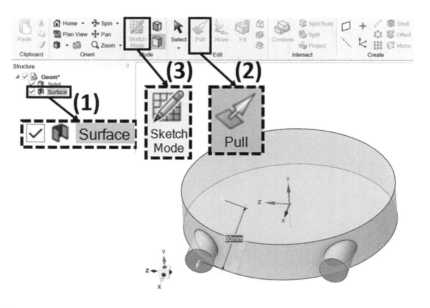

FIGURE 9.6

Now it is time to design the internal channel. Choose the *XZ* as the sketching plane and set your view perpendicularly to it, as in Figure 9.7. Using the **Line *(1)*, Construction Line *(2)* and Arc *(3)*** tools from the **Create** ribbon draw a shape shown in the picture. You can draw only half of the sketch and **Mirror** it by the *X*-axis *(4)*. The sweep trajectory is ready.

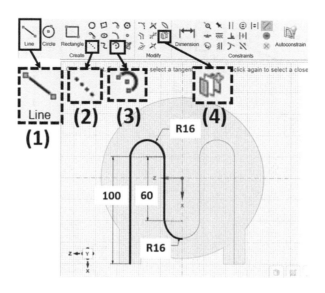

FIGURE 9.7

Tutorial 3 – Heat Receiver for a Solar Concentrating System

Now, prepare the shape which will be swept by the prepared trajectory. *Select New Sketching Plane* on the inlet/outlet face *(1)* as shown in Figure 9.8. Draw a *Circle (2)* with a diameter equal to 14 mm. Finally, switch the tab to the *Design* mode *(3)*.

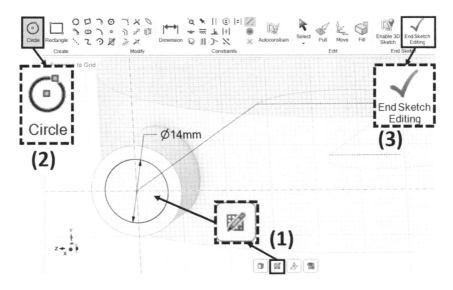

FIGURE 9.8

In this step create the geometry of the internal channel (Figure 9.9). Click on the newly created circular face *(1)* and the *Pull* tool *(2)*. Choose the *Sweep* option *(3)* and ensure that the *No merge* mode *(4)* is active. It prevents unification of the newly created body and the existing one. Now pick the trajectory to sweep along *(5)* and confirm the final choice by the Enter button. As a result, two bodies should be visible in the project *Structure (6)*. By clicking on them you could see that they are overlapping in the space in the internal channel where fluid should be located.

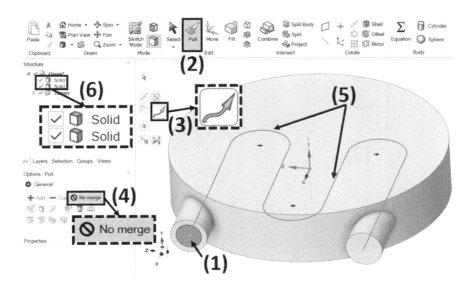

FIGURE 9.9

To remove the solid from the internal channel use the **Combine** tool *(1)* from the **Intersect** ribbon. Choose the receiver body *(2)* as the target object (as in Figure 9.10). Ensure that the **Cutter** *(3)* option is active and choose the internal channel body *(4)* as the cutter object. This operation will result in two completely separate bodies. For the final operations switch the tab to **Sketch Mode** *(5)*.

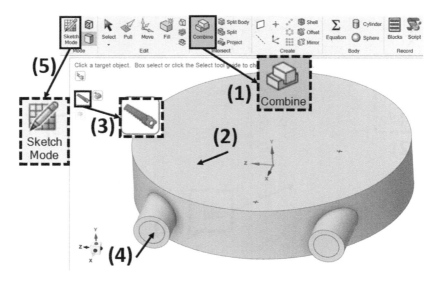

FIGURE 9.10

The upper side of the receiver collects concentrated solar irradiation. Due to the parabolic shape of the mirror, the amount of energy collected by the receiver varies with the distance from its center. Therefore, this surface should be divided into circular rings, where each ring represents one level of irradiation, as in Figure 9.11. To do this, **Select New Sketching Plane** *(1)* on the upper surface of the receiver and view the grid head-on *(2)*. Draw two circles *(3)* with the diameters of 40 mm and 136 mm. Finish editing your sketch *(4)*.

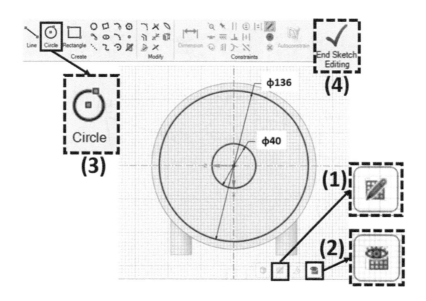

FIGURE 9.11

Tutorial 3 – Heat Receiver for a Solar Concentrating System

Two circular surfaces will be automatically created from the previous sketch, but they are overlapping. To solve this problem, imprint the previously prepared sketch into the solid body. Use the ***Project*** tool *(1)* from the ***Intersect*** ribbon. Click on the edges to imprint *(2)* and then select the ***Target Face (3)*** option. Now pick the face limiting the projection – in this case, it is the upper surface of the receiver. After this operation, the automatically created surfaces will be still visible in the project ***Structure***, as in Figure 9.12. ***Delete*** them *(4)*.

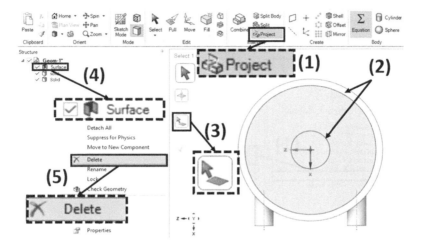

FIGURE 9.12

The final part of the geometry preparation is to create one part from the two bodies visible in the project ***Structure*** (Figure 9.13). Select both bodies *(1)* and choose the ***Move to New Component*** option *(2)* from the popup window. Check which body represents the receiver and which represents the fluid in the internal channel. You can do this by clicking on the V sign in the square located to the left of the body name. Then change the names of these bodies to solid and fluid, respectively *(3)*. The last step is to ***Share*** the topology *(4)* of the component.

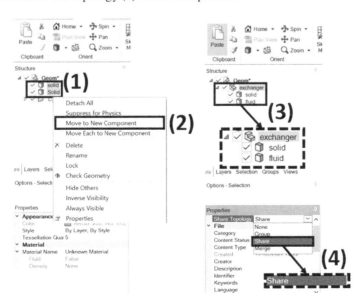

FIGURE 9.13

9.3 PREPROCESSING – MESHING

Select the *Mesh* module *(1)* from the *Component System* ribbon and drag it to the Ansys Workbench *Project Schematic (2)* as in Figure 9.14. Create an appropriate connection between the geometry and the mesh modules *(3)*. Double click the left mouse button (LMB) on the B3 cell *(4)* and open Mesh.

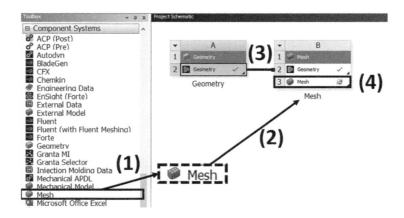

FIGURE 9.14

Find the *Geometry (1)* icon in the project tree and check if there is one component that consists of two bodies: solid receiver and fluid filling as in Figure 9.15. Then move to the *Mesh (2)* option. In the *Details of Mesh* window change the appropriate basic settings that are required for the CFD simulations: *Physics Preference – CFD (3)* and *Solver preference – Fluent (4)*. Generate the first mesh by clicking 1xRMB on the *Mesh (2)* and selecting the option *Generate Mesh (5)* from the popup window.

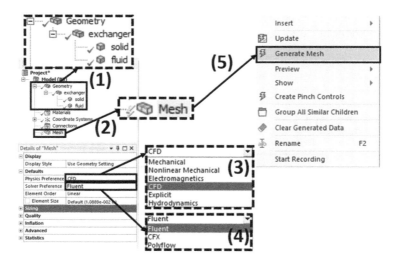

FIGURE 9.15

Find the *Quality* option in *Details of Mesh* and set the **Mesh Metric** parameter as *Skewness (1)*. It describes the difference between the shape of the cell and the shape of an equilateral cell with the same volume. The lower skewness, the better. The upper limit for skewness is 0.95 – higher skewed cells decrease accuracy and destabilize the solution. Rate the mesh quality based on the minimum,

Tutorial 3 – Heat Receiver for a Solar Concentrating System 143

maximum and average values of the analyzed parameter *(2)*. In the *Statistics* tab check the number of the mesh *Nodes* and *Elements (3)*. Compare the location of the mesh cells on the fluid and solid domains *(4)* as in Figure 9.16 – mesh topology is consistent because these two bodies create a single component with the shared topology. Nevertheless, the generated mesh is inadequate due to the size of the mesh elements.

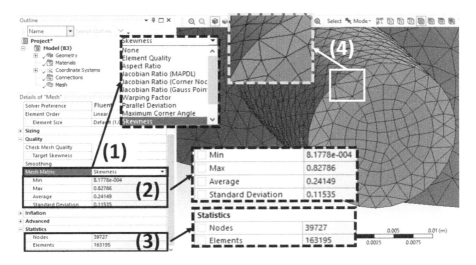

FIGURE 9.16

In the *Details of Mesh* change the *Element Size* value from default to *2 mm (1)*. Be careful and check your default unit of length. Generate the second mesh by clicking 1xRMB on *Mesh (2)* and selecting the option *Generate Mesh*. Check *Quality (3)* and *Statistic (4)* for the newly generated mesh. Now the mesh elements have more regular shapes and they are packed more densely *(5)* as in Figure 9.17.

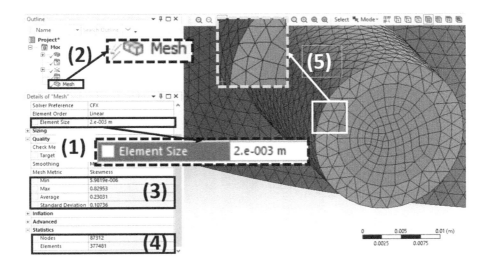

FIGURE 9.17

The fluid is flowing through the internal channel so the inflation layer should be generated (Figure 9.18). It increases the mesh density in the regions adjacent to the wall and allows to solve the boundary layer for turbulent flows. This function is dedicated only to the fluid domain, so select the *solid* domain *(1)* from the project tree and click 1xRMB. Choose the *Supress Body (2)* option from the popup window to hide the solid for physics. Use the *Volume Selection* mode *(3)* and click on the fluid body *(4)*. 1xRMB and choose option *Insert (5)* → *Inflation (6)*.

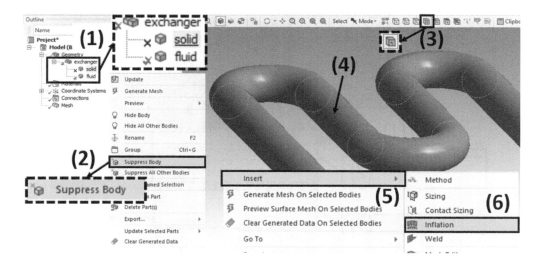

FIGURE 9.18

In the *Details of Inflation* window find the *Geometry (1)* option. In this row *1 Body* should be visible – the previously selected fluid domain. In the *Boundary (2)* row it is required to select all the surfaces that are in direct contact with the channel walls. Use the *Face Selection* mode *(3)* and select the side surfaces *(4)* as in Figure 9.19. The inlet and outlet surfaces must not be selected! Apply the selection *(2)*. Choose the *Inflation Option (5)* as the *First Layer Thickness* to precisely control the size of the first mesh element in inflation and set the *First Layer Height* to 0.25 mm. Leave other parameters unchanged.

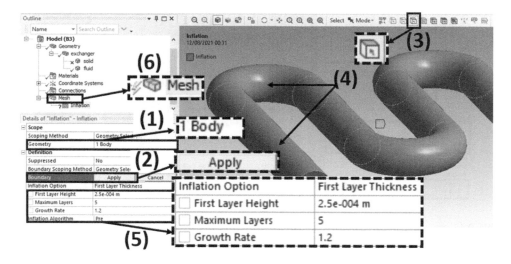

FIGURE 9.19

Tutorial 3 – Heat Receiver for a Solar Concentrating System

Restore the solid body by clicking 1xRMB on its name in the project tree *(1)*, then select *Unsuppress (2)* from the popup window list. *Generate* the third mesh for the whole component *(3)* as in Figure 9.20. Again, compare the mesh *Quality (4), Statistics (5)* and appearance *(6)* to the previous cases.

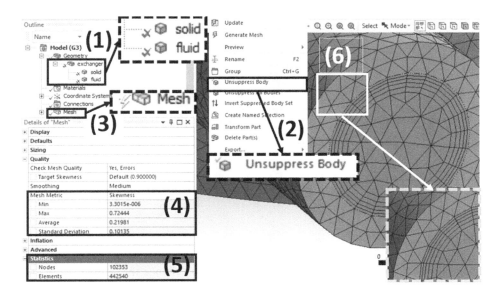

FIGURE 9.20

Activate the *Volume Selection* mode *(1)* and select the fluid domain *(2)* as in Figure 9.21. Click 1xRMB and from the list select *Insert (3)* → *Sizing (4)*. In the *Details of Body Sizing* window find the *Type* option and insert value *1.5 mm* in the *Element Size* row *(5)*. As the fluid domain is curved, activate the *Capture Curvature* option *(6)*. Now you can *Generate* the final mesh for the whole model *(7)*.

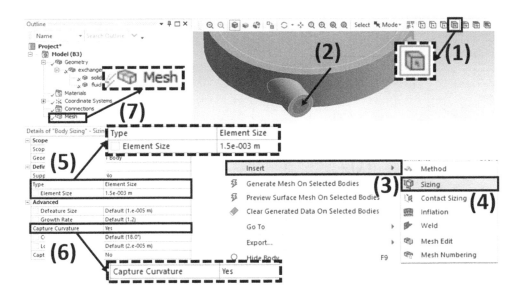

FIGURE 9.21

The final mesh is ready. Look at the mesh appearance *(1)* – pay special attention to the area closest to the channel walls (Figure 9.22). Evaluate mesh **Quality** *(2)* and **Statistics** *(3)*. In the **Mesh Metrics** window check the shape of the mesh elements *(4)* and their participation in the total number of mesh elements and quality *(5)*.

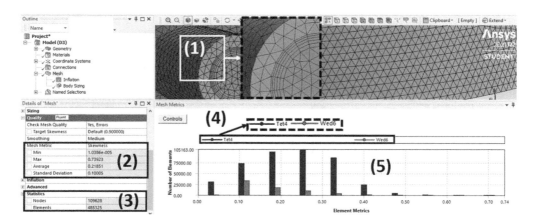

FIGURE 9.22

To locate the elements characterized by the worst quality, change the options of the displayed chart. Click **Controls** in the **Mesh Metrics** window and change the minimum value of **X-Axis in the popup** to display only the worst-quality elements. Change this parameter to the value defined by the formula: maximum skewness – 0.05. Change also the maximum value of **Y-Axis** to 10 (the number of the worst elements is low). There is no Apply button, just click **X** in the corner of the popup window *(2)*. Click on the bars displayed on the chart *(3)* and select them all. The worst-quality elements are now visible in the workspace *(4)* as in Figure 9.23. Now check another quality parameter **Orthogonal Quality**, which (in simplification) describes the equilaterality of the cell. Value equal to zero is the worst and means really skewed cells, whereas value one is the best and means fully equilateral cells (like a cube). Generally, Orthogonal Quality should not be lower than 0.05. To check the minimum, maximum and average value of this parameter, just select **Orthogonal Quality** parameter in the **Mesh Metric** row in **Quality** tab *(5)*.

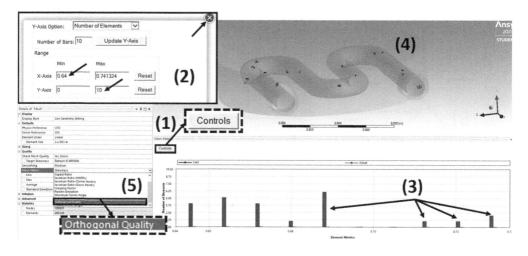

FIGURE 9.23

Tutorial 3 – Heat Receiver for a Solar Concentrating System

Named Selection is the function that allows to assign names to different objects to make them easier to find in the solver. Remember: to name a surface, use the *Face Selection* mode and to name a body use the *Volume Selection* mode *(1)*. Click *1xRMB* on the central circular surface *(2)* as in Figure 9.24 and then choose the *Create Named Selection (3)* option from the dropdown list. You can also just press N button from the keyboard. Then insert the name for the new *Named Selection* in the popup window *(5)*: *wall_i1* in this case and confirm by clicking *OK*.

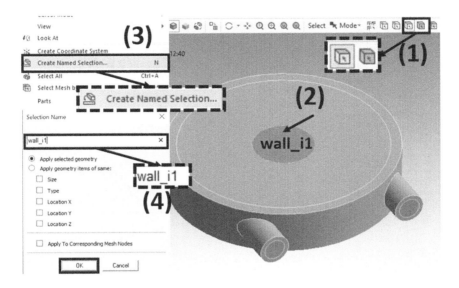

FIGURE 9.24

Repeat the same step and name other surfaces *(1)*: *wall_i2*, *wall_i3*, *wall_cold*, *wall_adaiabtic*, *inlet*, and *outlet* as in Figure 9.25. Give also a name to solid and fluid bodies *(2)*. Their names should appear in the project tree, below the *Named Selections* tab *(3)*. When you click on it, you can see in the workspace the relevant object highlighted in red.

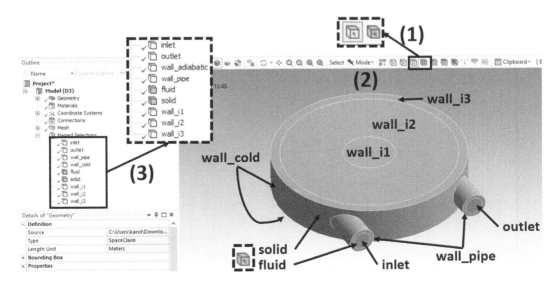

FIGURE 9.25

9.4 PREPROCESSING – SOLVER SETTINGS

Drag the *Fluent (1)* module to the *Project Schematic* in Workbench and set the appropriate connection *(2)* between the *Mesh* module (B3 cell) and *Fluent Setup* (C2) as in Figure 9.26. Click *1xRMB* on *B3* cell *(3)* and choose the *Update* option *(4)* to export the mesh to the solver. Start the solver by clicking *2xLMB* on the *C2* cell and in the popup window run the *Parallel* processing mode with the number of cores available on your computer.

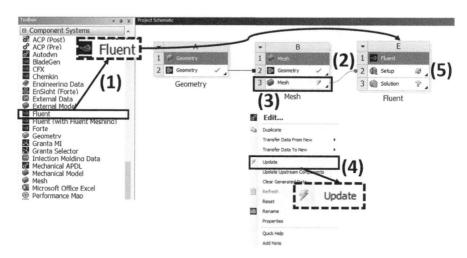

FIGURE 9.26

In the solver *Setup* find the *General* option *(1)* (Figure 9.27). Firstly, click *1xLMB Check (2)* button to check the details of the imported geometry and mesh. They will be displayed in the *Console (3)* window. Set the *Solver Type* as *pressure-based (4)* to analyze fluid flows with low velocity (< 200 m/s). To provide a steady state simulation choose the *Time* option as *Steady (5)*. The heat receiver during operation is tilted at an angle of 45°. To consider this in simulation, activate *Gravity* option and set the values of gravitational acceleration as: $X = 6.94 \, m/s^2$ and $Y = 6.94 \, m/s^2$ *(6)*.

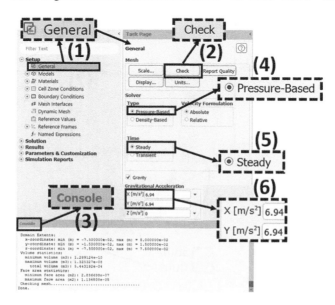

FIGURE 9.27

Tutorial 3 – Heat Receiver for a Solar Concentrating System   149

Move to the ***Models*** option *(1)* to select appropriate mathematical models of physical phenomenon occurring in the analyzed case (Figure 9.28). Firstly, activate the ***Energy Equation (2)*** for heat transfer. Do this by ticking the available option in the popup window *(3)* and confirm ***OK***. Secondly, choose the ***Viscous (4)*** option to choose a turbulence model for the fluid flow. In case of an inflation layer in the mesh, ***k-omega (5) the*** turbulence model in ***SST (6)*** mode is suitable.

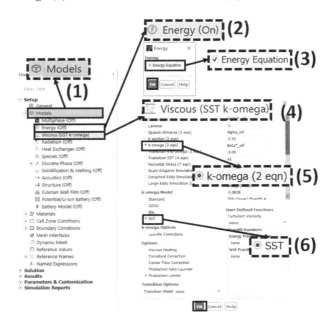

FIGURE 9.28

Now let's focus on ***Materials (1)***. By default, the air is set as fluid and aluminum as solid, as in Figure 9.29. In the studied case, the working fluid is thermal oil and the receiver is made of aluminum. Thermal oil is not specified in the Fluent database, so the new material should be defined manually. Start by clicking ***2xLMB*** on the ***air (2)*** material and change its name to thermal oil. It is necessary to describe density, specific heat and viscosity as temperature-dependent parameters. For these properties, select the ***polynominal*** option *(4)* from the dropdown list.

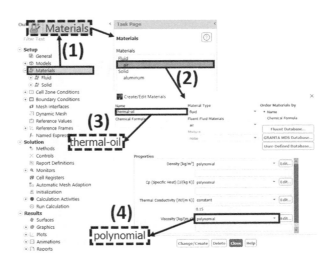

FIGURE 9.29

The temperature dependence of viscosity can be described by the equation:

$$v = 34.93062 - 0.4530667 \times T + 2.436633 \times 10^{-3} \times T^2 - 6.951656 \times 10^{-6} \times T^3$$

$$+ 1.10933 \times 10^{-8} \times T^4 - 9.386901 \times 10^{-12} \times T^5 + 3.290347 \times 10^{-15} \times T^6 \left[\frac{kg}{m\,s}\right],$$

so set the number of **Coefficients** as 7 *(1)* and insert their values *(2)*, starting from the constant term. A graphic interpretation of the function is also provided *(3)* as in Figure 9.30. Repeat the above steps with specific heat and density.

$$c_p = 1970.2 + 0.6274 \times T \left[\frac{J}{kg\,K}\right],$$

$$d = 1079.2 - 0.6223 \times T \left[\frac{kg}{m^3}\right].$$

The thermal conductivity is **Constant** and equal $\lambda = 0.15 \left[\frac{W}{m\,K}\right]$. Confirm all the changes *(4)*.

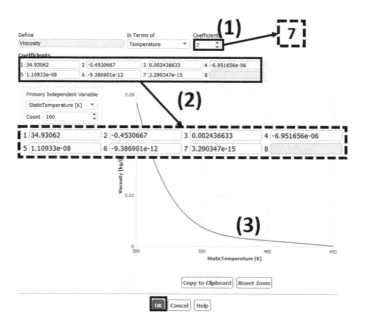

FIGURE 9.30

Find the **Cell Zone Conditions** *(1)* option (Figure 9.31) and check if the fluid body type is defined as *fluid (2)*. Then click **Edit** *(3)* and in the popup window set the fluid material as **thermal oil** *(4)*. Confirm changes by clicking **Apply** *(5)*. Repeat these steps with the solid domain (made of **aluminum**).

Tutorial 3 – Heat Receiver for a Solar Concentrating System

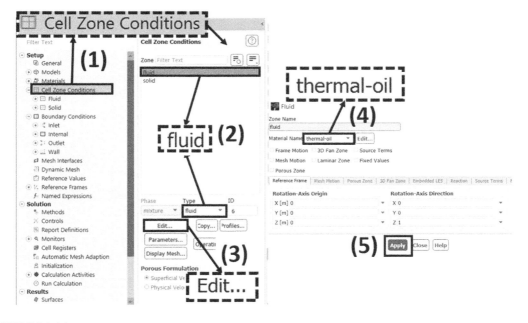

FIGURE 9.31

It's time to define Inlet, the first boundary condition. Click *2xLMB* on the ***Boundary Conditions (1)*** option and select ***Inlet (2)*** as in Figure 9.32. Change its ***Type*** from velocity-inlet to ***mass-flow-inlet (3)***. Open the details window by clicking *2xLMB* on the zone name *(2)*. In the ***Momentum*** tab insert inlet ***Mass Flow Rate*** value 0.05 kg/s *(4)* and set ***Direction Specification Method*** as ***Normal to Boundary (5)***. ***Turbulence*** should be described by ***Intensity*** (default value 5% is OK) and ***Hydraulic Diameter (6)***. In this project, the internal channel has got a 14 mm diameter and it's recommended to apply this value *(7)*. In ***Thermal*** tab set ***Total Temperature*** 288 K *(8)*.

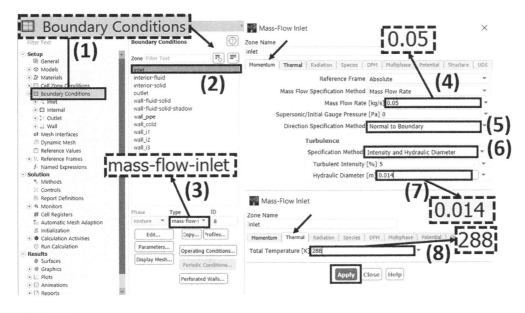

FIGURE 9.32

Being still in the *Boundary Conditions (1)*, move to the *Outlet (2)* as in Figure 9.33. Its default type is *pressure-outlet (3)*, which describes gauge pressure instead of outlet velocity. In the *Momentum* tab set the *Gauge Pressure* as 0 Pa *(4)* and define *Turbulence (5, 6)* as in the case of the Inlet boundary. In the *Thermal* tab, it is required to provide information about *Backflow Total Temperature*, if the reversed flow is expected. In this case, the backflow will not occur so leave the default value *(7)* and *Apply* all the changes.

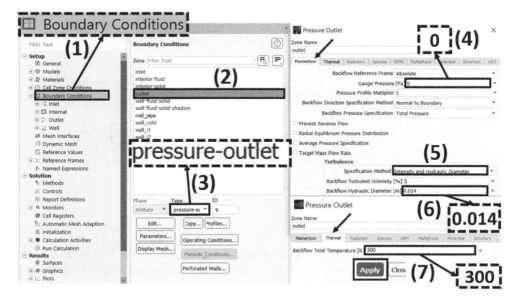

FIGURE 9.33

Now, move to the wall_adiabatic boundary condition *(1)* as in Figure 9.34. Its type is set by default as *wall (2)* with *Heat Flux (3)* equal to zero *(4)*. This is fine in the case of thermally insulated surfaces. Confirm these values by clicking *Apply*.

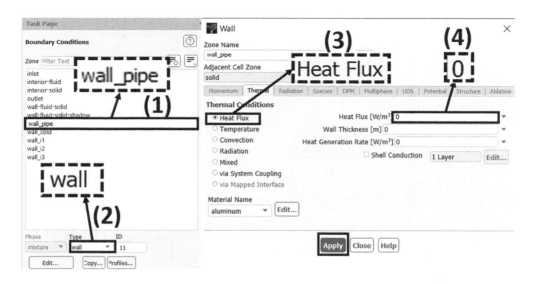

FIGURE 9.34

Tutorial 3 – Heat Receiver for a Solar Concentrating System   153

Wall_cold *(1)* is another wall – *Type (2)* boundary condition characterized by *Mixed (3)* heat transfer (Figure 9.35). Convection occurs with *25 W/(m² K) Coefficient (4)* and 298 K *Free Stream Temperature (5)*. The *External Radiation Temperature* is also 298 K *(5)* and the *External Emissivity* is 0.8 *(6)*.

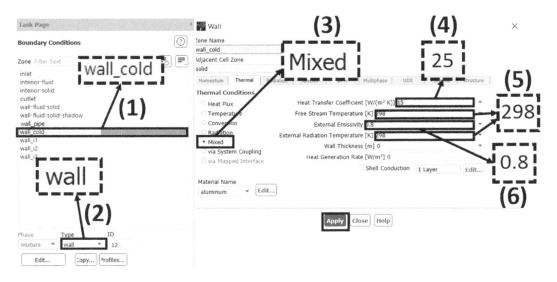

FIGURE 9.35

Boundary conditions on the receiver upper surface: *wall_ir1*, *wall_ir2* and *wall_ir3 (1)* are walls *(2)* with *Heat Flux* type heat transfer (Figure 9.36). The irradiance values are equal: *100,000 W/m²*, *70,000 W/m²*, and *25,000 W/m²*, respectively *(4)*.

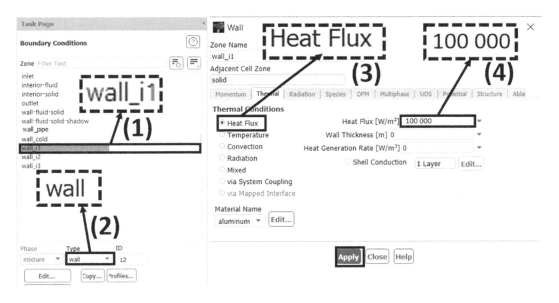

FIGURE 9.36

When the physics for the model is already determined, the next step is to define numeric *Methods* *(1)* of the solution process (Figure 9.37). Set the algorithm of pressure-velocity coupling as *Coupled* *(2)*, which ensures a stable computation process. Check, if *Spatial Discretization* for each parameter is based on *Second Order* or *Second Order Upwind (3)* schemes, which provide a more accurate solution and reduce the numerical diffusion.

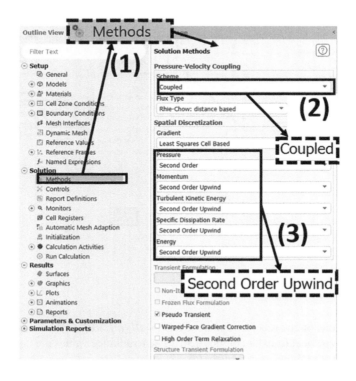

FIGURE 9.37

To supervise the computation process create at least one *Report definition (1)*, as in Figure 9.38. Click 2xLMB on this option and a popup window will open. Create a *New* one *(2)* by choosing from the dropdown list *Surface Report* with *Mass-Weighted Average* type *(3)*.

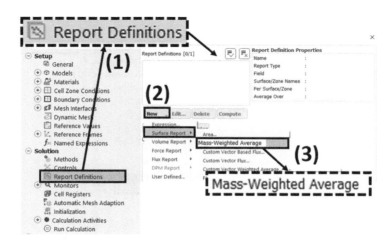

FIGURE 9.38

Tutorial 3 – Heat Receiver for a Solar Concentrating System

Change the Report Definition *Name* to temperature_outlet *(1)* and select *Temperature* as *Field Variable (2)*, as in Figure 9.39. In this case, there is only one option provided, *i.e. Static Temperature (3)*. As the name implies, this parameter should be calculated at the *outlet (4)*. Check if the *Report Plot* option *(5)* is active to monitor the changes in the outlet temperature in real time during calculations. Finally, confirm the settings with *OK*.

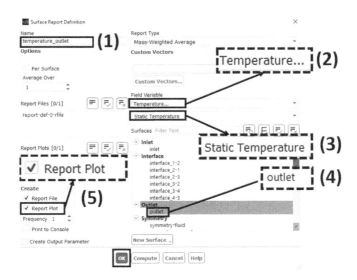

FIGURE 9.39

Expand the *Monitors* tab and select *Residual (1)* by clicking 2xLMB on it. In the newly opened window, all the parameters calculated during the solution are listed: *continuity, x, y, z-velocity, energy, k and omega* as in Figure 9.40. These parameters are calculated by the solver until the divergence between two subsequent iterations is lower than the defined *Absolute Criteria*. Lower the *energy* criteria from *1e-07* to obtain a more accurate solution *(2)*.

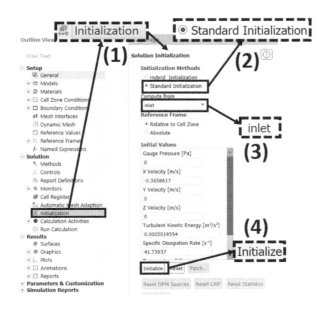

FIGURE 9.40

In each simulation, it is necessary to initialize the solution process (Figure 9.41). Find the ***Initialization*** option *(1)* and select the ***Standard Initialization (2)*** mode and compute the initial solution parameters from ***Inlet (3)***. Finally, click the ***Initialize*** button *(4)*. There won't be any changes or popup windows. Just move to the next step.

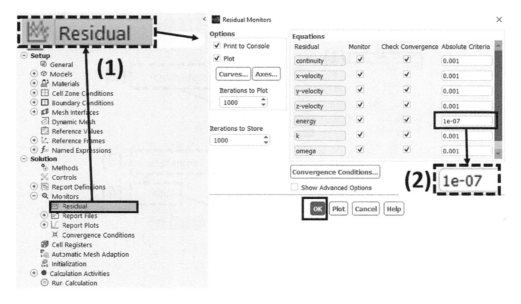

FIGURE 9.41

It is time to ***Run Calculation (1)***. But firstly, ***Check Case (2)*** to see if there are any recommendations regarding the solver settings. You can apply the suggestions or reject them. In the analyzed case no recommendations should be available *(3)*, as in Figure 9.42. Set the ***Number of Iterations*** as ***300 (4)*** – it should be enough to reach the convergence. Click ***Calculate*** button *(5)*. The level of calculation accuracy may be observed on the chart *(6)*. The curves which describe the error in the solution for each parameter separately should be heading down. The information that the calculation is complete will appear in the solver console and in the popup window.

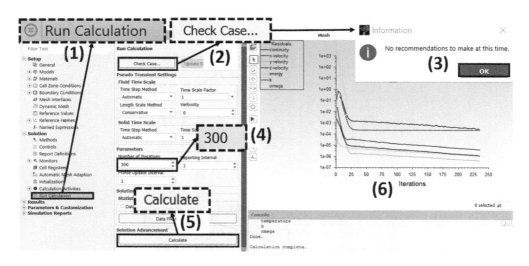

FIGURE 9.42

Tutorial 3 – Heat Receiver for a Solar Concentrating System   157

9.5 POSTPROCESSING

Find the **Results** module *(1)* in **Component Systems** and drag it to the **Project Schematic** as in Figure 9.43. Export **Solution** (*C3* cell) to the **Results** (*D2*) *(2)* and click *2xLMB* on the *D2* cell to launch the module.

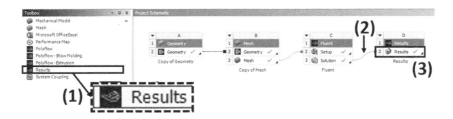

FIGURE 9.43

Select the **Contour** *(1)* tool and insert a **Name** *(2)*. In **Details of Contour** window select **All Domains** *(3)* and specify **Locations** as all external surfaces. To do this, click on the three dots icon *(4)* and select the necessary surfaces, as in Figure 9.44. Confirm your choice by clicking *OK* or Enter. Then choose **Variable** which will be displayed: **Temperature** *(5)* and its **Range: Global** *(6)*. In the *# of Contours* cell increase the number of temperature isolines to *50 (7)*. Confirm the settings by *Apply* and enjoy the first contour. Create the second contour – visualize the **Wall Heat Flux** on the upper surface of the receiver.

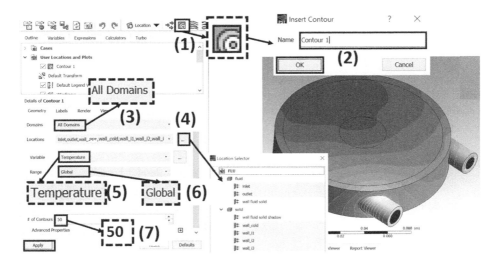

FIGURE 9.44

Deactivate the generated temperature distribution by clicking a tick in the project tree window *(1)*. Now generate a section plane that will allow visualizing data inside the model geometry. Open **Location** *(2)* and select **Plane** *(3)* from the dropdown list. *Name* your section *(4)* and define its details: plane in the *ZX (6)* orientation with *Y* equal 0 *(7)*. This will result in a cross-section through the middle of the receiver, as in Figure 9.45. By changing this parameter you can easily move the plane up and down. Check if **Plane Type** is **Slice** *(8)* and confirm your selections. Using the procedure described previously, add another temperature contour and locate it on the newly created plane. This time, select **Domains** as **Fluid** – the temperature distribution will be displayed only in the fluid body. Set **Range** as **Local**.

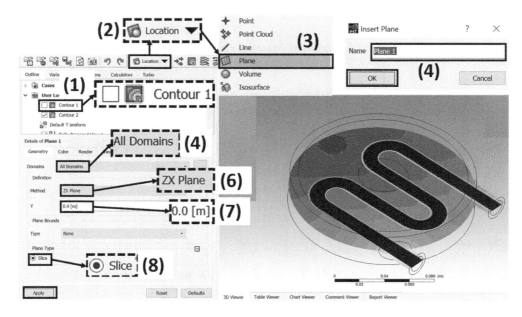

FIGURE 9.45

To calculate thermal power *(P)* of the heat receiver you have to define all the necessary factors, *i.e*: mass flow *(m)*, specific heat *(cp)* and temperature difference *(dT)* between inlet and outlet. For this purpose use the ***Expression (1)*** tool (Figure 9.46). Let's start with the mass flow. Define the ***Name*** of the factor *(2)* as m and in ***Details of Expression*** insert: ***massFlow()@inlet***. Confirm *(4)* and the current value of the mass flow will appear in the ***Value*** cell *(5)*. Use this procedure to add the remaining expressions:

dT: areaAve(T)@outlet-areaAve(T)@inlet
cp: 1970.2 [J*kg^–1*K^–1] –0.622[J*kg^–1*K^–2] *volumeAve(T)@fluid
Create the final expression and calculate power:
P: m * cp * dT

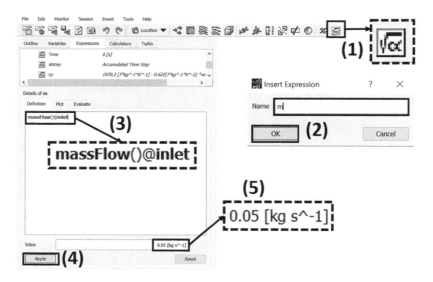

FIGURE 9.46

Tutorial 3 – Heat Receiver for a Solar Concentrating System

9.6 ADDITIONAL EXERCISE

The distribution of irradiance is one of the factors that have the biggest influence on heat generation in a solar receiver. In case 2 analyze the same receiver as in case 1, but with different distribution of the incident energy (Figure 9.47). Case 2 is an example of a receiver located closer to the focal point of the mirror than the receiver in case 1. The values of irradiance are higher, but at the same time, the area covered by highly-concentrated radiation is significantly reduced.

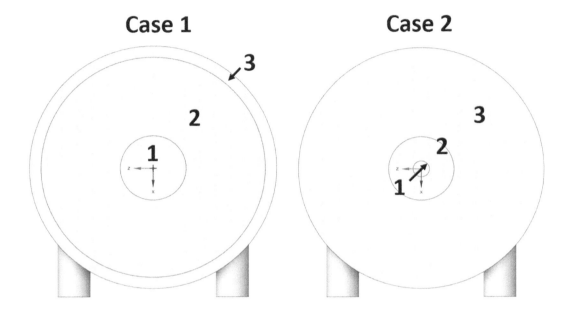

FIGURE 9.47

To compare the thermal power of the receiver in both cases you have to:

(1) in the **Geometry** module change the size of the circles that set the boundaries of *solar irradiation* levels,
(2) in the **Mesh** module assign new surfaces to the named selections: **wall_ir1**, **wall_ir2**, **wall_ir3**,
(3) in the **Fluent** module change the values of heat flux for **wall_ir1**, **wall_ir2**, and **wall_ir3** boundary conditions.

TABLE 9.1
Details of Irradiation Distribution for Cases 1 and 2

	Case 1		Case 2	
Zone	Radius [mm]	Irradiance [W/m²]	Radius [mm]	Irradiance [W/m²]
1	20	100,000	5	220,000
2	68	70,000	20	170,000
3	75	25,000	75	20,000

Part III

Photoelectric-Conversion-Based Technologies

10 Theoretical Background

10.1 DEVELOPMENT OF TECHNOLOGY

In Chapter 6, it has already been explained how to use solar energy to generate electricity by producing heat that can be used to run heat engines connected to generators. This method requires two stages – heat production and heating of the operating medium in the engine or turbine system – and employment of thermal solar collectors (especially with certain concentration technology). Photovoltaic devices are characterized as the ones that can convert sunlight directly into electricity, without the requirement to use any moving components or operating medium.

Internal photovoltaic effect that is the basis of a photovoltaic cell operation is the phenomenon of excitation of the electrons in the valence band by photons with sufficiently high energy (according to the quantum theory) and the movement of such electrons to the conduction band in the semiconductor material. The final consequence of the above described effect is generation of the electromotive force.

Photoelectric effect was discovered in 1839 by a French scientist, Alexandre Edmond Becquerel. He observed an increase in the conductivity of a lighting system when he was carrying out an experiment with electrodes and an electrolyte. Becquerel was able to examine the impact of different light wavelengths. The best results were observed for the blue and ultraviolet range [1]. The first selenium photovoltaic cell was created by Charles Fritts in 1883. It was in the form of a thin selenium layer covered by gold foil. The efficiency of the manufactured cell reached around 1% [2].

Although the first PV cell based on principles similar to those of the classic modern cells was already constructed in the 1920s, practical applications of photovoltaics appeared only in the early 1970s [3]. In 1962 Bell Telephone Laboratories equipped the TELSTAR satellite with a set of numerous small PV arrays. The "magic value" of 10% efficiency of commercial PV cells was first exceeded in 1959 (Hoffman Electronics).

Through decades photovoltaic technologies have been significantly developed. At the beginning of the 21st century companies around the world were competing to achieve the highest cell efficiency or the longest distance covered by the PV powered vehicles. For example, in 2001 NASA constructed the "Helios" photovoltaic plane that was able to fly at an altitude greater than 18 miles.

Today, the availability of advanced-technology-based PVT cells is wide enough to allow student scientific teams to compete in the same way as the biggest market players used to do twenty years earlier. Figure 10.1 presents examples of two student projects run at the AGH University of Science and Technology (Krakow, Poland): "AGH Solar Boat" [4] and "AGH Solar Plane" [5]. Both teams won prizes in frames of international events and contests.

FIGURE 10.1 Unique vehicles powered by photovoltaic modules constructed by student teams from the AGH University as part of the projects: (a) "AGH Solar Boat" and (b) "AGH Solar Plane".

10.2 STATISTICAL DATA

Figure 10.2 shows the installed capacity of photovoltaic technologies in the top ten countries around the world in 2021. The first place in this ranking had belonged continuously since 2015 to China, with 329 GW installed capacity in 2021 [6]. The USA and Japan took second and third place, respectively, with about 94 GW and 74 GW of installed photovoltaic. These three countries shared more than half of the world's installed photovoltaic capacity in 2021. Germany was the world's leader in this field until 2014, but then the investments slowed down. In 2021 Germany took 4th place with 58 GW of installed PV capacity. All the remaining countries had less than 50 GW of installed capacity, for *e.g.* India had 49 GW and Italy 23 GW. Australia and South Korea had about 19 GW of photovoltaic, whereas Vietnam and France had 17 GW and 15 GW, respectively.

Figure 10.3 presents the installed capacity of photovoltaic technologies in the world during the 10-year period, from 2012 to 2021. Generally, in the analyzed period the installed capacity increased by a factor of eight: from 102 GW in 2012 to 843 GW in 2021, which represents 27.5% of the global

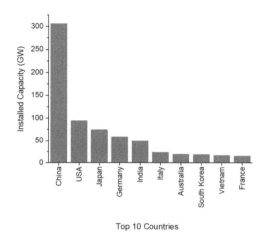

FIGURE 10.2 Top ten countries with the highest installed capacity of photovoltaic technologies in 2021. (Based on Ref. [7].)

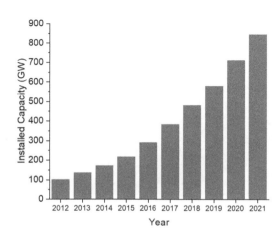

FIGURE 10.3 Cumulative installed capacity of photovoltaic technologies in the world. (Based on Ref. [7].)

Theoretical Background

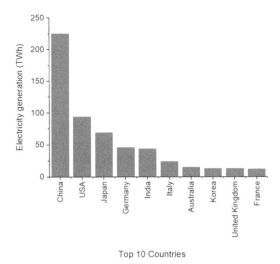

FIGURE 10.4 Top ten countries with the highest electricity generation from solar photovoltaic technologies in 2019. (Based on Ref. [7].)

installed capacity in RES. Between 2012 and 2021, there was a constant increase in the annual net additions of installed photovoltaic capacity. The most significant addition was noted in 2021 and it was equal about 113 GW.

The growth of installed photovoltaic capacity is strictly related to the lower prices of PV modules and reduced costs of the system equipment. The global average cost of utility-scale photovoltaic installation gradually declined over the ten-year period to finish at just under 883 USD/kW in 2020. It is worth mentioning that the total installed cost calculated for 2020 was 12% lower, compared to 2019 and 70% lower than in 2012. There is a wide range of photovoltaic technologies nowadays, nevertheless, the first-generation solar cells: crystalline silicon still dominate the market with 85%–90% share. The second-generation thin-film solar cells represent only about 0%–15% of the market, whereas the third-generation ones have less than 1% [8].

The bar chart presented in Figure 10.4 shows the top ten countries with the largest electricity generation from photovoltaics in 2019. Generally, solar PV with 679 TWh of produced energy is the third-largest renewable electricity technology, behind hydropower and wind. In the analyzed rank, the first position belongs to China, with nearly 225 TWh of generated energy, which stands for 33% of the global energy production from photovoltaics. This significant share is caused by the dynamic development of industry and therefore, the increased energy demand. It can be seen that USA and Japan, respectively second and third in this rank, have smaller production than China. In Australia, Korea, The United Kingdom and France, the energy production from photovoltaics is on a similar level, about 12 TWh. It is worth noting that the European countries included in this rank: Germany, Italy, The United Kingdom and France have high energy generation due to the looming policy deadlines regarding environmental protection.

10.3 CLASSIFICATIONS AND CHARACTERISTICS

The standard photovoltaic cell, which in fact is based on the *p-n* junction, consists of a couple of key-components, such as an emitter (*n*-region), substrate (*p*-region), anti-reflecting layers and electrical contacts allowing to close the external circuit between the anode and the cathode. Schematics of an individual cell, with the components listed and described, can be found in literature sources, like [2]. PV cells are connected in a module and depending on the connection type (in series or parallel) their resultant parameters (output voltage and current) differ. Modules are connected into strings that make PVT panels. Connecting modules in series results in an increase in total voltage. This kind of solution is preferred due to relatively low transmission losses; however, it cannot be applied in the case of some simple charging controllers. It comes from the requirement to adjust the controller output voltage to the battery voltage. Alternatively, the parallel connection which increases current can be applied.

Among many different benefits of photovoltaics listed in the literature, the most common is the PV's possibility to operate for a relatively long time without high maintenance costs, a low environmental impact, the possibility to reduce emissions and reducing peak loads. Of course, as in the case of the solar thermal collectors, real issues of the technology application depend on many factors, starting with the geographical location and weather conditions, and finishing on the economic barriers (for example, unclear regulations regarding the power pricing or just a lack of them) [2,3].

As it has already been emphasized, solar radiation energy can be used both for heat and electricity production. In certain areas both forms of energy are usually required, so thermal solar collectors and PV modules can be considered there. It is reasonable to take into account combined heat and power generation to reduce the area that is needed to construct the solar energy installation. It allows to improve the overall energy efficiency utilization [9–11].

As it is easy to guess, the first ideas of the hybrid photovoltaic – thermal systems (PVT) are as old as the first commercial PV cells. The beginnings of the studies date to the 1970s, when the first fluid-based device was proposed (1976). Although half a century has passed since then, it cannot be said that PVT technology is mature today. There are still some issues requiring deeper analyses and many valuable results have been recently reported in literature sources [12–14].

The development of this technology still requires a lot of effort, but its advantages are obvious [9]. First of all, as it has already been mentioned, it allows not only to utilize the space much better than individually installed thermal collectors and PV panels but also to keep the architectural uniformity of the roof or facade [15]. Furthermore, the installation cost of the hybrid system is relatively low, compared to individual ones. Of course, the total efficiency of the energy conversion in a cogeneration system is much higher than that obtained from classic systems.

Both in case of the PV cells and PVT modules numerous different solutions can be listed. In general, it can be concluded that the main difference between cells comes from the materials used for their production. Classification presented in Ref. [2] covers four main generations of PV cells. A detailed division into subsequent ones has been presented in Figure 10.5 (each block includes information about the cell efficiency range).

Regarding the first-generation cells, it can be noticed that usually different kinds of silicon crystallites are used for their production. Silicon-based PV cells are quite efficient and characterized by a long life. The material is easily available and cheap. When it comes to polycrystalline cells, their production process is also quite inexpensive. As it can be seen in Figure 10.5, Gallium Arsenide cells are the most efficient in the first-generation class. Moreover, these cells are less temperature sensitive [2].

Based on the information on the second-generation cells it can be concluded that they are not much more efficient. In their design, the center of gravity moved rather to the reduction of production costs and material consumption than to achieving higher efficiency compared to the first generation. Additionally, materials like CdTe/Cds provide higher absorption ability.

The third generation of cells adds functionality to the efficiency and economical issues. In simple terms, it can be said that the third-generation cells are lighter and sometimes (like in the case of organic or perovskite cells) flexible, which significantly extends the range of their potential applications. There is also no doubt that some of the third-generation cells are high efficiency ones (like "MJ" cells).

Manufacturers of fourth-generation cells apply nanotechnology to combine organic and inorganic nanoparticles and nanomaterials. This can result in further progress in functionality (printability, depending on requirements stiffness or flexibility, strength others…).

Theoretical Background

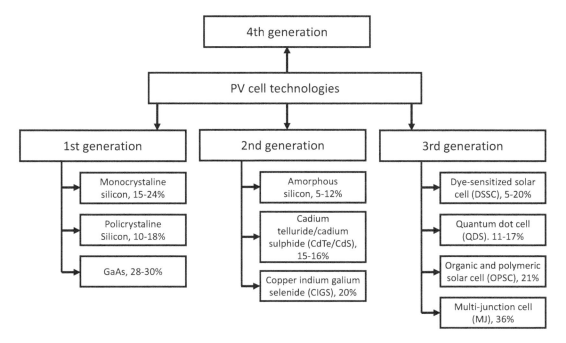

FIGURE 10.5 General classification of PV cell technologies. (Based on Ref. [2].)

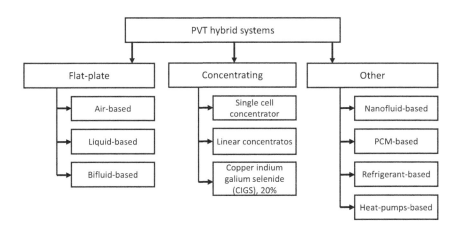

FIGURE 10.6 General classification of PVT technologies. (Based on Ref. [9].)

Figure 10.6 presents a simplified classification of the hybrid PVT technologies. The first group includes devices characterized by the use of different kinds of operating medium but always based on the flat-plate absorber technology. As there is no need for the application of a concentrating element (like a dish or lens) they are relatively cheap, especially when it comes to air-based systems. Of course, concentration of the solar radiation incident on the PV cell improves the electrical output of the PV cell, however, depending on the concentration technology, the cell is heated up more or less significantly, which also negatively affects the energy conversion efficiency. Both these facts make the hybrid PVT receivers a really attractive solution for the concentrating systems. An additional way to improve the efficiency of PVT systems is the application of a working medium offering heat capacity (but also other properties, like conductivity) higher than that of common operating fluids, like air, water, or even glycol solution. Some systems use innovative materials, such as nanofluids, phase change materials or low-boiling mediums, which allow to extend the range of system functionalities.

Despite many differences coming from materials and construction, both classic PV modules and hybrid PVT collectors have certain components that are common regardless of the specific device technology. Thus, general description of PV and PVT construction is given below.

10.3.1 Photovoltaic Modules and Systems

Photovoltaic panel is a device that converts solar radiation into direct current electricity, using solar modules (Figure 10.7) composed of 60–72 solar cells. Single cells are connected electrically by busbars in series and parallel to achieve the desired voltage and amperage. Then they are laminated with transparent elastomeric polymer, called EVA, which holds them in position and also bonds to the glass cover and polymer Tedlar backsheet. The tempered glass cover and EVA film protect PV cells from unfavorable weather conditions, moisture, dust and mechanical damage. The front layers of the panel must provide a high degree of light permeability so they are transmissive and covered by antireflective coating. The whole panel is enclosed in a metal frame providing the construction support. On the back of the panel, the junction box with electrical connectors is usually located [16].

Figure 10.8 presents the scheme of a 7.5 kWp photovoltaic system located on the roof of one of the buildings belonging to the AGH University of Science and Technology (Krakow, Poland). The system is divided into three independent arrays. Two of them consist of polycrystalline panels facing East and West under a 10° tilt angle. One array includes two strings with seven panels each. In the third string, 12 thin-film panels are mounted. All of them face the South under a 7.5° tilt angle.

Each array is equipped with a separate inverter which enables its independent operation. The electrical energy is transferred to the switch box and then to the electrical grid. The power obtained from this installation provides approximately 2.5% energy demand for the above mentioned building [17].

10.3.2 Hybrid (Photovoltaic – Thermal) Collectors

As it has already been said, a hybrid photovoltaic thermal (PVT) module is a device that combines two solar technologies: photovoltaic panels and thermal collectors [18]. This makes it possible to produce usable thermal and electrical energy. The construction of a PVT panel (Figure 10.9) often

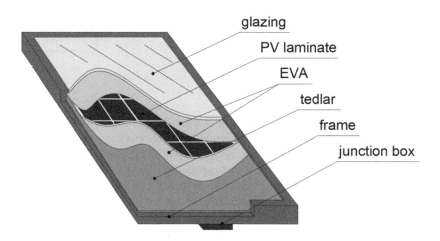

FIGURE 10.7 Construction of a photovoltaic module.

Theoretical Background

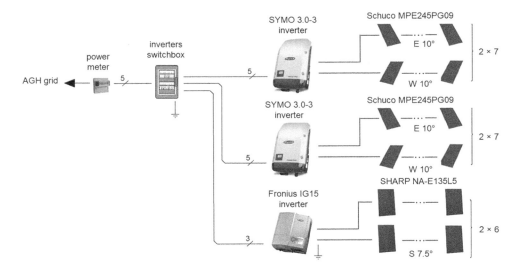

FIGURE 10.8 Scheme of the photovoltaic facility located on the roof of the AGH University building.

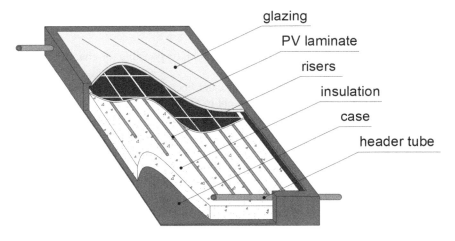

FIGURE 10.9 Construction of a hybrid photovoltaic – thermal solar collector.

includes a glass cover transparent to solar radiation, which provides better thermal insulation and protects the photovoltaic module. Nevertheless, like in the case of solar thermal collectors (see Chapter 6), there are also unglazed constructions. The role of the thermal absorber is played by the layer of PV cells [19]. They are encapsulated into the EVA polymer which provides hermetic operating conditions and holds cells in the right position. Solar radiation passes through the transparent layers of glass and EVA and hits the solar cells. Only a fraction of the incident energy is converted into electricity. The remaining part of the radiation is converted into heat, which adversely affects their operation. This heat is transferred through the polymer backsheet of the PV panel to the thermal part of the hybrid collector. The connection between the photovoltaic and the thermal part of the hybrid collector lays a vital role in the heat transfer process [20]. The heat exchanger consists of a metal (copper or aluminum) sheet and channels welded underside. The piping is filled with working fluid which carries away the heat and reduces the temperature of the PV cells. The bottom and sides of the PVT collector are usually covered with insulation to reduce thermal losses.

Classic PVT systems tend to use water since it has better thermal properties compared to air. The water based PVT system has higher electrical and thermal efficiencies compared to the air-based PVT system. Glazing is suitable for increasing thermal efficiency but it decreases electrical efficiency of the system. An Unglazed PVT system is favored if the high electrical output is required, while glazing is preferred when high thermal or overall system efficiency is in demand. However, single glazing is more preferable than double glazing. The type of absorber plays a great role in exploiting solar energy, where aluminum and copper are the commonly used materials in solar thermal absorbers.

Hybrid PVT panels are commonly used in locations where roof-panel spacing is limited. The temperature of the heat strongly depends on the construction of a specific collector and varies from 20°C to 100°C. Depending on the temperature, the heat may be used for domestic heating and cooling applications, drying, heating swimming pools or as a heat pump source [19,21]. The produced electricity usually supports the above described processes [22]. Water heating is relatively the most common application that exploits PVT systems. Generally, PVT systems are suitable for applications that require thermal energy more than electrical energy. In all applications, the performance of PVT systems is better than that of thermal systems and PV systems working alone.

10.4 FUNDAMENTALS OF ENERGY CONVERSION AND BALANCE

10.4.1 Photovoltaics

As it has already been stated, photovoltaic devices perform direct conversion of solar radiation into electricity. The main ideas of PV conversion are as follows:

- creation of a material with an internal electric field, responsible for charge separation and transportation to the electrodes,
- delivery of solar radiation (photons) of sufficient energy to transport an electron from its basic level to the conduction band (equivalent to an electron-hole pair generation).

A description of these phenomena (in certain simplification) is presented in the further part of the subchapter.

Electric behavior of materials depends on the band gap between the already mentioned valence band and conduction band (Figure 10.10). If this gap is equal zero (or very small), *e.g.* metals – then we have good current conductors, because in normal conditions electrons are present in the conduction band. For insulators, the value of the band gap is high and it is impossible to transfer electrons from the valence band to the conduction band. For semiconductors, this gap is relatively small and *e.g.* solar radiation is able to transfer electrons to the conduction band. A simplified structure of *pn*-junction is presented in Figure 10.11.

Two layers: the so called *n*-layer with doped electrons (negative extra charge is present) and the *p*-layer with doped holes (positive extra charges) create a *p-n* junction. Between these layers, an internal layer with depleted charge density is created as a result of charge diffusion. This layer is very thick but effectively separates two n and *p* zones with effective negative and positive charges. In consequence, an internal electric field is created, which is responsible for the separation of the positive and negative charges (electron-hole pair) and for the transport of electrons and holes to the opposite electrodes due to the acting electric forces.

Considering the spectrum of incident solar radiation, only the photons with an energy equal to or larger than the band gap can participate in the photovoltaic effect (equation 10.1):

$$h\nu = \frac{hc}{\lambda} \geq E_g \qquad (10.1)$$

where E_g – bandgap, λ – wavelength, c – light speed, and h – Planck constant.

Theoretical Background

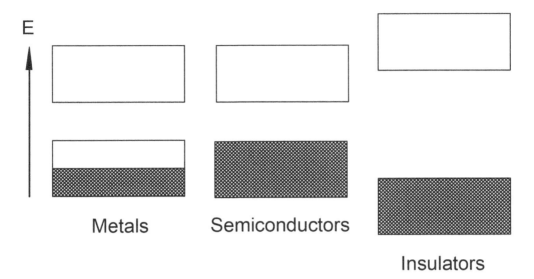

FIGURE 10.10 Electric structure of materials.

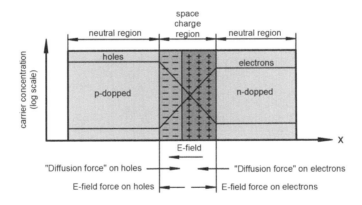

FIGURE 10.11 A simplified structure of a p-n junction. A junction is the region with a space charge (depletion region). An electric field resulting from the non-constant charge density is indicated, as well as the forces acting on the electrons and holes generated by photons.

The range of useful photon energy for a typical spectrum of solar radiation and bandgaps of photovoltaic materials (mostly silicon) are visible in Figure 10.12a (wavelengths below λ).

Therefore, only part of the photons can contribute to the overall photovoltaic effect.

Another limitation is related to the properties of the crystal structure of photovoltaic materials - *i.e.* we can assume, with some approximation, that one photon fulfilling condition (10.1), despite its energy can generate only one electron-hole pair. It means that its surplus of energy (over E_g) is wasted (spent on heat generation). Therefore, the efficiency of the so called single-junction PV cell is strongly limited. Figure 10.12a illustrates this limitation. The parameter describing this effect is called spectral response.

It is visible that maximum efficiency is reached only for the photon energy equal to the bandgap width. The values indicated by the "ideal cell" line result from the condition that surplus energy

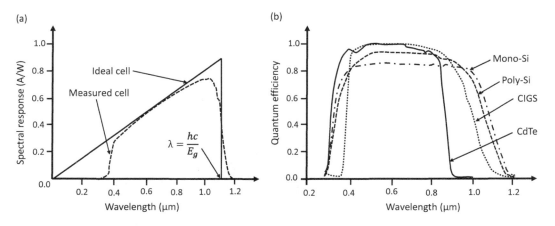

FIGURE 10.12 Spectral response of the solar cell (a) and typical quantum efficiency (b).

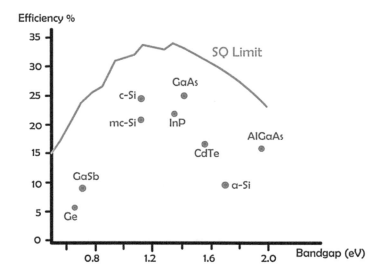

FIGURE 10.13 Efficiency of solar cells in function of a bandgap.

cannot be used for the photovoltaic effect. It does not reach 1 due to the so called quantum efficiency. The real curve is marked as a "Measured cell". The effect of quantum efficiency is presented in Figure 10.12b.

This effect is responsible for the real efficiency of a solar cell as given in Figure 10.13:

It is visible that the above mentioned effect has a strong influence on efficiency and there exists an optimal value of a bandgap (*ca.* 1.4 eV) corresponding to the efficiency of *ca.* 26%. Silicon material is quite close to this optimal value (1.12 eV). The curve presenting the theoretical limit of the efficiency is called the Shockley-Queisser limit.

In solar cells electron-hole pairs are separated due to opposite charges: electrons are transported to the top layer and finally reach the negative electrode, and positive holes are transported in the opposite direction, to the bottom, where the positive electrode is situated. Electrons are collected by the top electrode (negative) which is in the form of metal strips, wires, *etc.*, to let through the incident light, while the positive electrode has the form of metal foil.

Theoretical Background

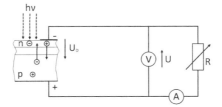

FIGURE 10.14 Basic model of an electric circuit with a solar cell (positive and negative electrodes are indicated together with the charge separation by the electric field).

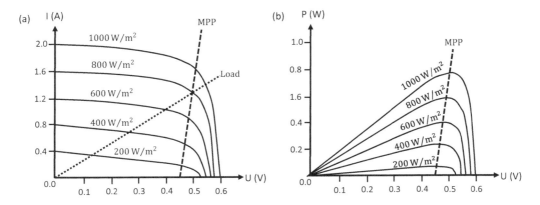

FIGURE 10.15 (a and b) Basic characteristics of a solar cell for varied intensity of solar radiation, some special points like: open circuit, shortcut and maximum power point (MPP) are indicated.

The essential properties of a solar cell can be described by its characteristics, which determine the conditions of the transfer of energy generated in a solar cell to the electric circuit and load. Let us consider the following model of a circuit (Figure 10.14):

Considering charge generation and its delivery to the electric circuit we obtain the following characteristics (Figure 10.15), known as current-voltage and power-voltage:

It is visible that for special cases, like (a, b) shortcut circuit and open circuit, the efficiency of power transfer from the cell to the circuit is equal to zero (according to the well-known formula that the value of electric power is calculated by the multiplication of voltage by current intensity).

This means that there should exist a maximum of the power function – and it is indicated in Figure 10.15 as MPP (maximum power point). It means that the maximum efficiency of power transfer can be reached only if we use the specially selected value of the load, according to equation 10.2, (notation as in Figure 10.14):

$$P_L = I^2 R \qquad (10.2)$$

This function reaches its maximum for $R = R_{int}$ and then the maximum power $P_L^{max} = \dfrac{U^2}{4R_{int}}$.

Another very important factor describing the property of a solar cell is the Fill Factor (FF), the idea of which is illustrated in Figure 10.16.

The following equation describes FF:

$$FF = \dfrac{P_L^{max}}{P_T} = \dfrac{I_{MP} \cdot V_{MP}}{I_{SC} \cdot V_{OC}} \qquad (10.3)$$

where indexes denote SC – shortcut circuit and OC – open circuit (see also Figure 10.15)

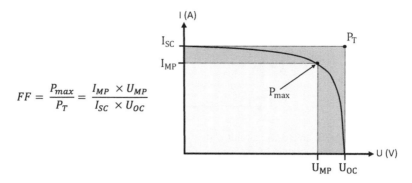

$$FF = \frac{P_{max}}{P_T} = \frac{I_{MP} \times U_{MP}}{I_{SC} \times U_{OC}}$$

FIGURE 10.16 Fill factor idea as a relation of the plotted rectangular surface.

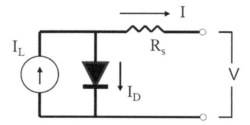

FIGURE 10.17 The four-parameter model of a solar cell represented as a diode and a serial resistor.

The Fill Factor is essentially a measure of quality of the solar cell. It is calculated by comparing the maximum power possible to obtain P_L to the theoretical power P_T. FF can also be interpreted graphically as the ratio of the rectangular areas (area of the smaller dot-line box to the larger area of dot-line box).

All the electric properties of solar cells can be described and modeled using several electric models. The simplest one is called the four-parameter model and can be illustrated by the following diagram (Figure 10.17).

The solar cell is represented as the current source I_L (with current intensity controlled by solar radiation), parallel diode D and a serial resistor R_s.

Even this simple model can explain several properties of a solar cell with quite satisfactory accuracy. A detailed description and solution of the model are presented in the next subchapter to illustrate the thermal properties of solar cells.

10.4.2 Thermal Photovoltaics (PVT)

Due to the negative influence of temperature on the operation of solar cells, especially on the power output, the idea of thermophotovoltaics (PVT) appeared as a possible solution to this problem. In this subchapter, the problem will be defined and a suggested solution by PVT presented.

The influence of temperature on the performance of solar cells can be illustrated in the following pictures (Figure 10.18).

It is visible that the loss of efficiency occurs with the temperature increase. This effect varies for different materials, but with some simplification, the average values are ~ 0.5% / K. It means that a temperature increase by 50 K can lead to the reduction in power generation by 25%. The temperature

Theoretical Background

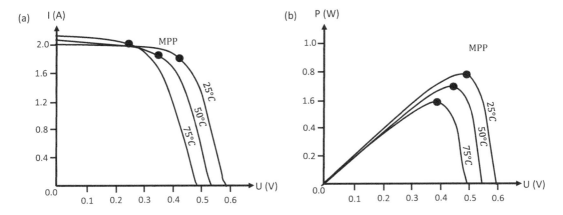

FIGURE 10.18 (a and b) Temperature effect of solar cell performance.

increase inside a solar cell structure is a common and difficult-to-reduce effect, as a solar cell converts most of the radiation energy to heat.

A theoretical description of this effect can be performed using the four-parameter model presented in Figure 10.18 as the property of the semiconductor diode depends on temperature.

Thermal behavior of a solar cell can be modeled assuming that:

- diode current depends on the temperature according to the formula:

$$I_D = J_0 \cdot \left[\exp\left(\frac{e}{nkT_c} \right) - 1 \right] \quad (10.4)$$

where J_0 is the reverse saturation current, e is the charge of the electron, n is the diode ideality factor, k is the Boltzmann constant and T_c is the cell temperature in absolute scale.

Reverse saturation current density, J_0, is a measure of the leakage (or recombination) of minority carriers across the p–n junction in reverse bias. This leakage is a result of carrier recombination in the neutral regions on either side of the junction and, therefore J_0, primarily controls the value of V_{OC} in the solar cells. The minority carriers are thermally generated; therefore, J_0 is highly sensitive to temperature changes.

Using the reverse saturation current, it is possible to assess the material properties and details of the construction of the p-n junction. A function of the thermal equilibrium between the concentrations of the minor carriers of the p-n junction is the following:

$$J_0 = qn_i^2 \cdot \left(\frac{D_n}{L_n N_A} + \frac{D_p}{L_p N_D} \right) \quad (10.5)$$

where:
n_i is the intrinsic carrier concentration in the semiconductor material $q_n = e \cdot n_i$;
D_n and D_p-diffusivity of the hole and electrons of the material of type n and p;
L_n and L_p-diffusion length hole and electrons of type n and type p respectively;
N_a and N_d-concentration of acceptors in the p-type region and concentration of donors in the
 n-type region.

These parameters are generally a function of the cell temperature, describing thermal behavior of the solar cell.

J_o is strongly determined by the proportionality to n_i^2 and n_i can be represented as:

$$n_i^2 = N_c N_v \exp\left(\frac{-E_g}{kT}\right) = 4\left(\frac{2\pi kT}{h^2}\right)^3 \cdot m_e^{\frac{3}{2}} m_h^{\frac{3}{2}} \exp\left(\frac{-E_g}{kT}\right) \quad (10.6)$$

where N_c and N_v are the effective density of states in the conduction band and valance band, and m_e, m_h are the effective mass of electron and hole, respectively.

Combining equations 10.5 and 10.6, the expression for J_0 can be written in terms of temperature and bandgap energy in general form as:

$$J_0 = CT^3 \exp\left(\frac{-E_g}{kT}\right) \quad (10.7)$$

giving general dependence on temperature. As visible in the above equation, doping and the material parameters of solar cells are combined in one constant C.

This expression can be used for further calculations of the temperature coefficient $1/J_0$, dJ_0/dT, dV_{OC}/dT, FF and its temperature dependence, solar cell efficiency and others. For details, see *e.g.* [23].

It is also necessary to include temperature dependence of the bandgap in semiconductors. $E_g(T)$ can be described as [23]:

$$E_g(T) = E_g(0) - \frac{\alpha T^2}{T + \beta} \quad (10.8)$$

where $E_g(0)$ its value at 0 K, α and β are constants. For example, for silicon semiconductors: $E_g(0) = 1.1557$ eV, $\alpha = 7.021$ (eV/K) 10^{-4} and $\beta = 1108$ K.

It is visible that the bandgap width decreases with the temperature growth. The bandgap is responsible for the V_{OC} and this effect is visible in Figure 10.18.

In this context the idea of PVT is very simple – let us apply a solar collector to remove heat from the silicon structure of PV, as a result decrease its temperature and finally use this extracted heat for our purposes. What we achieve in effect is a hybrid device acting as a cogeneration unit.

To describe the thermal properties of the device, we can apply the modified formula for the collector efficiency [24]:

$$f_{TPV} = \frac{\frac{1}{U_L}}{W\left\{\frac{1}{U_L[D+(W-D)\cdot F]} + \frac{1}{C_b} + \frac{1}{\pi\alpha D_i}\right\}} \quad (10.9)$$

where F is given by equation 10.7 with modified coefficient m, according to equation 10.10:

$$m = \frac{U_L}{\lambda_p \delta_p + \lambda_{PV} \delta_{PV}} \quad (10.10)$$

where δ_p is the absorber thickness, λ_p is the absorber thermal conductivity (as for thermal collector) and additionally, δ_{PV} is the PV panel thickness and λ_{PV} is the PV thermal conductivity. Comparing

Theoretical Background

equations 10.4 and 10.10, it is visible that an additional term $\lambda_{PV}\delta_{PV}$ responsible for thermal part of the device is added.

U_L is the overall loss coefficient for PVT and is expressed as follows:

$$U_L = U_t + U_b + U_e \qquad (10.11)$$

where U_b is the loss coefficient of the bottom, U_t is the top coefficient and U_e is the edge coefficient. The top coefficient is given by:

$$U_t = \cfrac{1}{\cfrac{N}{\cfrac{c}{T_{Pm}\left(\cfrac{T_{Pm}-T_a}{N+f}\right)^{e+\frac{1}{h_k}}}}} + \cfrac{\sigma(T_{Pm}+T_a)\cdot(T_{Pm}^2+T_a^2)}{\cfrac{1}{\epsilon_p + 0.00591Nh_k} + \cfrac{2N+f+0.133\epsilon_p}{\epsilon_{PET}} - N} \qquad (10.12)$$

where: N is the number of glass covers, T_{Pm} average temperature of the PV module, ϵ_p emittance of the PV module, ϵ_{PET} emittance of the transparent cover of the PV cell h_k, the assumed value of coefficient e is 0.33, c and f are empirical functions, and can be expressed as:

$$c = 365.9 \cdot (1 - 0.00883 \cdot \beta + 0.0001298 \cdot \beta^2) \qquad (10.13)$$

$$f = (1 - 0.04 \cdot h_k + 0.000 \cdot h_k^2) \cdot (1 + 0.091 \cdot N) \qquad (10.14)$$

where β is incidence angle of solar radiation or collector slope.

The expression for solar cell temperature can be calculated as follows:

$$T_c = \frac{\eta_0 S + U_{tc,a} T_a + h_{c,p} T_p}{U_{tc,a} + h_{c,p}} \qquad (10.15)$$

where indexes tc, p and c, p are related to the PV device plate.

Taking into account the relation presented above, the electrical efficiency η_{el} of a PV module is:

$$\eta_{el} = \eta_c \left[1 - 0.0045(T_c - T_{ref})\right] \qquad (10.16)$$

and the thermal efficiency of the PVT collector is given by:

$$\eta_{th} = F_R \left[(\tau\alpha)_{PV} - U_L \left(\frac{T_i - T_a}{G}\right)\right] \qquad (10.17)$$

According to Ref. [24] presenting an experimental study, the thermal and electrical efficiency is as follows:

It is visible that in comparison to the system with no cooling ($\dot{m} = 0$) an increase in electrical efficiency is observed for increasing flow. However, electrical efficiency also depends on the intensity of solar radiation (for larger intensities is lower). For thermal efficiency situation is similar. The optimal value of the flow can be seen in Figure 10.19a.

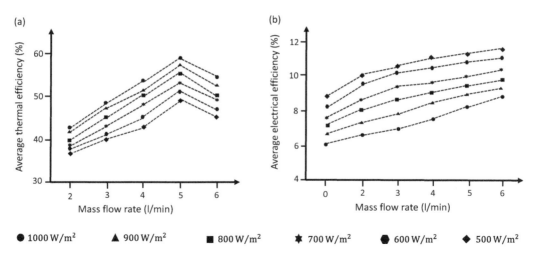

FIGURE 10.19 (a and b) Thermal and electrical efficiency of PVT in function of mass flow $\dot{m}$.

REFERENCES

1. Photovoltaic effect, https://www.sciencedirect.com/topics/engineering/photovoltaic-effect (last access: 14.04.2022).
2. P. Sharma, P. Goyal, Evolution of PV technology from conventional to nano – materials, *Materials Today: Proceedings*, 28 (2020), 1593–1597.
3. T. B. Johansson et al., *Renewable Energy. Sources for Fuels and Electricity*, Island Press, Washington, 1993.
4. Construction of the solar boat at AGH University, http://www.aghsolarboat.pl/ (last access: 11.06.2023).
5. Construction of the solar plane at AGH University, http://solarplane.agh.edu.pl/ (last access: 11.06.2023).
6. M. V. D. Hoeven, Energy and climate change, world energy outlook special report, IEA; 2015, https://www.iea.org/reports/energy-and-climate-change (last access: 14.04.2022).
7. Statistics Data of the International Renewable Energy Agency, www.irena.org (last access: 07.05.2022).
8. M. V. D. Hoeven, Technology roadmap: Solar photovoltaic energy. International Energy Agency IEA, https://www.iea.org/reports/technology-roadmap-solar-photovoltaic-energy-2014 (last access: 14.04.2022).
9. Y. Riffonneau et al., Optimal powerflow management forgrid connected PV systems with batteries, *IEEE Transactions on Sustainable Energy*, 2 (3) (2011), 309–320.
10. A. Herez et al., Review on photovoltaic/thermal hybrid solar collectors: Classifications, applications and new systems, *Solar Energy*, 207 (2020), 1321–1347.
11. C. S. Rajoria et al., Development of flat-plate building integrated photovoltaic/thermal (BIPV/T) system: A review, *Materials Today: Proceedings*, 46 (11) (2021), 5342–5352.
12. M. Herrando et al., The challenges of solar hybrid PVT systems in the food processing industry, *Applied Thermal Engineering*, 184 (2021), 1–15 (116235).
13. E. Yandri, Development and experiment on the performance of polymeric hybrid photovoltaic thermal (PVT) collector with halogen solar simulator, *Solar Energy Materials and Solar Cells*, 201 (2019), 1–11 (110066).
14. I. Guarracino et al., Systematic testing of hybrid PV-thermal (PVT) solar collectors in steady-state and dynamic outdoor conditions, *Applied Energy*, 240 (2019), 1014–1030.
15. H. A. Zondag et al., The yield of different combined PV-thermal collector designs, *Solar Energy*, 74 (2003), 253–269.
16. C. Lupangu, R. C. Bansal, A review of technical issues on the development of solar photovoltaic systems, *Renewable and Sustainable Energy Reviews*, 73 (2017), 950–965.
17. M. Żołądek, K. Sornek, K. Papis, R. Figaj, M. Filipowicz, Experimental and numerical analysis of photovoltaics system improvements in urban area, *Civil and Environmental Engineering Reports*, 28 (2018), 13–24.

18. C. Lamuatou, D. Chemisana, Photovoltaic/thermal (PVT) systems: A review with emphasis on environmental issues, *Renew Energy*, 105 (2017), 270–287.
19. F. Shan, F. Tang, L. Cao, G. Y. Fang, Performance evaluations and applications of photovoltaic-thermal collectors and systems, *Renewable and Sustainable Energy Reviews*, 33 (2014), 467–483.
20. D. Das, P. Kalita, O. Roy, Flat plate hybrid photovoltaic- thermal (PV/T) system: A review on design and development, *Renewable and Sustainable Energy Reviews* 84 (2018), 111–130.
21. A. Sharma, C. R. Chen, N. V. Lan, Solar-energy drying systems: A review, *Renewable and Sustainable Energy Reviews*, 13 (2009), 1185–1210.
22. A. Kumar, Historical and recent development of photovoltaic systems: A review, *Renewable and Sustainable Energy Reviews*, 102 (2019), 249–265.
23. P. Singh, N. M. Ravindra, Temperature dependence of solar cell performance – an analysis, *Solar Energy Materials & Solar Cells*, 101 (2012), 36–45.
24. N. Tamaldin, M. A. M. Rosli, F. A. Sachit, Theoretical study and indoor experimental validation of performance of the new photovoltaic thermal solar collector (PVT) based water system, *Case Studies in Thermal Engineering*, 18 (2020), 100595.

11 Tutorial 4 – Photovoltaic Farm

11.1 EXERCISE SCOPE

The goal of this tutorial is to present the impact of weather conditions on the operation of photovoltaic panels. This tutorial takes into consideration the position of the sun in the sky, cloudiness and wind velocity. As shown in Figure 11.1, the geometry consists of two photovoltaic panels located one behind the other, which makes it possible to predict the shadowing effect. Moreover, the panels are located in an enclosure, so the cooling effect of wind may be observed.

As you could find out from the theoretical background (see Chapter 10), the operation of a photovoltaic panel is strongly dependent on temperature. This model helps to estimate the influence of the shadowing effect and natural cooling on the temperature distribution of the PV panels. This tutorial presents the methodology of:

- creating an enclosure around 3D objects and naming surfaces in *Ansys SpaceClaim DirectModeler*,
- generating a polyhedral mesh and evaluating its quality in the *Fluent Meshing* module,
- applying the *Surface-to-Surface* radiation model with the additional *Solar Load* sub-model, allowing to take into account panel location (geographical coordinates) and specified time of the year (date and hour),
- displaying temperature contours and flow patterns around the PV panels.

Pay special attention to the description of the boundary condition representing the ground.
All of the images included in this tutorial use courtesy of Ansys, Inc.

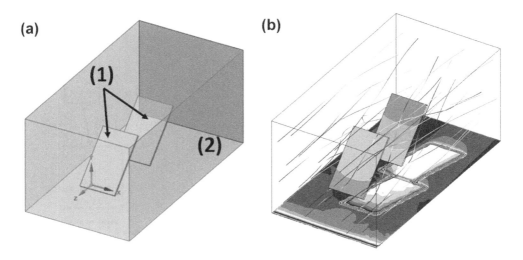

FIGURE 11.1 (a) Geometry considered in the exercise and (b) temperature distribution and wind streamlines obtained from the simulation.

DOI: 10.1201/9781003202226-14

11.2 PREPROCESSING – GEOMETRY

Drag the *Geometry* module *(1)* to the Ansys Workbench *Project Schematic* as in Figure 11.2. To open the geometry editor, double click the left mouse button *(LMB)* on the *A2* cell *(2)*.

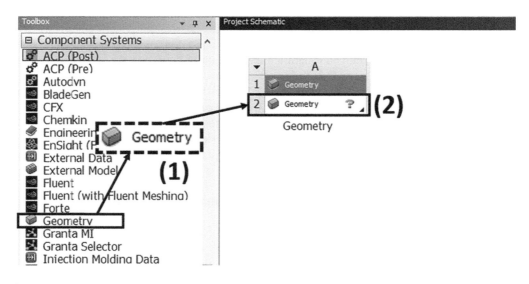

FIGURE 11.2

The *Sketch* tab will open (Figure 11.3). Firstly, choose *XY* as the sketching plane *(1)*. Secondly, click *LMB* on the *Plan View* *(2)* to view the *XY* grid head-on. Draw the PV panel using the *Rectangle* tool from the *Create* ribbon *(3)*. Start drawing from the coordinate system origin *(4)* and set the dimensions as shown in the picture. Finish your work by choosing the *End Sketch Editing* option *(5)*.

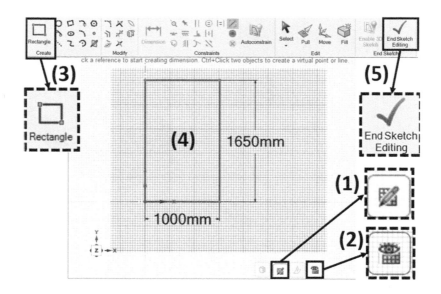

FIGURE 11.3

In the *Design* tab, click on the *Pull* tool *(1)* from the *Edit* ribbon, which will allow you to create a 3D object from the previously prepared sketch. Select a rectangular surface created from the sketch and then pull the sketch backward (along Z-axis) as in Figure 11.4. Set the thickness to *50 mm (2)*. After that, use the *Select* option to select the whole 3D object *(3)*. Next use the *Move (4)* tool to rotate the panel by an angle. Select the rotation direction by clicking on the appropriate triade axis *(5)* and set the rotation angle to *315° (6)*, so that the panel slope will be 45°.

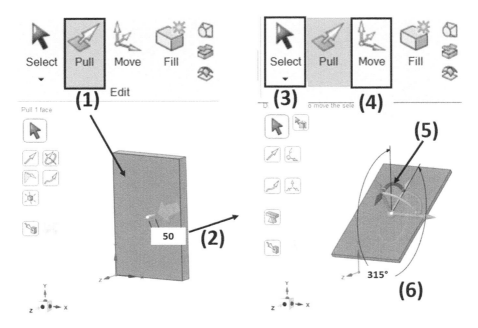

FIGURE 11.4

The geometry of one PV panel is ready, so it is time to copy it and create the second panel. *Select (1)* the whole geometry and then use *Ctrl+C* and *Ctrl+V* shortcuts. After the pasting operation, there will be no change in the project's appearance because these two bodies overlap. In the project *Structure* hide one of the bodies *(2)*. Then use the *Move (3)* tool to shift the visible geometry *1,750 mm* backward (along Z-axis) as in Figure 11.5 *(4)*.

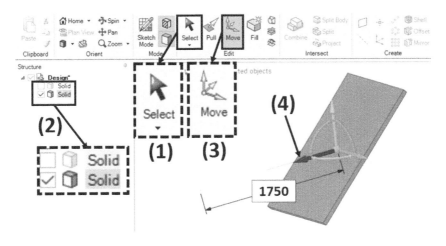

FIGURE 11.5

Make all the bodies visible *(1)* and create an ***Enclosure*** *(2)* around them, which will separate the control volume. From the ***Create Options*** choose the external shape of the enclosure as ***Box*** and deactivate the ***Symmetric dimensions*** *(3)* option. Then select both panels *(4)* to indicate the bodies which should be placed in an enclosed space (Figure 11.6).

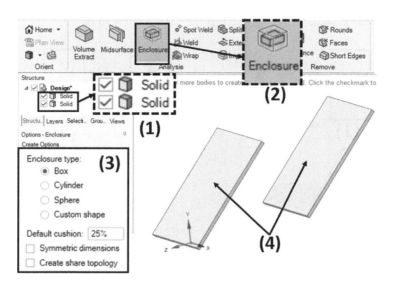

FIGURE 11.6

The default enclosure will be created around the selected bodies. Change its dimensions as shown in steps *(1)* and *(2)* in Figure 11.7. To make this task quicker, you can use the *Tab* key to switch between individual dimensions. After that, confirm your settings by clicking the green tick *(3)* or the ***Enter*** button to create the final enclosure.

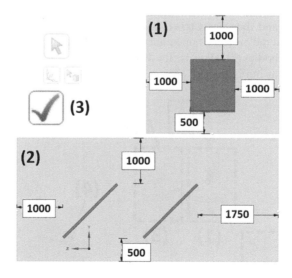

FIGURE 11.7

Note that, due to the limitations of the Student license, the generated enclosure is relatively small. Generally, the dimensions of enclosure should be calculated based on the characteristic length of the panels L and the appropriate multiplier.

Tutorial 4 – Photovoltaic Farm

Now create one component from the bodies visible in the project **Structure** (Figure 11.8). Select bodies representing both panels and fluid (enclosure) *(1)*, **click 1xRMB** and choose the **Move to New Component** option *(2)* from the popup window. All three bodies will be placed in a newly created component *(3)*. Change the **Share Topology** option from **None** to **Share** *(4)*. Finally, delete the empty component named Enclosure *(5)*.

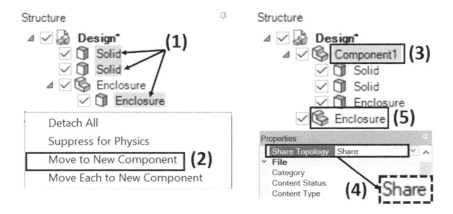

FIGURE 11.8

Using the **Select** *(1)* option choose the front face of the geometry *(2)* as in Figure 11.9. Then move to the **Groups** tab *(3)* to create **Named Selection**. Assigning names to different surfaces makes them easier to find in the mesh and solver modules. Click on the **Create NS** option *(4)* and insert the surface name as *inlet* *(5)*.

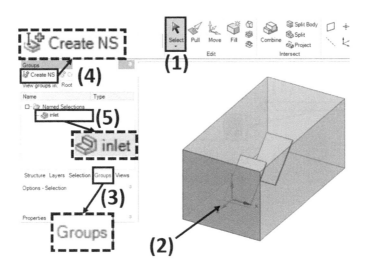

FIGURE 11.9

Repeat the same step and name other surfaces as in Figure 11.10: **outlet**, **wall_air** (remaining side and top walls of the enclosure) and *ground* (surface located at the bottom). Then hide the body representing the enclosure to see only two panels. Name the upper surface of the first panel **absorber_1**, and the upper surface of the second panel **absorber_2**. In the last step, all the remaining surfaces on the sides and at the bottom of both panels should be named **wall_pv**. Remember to show all the bodies after finishing the **Named Selection** operation.

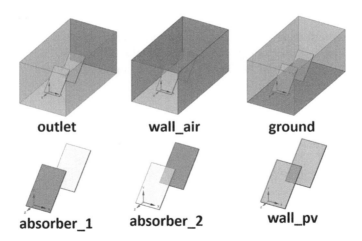

FIGURE 11.10

11.3 PREPROCESSING – MESHING

Select the ***Fluent (with Fluent Meshing)*** module *(1)* from the ***Component System*** ribbon and drag it to the Ansys Workbench ***Project Schematic*** *(2)* as in Figure 11.11. Create an appropriate connection between the geometry and the mesh modules *(3)*. Double click the left mouse button (LMB) on the ***Mesh*** cell *(4)* to open ***Fluent Meshing***. Set the number of ***Meshing Processes*** *(5)* to four and ***Start*** *(6)* the software. Student license allows to use four processors in parallel mode and it is enough to provide a smooth meshing process.

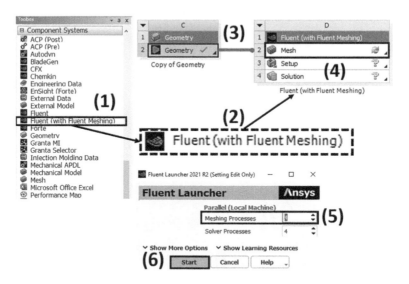

FIGURE 11.11

In the ***Fluent Meshing*** module, the process of mesh generation is structured as in Figure 11.12, so the user is guided through the individual steps. Find the ***Workflow*** *(1)* tab and choose the ***Import Geometry*** *(2)* option. In the details window *(3)* the default location of the geometry file is shown. Confirm the selection by clicking the ***Import Geometry*** *(4)* button. The geometry will be loaded to the meshing module and then displayed in the workspace.

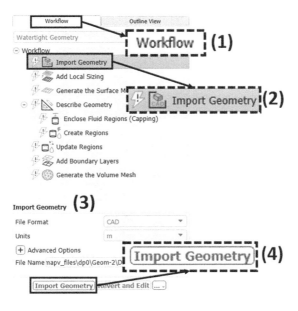

FIGURE 11.12

The next step is dedicated to *Local Sizing (1)*. Confirm that the local sizing will be added *(2)* and select the *Size Control Type* as *Body Size (3)* with *Target Mesh Size* set to 0.07 m *(4)*. This sizing should be located in the fluid domain. Select the body represented by the enclosure name *(5)*, as in Figure 11.13. Finally, confirm your settings and click *Add Local Sizing (6)*. After that, some *Size Boxes* will be drawn and displayed in the workspace to visualize the target size of the mesh elements. Repeat the steps described in this section and add another *Body Sizing* in the solid bodies which represent the PV panels. In this case, set the target mesh size to 0.02 m to provide a minimum three-layered mesh in the solid domain.

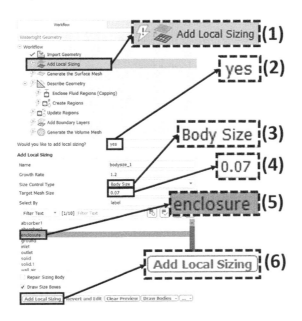

FIGURE 11.13

Find the ***Generate the Surface Mesh (1)*** option (Figure 11.14) and in the details window, set the ***Size Function*** as ***Proximity (2)***. In the analyzed geometry there are no rounded elements, so the ***Curvature*** option is unnecessary. ***Generate the Surface Mesh (3)*** and check the general mesh appearance in the project workspace.

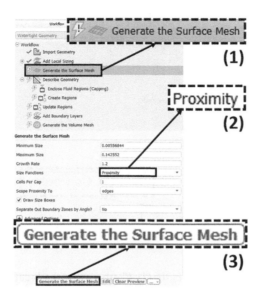

FIGURE 11.14

Before the volume mesh generation, the imported geometry has to be described in a more detailed way (Figure 11.15). Move to the ***Describe Geometry (1)*** option and specify that the ***Geometry Type*** consists of ***both fluid and solid regions and/or voids (2)***. Leave unchanged the remaining options – the defaults are well-suited to the analyzed case. Confirm these initial settings *(3)* and move to the ***Update Boundaries*** subsection. The displayed table contains all the surfaces named in the geometry module. It is time to define their ***Boundary Type***. Make sure that the ***inlet*** is set as ***velocity-inlet***, *the outlet* as ***pressure-outlet*** and all the remaining named selections are ***wall-type***. Finally, ***Update Boundaries (6)***.

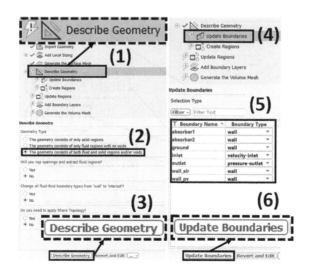

FIGURE 11.15

Now, *Create Regions (1)* by setting one fluid region *(2)* in the analyzed case as in Figure 11.16. After confirming this setting *(3)*, all regions have to be updated *(4)*. Choose the proper *Region Type (5)* for all the bodies created in Space Claim: the enclosure represents the *fluid* body, whereas two panels are made of the *solid*. Again, confirm the introduced changes *(6)*.

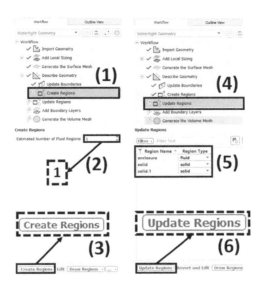

FIGURE 11.16

Now, Add *Boundary Layers (1)* to make the mesh finer in the selected near-wall regions (Figure 11.17). Select the *Offset Method Type* as *smooth-transition (2)* with four layers *(3)*. Settings for the *Transition ratio* and *Growth rate* leave unchanged. Specify that the inflation should be generated only in *fluid-regions (4)* on *solid-fluid-interfaces (5)*. This boundary layer will be created around PV panels *(6)*. Then add the second inflation above the ground surface. Use the same settings as previously, except for the boundary layer location – this time it should grow only in the selected-zones (ground).

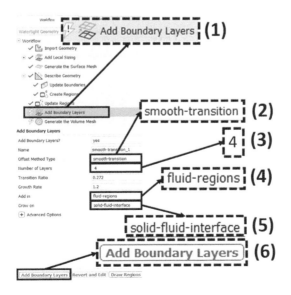

FIGURE 11.17

In the final step, ***Generate the Volume Mesh (1)*** with the ***polyhedral*** method ***(2)*** as in Figure 11.18. In this type of mesh, polygonal elements are formed around each node in the tetrahedral mesh. The great advantage of the polyhedra is the fact that a relatively low number of iterations is required to achieve the specified level of convergence. In the volume mesh options, it is also possible to change the ***Global cell length***, but size limits for all bodies have already been inserted as ***Local Sizing*** (Figure 11.13). Generate the mesh *(3)*.

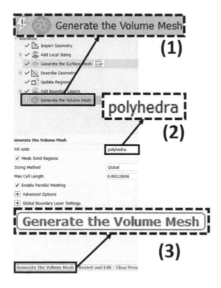

FIGURE 11.18

General information about the generated mesh will be displayed in ***Console (1)*** as in Figure 11.19. You can find there data on the mesh quality and the number of elements *(2)*. To find more about the generated mesh, find the ***Report*** tab and select ***Report → Mesh Size…***. To take a closer look at the generated mesh, ***Insert Clipping Planes (3)***. To see the cross-section of the volume mesh, activate the ***Draw Cell Layer (4)*** option. It is possible to change the section planes (switching between ***Limit in X, Y*** or ***Z***) and also move them with triad.

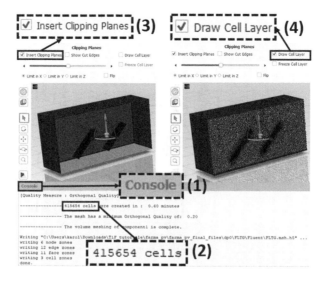

FIGURE 11.19

Take a closer look at the general mesh appearance, as in Figure 11.20. Examine the shape of the mesh elements and find the regions with the inflation layer around the PV panel *(1)* and close to the ground *(2)*.

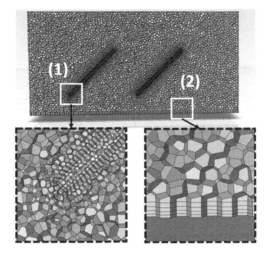

FIGURE 11.20

To analyze the quality of the generated mesh in a more detailed way, find the **Report** tab *(1)* and choose the **Cell limits…** option from the dropdown list. Select the **Quality Measure** as **Orthogonal Quality** *(3)* and all **Cell Zones** *(4)*, where the quality will be examined. Orthogonal Quality describes the similarity between the shape of the analyzed cell and the shape of an equilateral cell of the same volume. The bigger the similarity is, the better. Generally, mesh elements should have orthogonal quality higher than 0.05- cells of lower values decrease the accuracy and destabilize the solution. To display values, click **Compute**. Analyze *(5)* the minimum, maximum and average values of the selected quality parameter (Figure 11.21). **Quality Measure** tool is also useful when the external mesh is loaded and it is necessary to rate its quality.

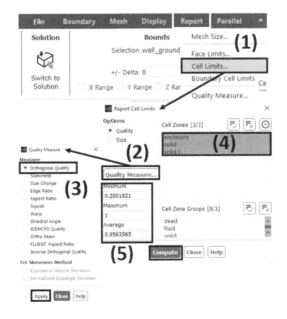

FIGURE 11.21

You can also display a chart presenting the number of cells of the specified quality. To do so, use the ***Display → Plot → Cell distribution (1)*** option. In a new window, specify the ***Quality Measure (2)*** and also the ***Cell Zones*** which will be analyzed ***(3)***. In the ***Axes…*** option, change the ***Y-axis*** type to the logarithmic scale ***(5). Plot (6)*** the chart and examine the participation of the mesh elements in the mesh quality (Figure 11.22).

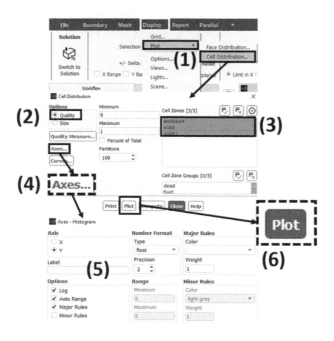

FIGURE 11.22

11.4 PREPROCESSING – SETTING SOLVER

Find ***Switch to Solution*** option ***(1)*** to start the Fluent solver as in Figure 11.23. In the popup window confirm your choice by clicking ***Yes (2)***. Note, when using ***Fluent Meshing*** it is not necessary to go back to the ***Ansys Workbench*** window to launch the solver.

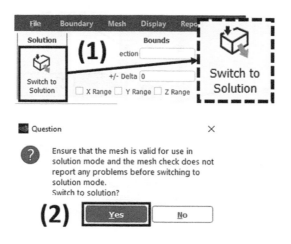

FIGURE 11.23

Tutorial 4 – Photovoltaic Farm

In solver **Setup** find the **General** option *(1)*. Firstly, click the 1xLMB **Check** *(2)* button to check the geometrical details of the created geometry and mesh. They will be displayed in the **Console** *(3)* window as in Figure 11.24. Set the **Solver Type** as pressure-based *(4)* to analyze fluid flows with a low velocity (<200 m/s). To provide a steady-state simulation choose the **Time** option as **Steady** *(5)*.

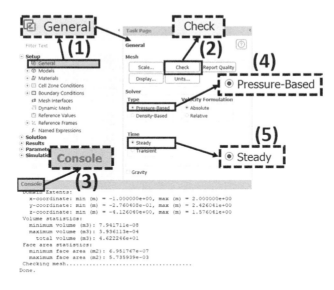

FIGURE 11.24

Move to the **Models** option *(1)* to select appropriate mathematical models of the physical phenomena occurring in the analyzed case: heat transfer, turbulent fluid flow (Figure 11.25) and solar radiation (Figure 11.26). Firstly, activate the **Energy Equation** *(2)* for heat transfer. Do this by ticking the available option in the popup window *(3)* and confirm **OK**. Secondly, choose the **Viscous**

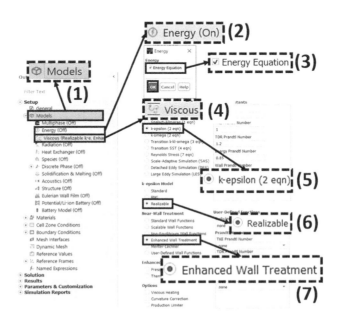

FIGURE 11.25

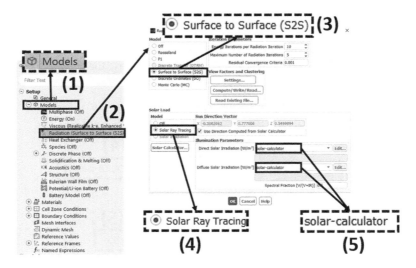

FIGURE 11.26

(4) option to choose the **k-epsilon** *(5)* turbulence model in the **Realizable** *(6)* mode with **Enhanced Wall Treatment** *(7)* functions. They effectively reflect thermal effects occurring in near-wall region.

Still in the **Models** *(1)* tab, move to the Radiation (2) tab. Activate the **Surface to Surface (S2S)** (3) model, which takes into account only walls - the domain material (air) is treated as fully transparent (Figure 11.26). Activate the **Solar Ray Tracing (4)** method and use *solar calculator* to state the value of **Direct** and **Diffusive Solar Irradiation (5)**.

Activate the **Solar Calculator (1)** to specify the details of solar radiation (Figure 11.27). In the **Global Position** section *(2)*, enter the appropriate geographic coordinates of Krakow (**Longitude** 20°, **Latitude** 50°) and **Timezone** (+0 GMT). In the **Date and Time (3)** section, select the appropriate date as *21st June* and time *12:00*. Based on these data, the angular height and azimuth of the Sun will be calculated. Then define the position of the created geometry (and mesh) in relation to the directions

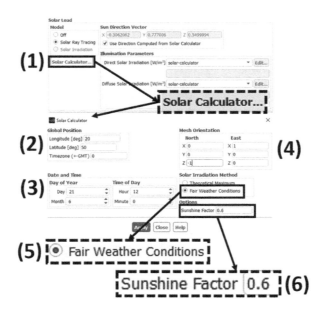

FIGURE 11.27

Tutorial 4 – Photovoltaic Farm

of the world *(4)*. The panels are facing south, so **North** is located in **Z** direction and **East** in **X** direction. To calculate the maximum theoretical energy flux, select **Fair Weather Conditions** in section *(5)* and set the **Sunshine Factor** *(6)* as **0.6**, which means that 40% of the sky is covered by clouds.

Now let's focus on **Materials *(1)***. By default, the air is set as fluid and aluminum as solid. In the studied case, the flowing fluid is air, the panels are made of numerous materials (silicon, glass, aluminum, *etc.*) and the bottom wall represents the ground. Both, the panels and the ground materials are not specified in the **Fluent database**, so the new materials should be defined manually (Figure 11.28). Start by clicking **2xLMB** on the **aluminum *(2)*** material and change its name to panel *(3)*. It is necessary to describe **density *(4)*, specific heat *(5)*** and **thermal conductivity *(6)*** as constant parameters according to Table 11.1. A description of PV panels with only one material is a huge simplification, but sufficient for the purpose of this analysis.

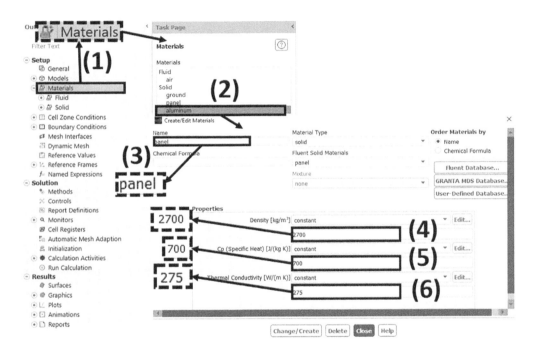

FIGURE 11.28

TABLE 11.1
Material Properties

	Density, kg/m³	Specific Heat, J/(kg K)	Thermal Conductivity, W/(m K)
Panel	2700.0	700.0	275.0
Ground	1600.0	1600.0	1.5

Repeat the steps described above and create the second missing material representing the ground.

Find the **Cell Zone Conditions** (Figure 11.29) option *(1)* and check if the enclosure is defined as *fluid (2)*. Then click **Edit** *(3)* and in the popup window set the fluid material as air *(4)*. Confirm changes by clicking **Apply**. Repeat these steps with the solid bodies representing PV panels. Remember to set their material as a newly defined panel.

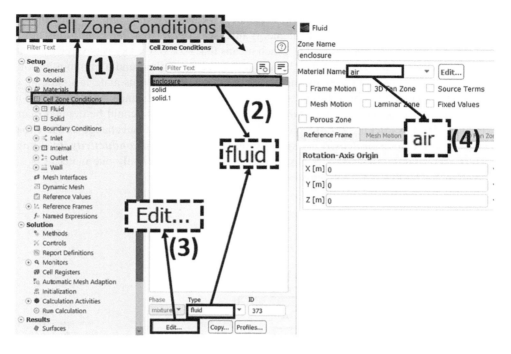

FIGURE 11.29

It's time to define Inlet, the first boundary condition (Figure 11.30). Click *2xLMB* on the *Boundary Conditions (1)* option and select the *Inlet (2)* defined as *velocity-inlet (3)*. Open the details window by clicking 2xLMB on the zone name *(2)*. In the *Momentum* tab insert *Velocity Magnitude* 0.05 m/s *(4)*. In the *Thermal* tab set *Temperature* of flowing air as *300 K (5)*. Move to the *Radiation* tab and deactivate the participation of inlet in *Solar Ray Tracing* and *View factor Calculation*. However, be aware that surfaces excluded from the View factor calculations emit heat radiation as a body with a temperature of *300 K*.

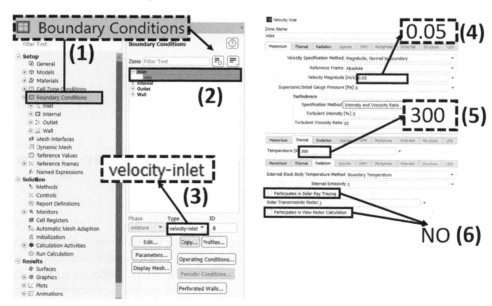

FIGURE 11.30

Tutorial 4 – Photovoltaic Farm   197

Being still in the ***Boundary Conditions (1)***, move to the ***Outlet (2)***. Its default type is ***pressure-outlet (3)***, which describes gauge pressure instead of outlet velocity (see Figure 11.31). In the ***Momentum*** tab set the ***Gauge Pressure*** to *0 Pa (4)* and in the ***Thermal*** tab leave the default ***Backflow Total Temperature (3)*** as the reversed flow is not expected. Like the inlet, the outlet is not participating in the radiation simulation *(6)*.

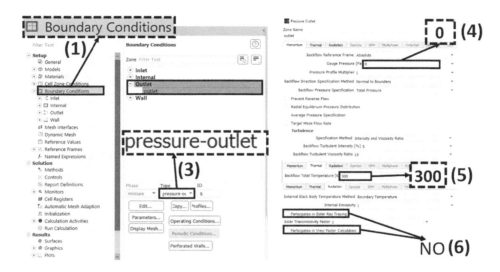

FIGURE 11.31

Now, move to *absorber1* boundary condition *(1)*. Its type is set by default as ***wall (2)*** with ***No Slip*** shear *(3)*, and the ***Coupled (4)*** thermal condition (as in Figure 11.32). Change the material type to ***panel (5)*** and in the ***Radiation*** tab confirm the participation of this wall in both: ***Solar Ray Tracing*** and ***View factor Calculation (6)***. Copy these settings to the second absorber. Absorber walls are accompanied by shadow walls, which are the imprints of the original surfaces into fluid domain. From the practical point of view, the perfect heat transfer occurs between them.

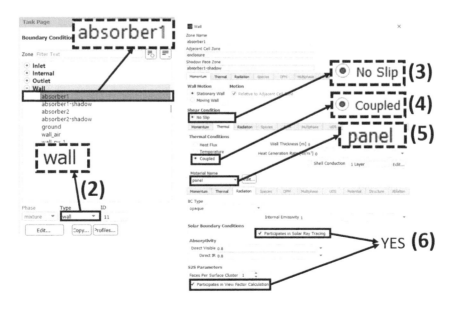

FIGURE 11.32

Wall_air *(1)* is another wall – *Type (2)* boundary condition. It is characterized by lack of shear, so change the **Shear Condition** from *No Slip* to *Specified Shear (3)* with *X, Y* and *Z components* equal to zero *(4)* as in Figure 11.33. In the **Thermal** tab, set the *Heat Flux* value to *0* to imitate the adiabatic condition. In the **Radiation** tab deactivate the **Ray-Tracking** and *View Factors* options, as the *wall-air* is not participating in radiation *(6)*.

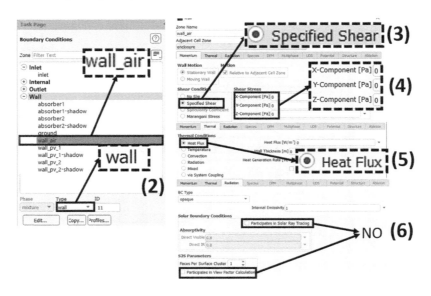

FIGURE 11.33

Ground *(1)* is *a* wall-*Type (2)* boundary condition with *No Slip (3)* option in the **Momentum** tab and **Temperature** *(4)* in the **Thermal** tab. The temperature 283 K *(5)* represents the ground temperature at the depth where the effect of solar radiation is negligible (Figure 11.34). In this case, the thermal penetration depth is equal to *3 m (6)*. The ground takes part in solar radiation, so the **Solar Ray Tracing** and **View factor Calculation** options have to be activated *(6)*.

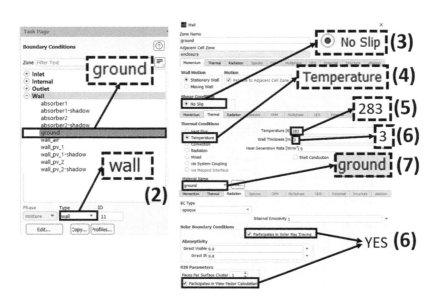

FIGURE 11.34

Tutorial 4 – Photovoltaic Farm

Boundary conditions on the panels side and bottom surfaces: wall_pv1, and wall_pv2 *(1)* are walls *(2)* with ***No Slip*** *(3)* and ***Coupled*** *(4)* heat transfer (Figure 11.35). They are made of panel material *(5)* and do not participate in solar radiation *(6)*.

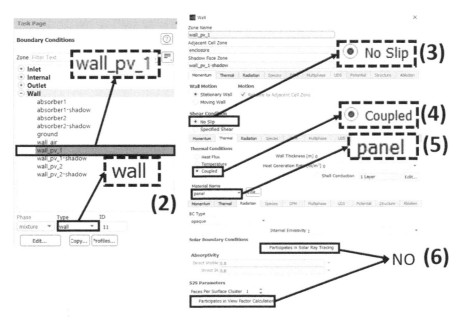

FIGURE 11.35

To solve the radiation equation simultaneously with the other ones, it is necessary to generate a file with information about ***View Factors***. They describe the interactions between the surfaces involved in the radiation. Generate them by clicking the ***Compute/ Write/ Read (1)*** option in ***View Factors and Clustering*** section (see Figure 11.36). Use the default file location to save the generated file.

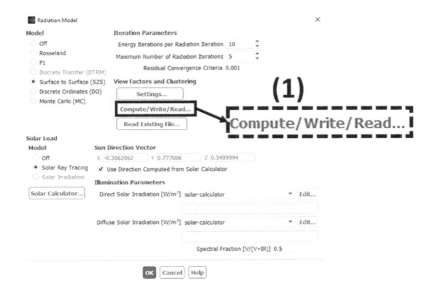

FIGURE 11.36

To supervise the computation process create at least one **Report definition** *(1)*, like velocity at the outlet or temperature at a specified surface (Figure 11.37). Click *2xLMB* on this option and a popup window will be opened. Create a *New* report *(2)* by choosing from the dropdown list *Surface Report* with *Area-Weighted Average* type *(3)*.

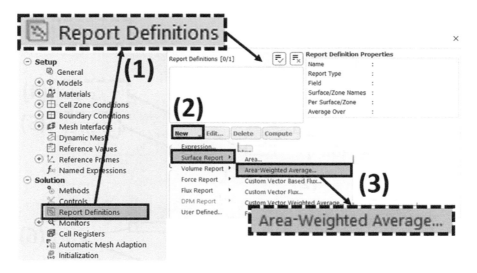

FIGURE 11.37

Change the Report Definition *Name* to t-absorber *(1)* and select *Static Temperature* as *Field Variable (2)*. This parameter should be calculated at the *absorber1 (3)* surface. Moreover, activate the *Print to Console* option *(4)* to monitor changes in the absorber temperature in real time during calculations, and confirm settings with *OK* (Figure 11.38).

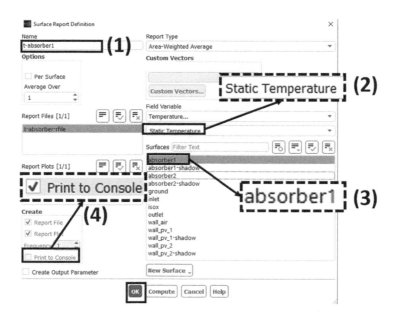

FIGURE 11.38

Tutorial 4 – Photovoltaic Farm

When the physics for the model is already determined, the next step is to define numeric *Methods* *(1)* of the solution process as in Figure 11.39. Set the algorithm of *pressure-velocity coupling* as *Coupled (2)*, which ensures a stable computation process. Check whether *Spatial Discretization* for each parameter is based on *Second Order* or *Second Order Upwind (3)* schemes. The latter provides a more accurate solution and reduces the numerical diffusion. Deactivate the *Pseudo Transient* option *(4)*, which is a form of under-relaxation for steady-state cases.

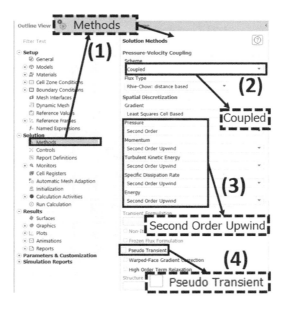

FIGURE 11.39

In each simulation, it is necessary to initialize the solution process as in Figure 11.40. Find the *Initialization* option *(1)* and select the *Standard Initialization (2)* mode. Then compute the initial solution parameters from *Inlet (3)*. Finally, click *Initialize* button *(4)* and move to the next step.

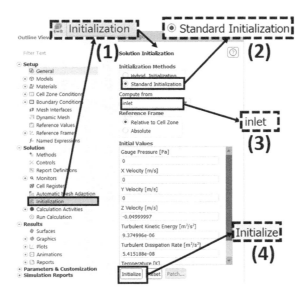

FIGURE 11.40

It is time to **Run Calculation (1)** as in Figure 11.41. Set **Number of Iterations** to 100 *(2)* – it should be enough to reach the convergence. Click the **Calculate** button *(3)*. During calculations observe the residuals and the average temperature calculated at the absorber surface. The curves which describe the error in the solution for each parameter separately should be heading down. The information that the calculation is complete will appear in the solver console and in the popup window.

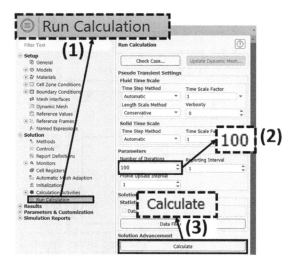

FIGURE 11.41

11.5 POSTPROCESSING

Create a visualization of the results in a Fluent- built-in post-processor as in Figure 11.42. Find the **Results (1)** tab and move to **Graphics → Contours** (2). In a new window, you may insert a new **Name (2)**. In the **Contours of** section, select **Temperature – Static Temperature** as the displayed parameter. In **Options (4)** activate the **Filled** way of contour appearance and **Global Range, Auto Range** for temperature. Specify the locations where the contour will be displayed at **Surfaces** of **absorber1, absorber2** and **ground (5)**. Confirm your choices by clicking **Save/Display**. Pay attention to the shadow shape and orientation. As you can see the temperature distribution on the absorbers is nearly uniform due to the use of a global range of values.

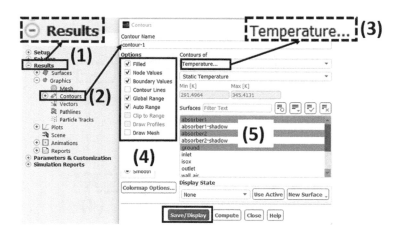

FIGURE 11.42

Tutorial 4 – Photovoltaic Farm

To make the visualization clearer, create the second contour of temperature. This time deactivate the **Global Range** and leave the **Auto Range** option *(1)* as in Figure 11.43. Display this contour only at the absorber surfaces *(2)* to observe temperature gradients. Which absorber has got higher values of temperature?

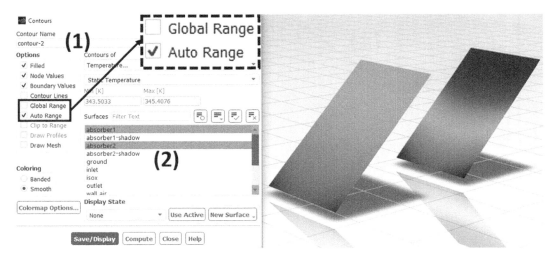

FIGURE 11.43

Generate the third contour, this time display the velocity *(1)* distribution at the geometry cross-section as in Figure 11.44. To create the section plane use the **New Surface → Iso-Surface** option *(2)*. Set the **Surface of Constant** based on **Mesh X-Coordinate *(3)*** with X value equal to *0.5 m (4)*. Then select the newly created plane as the location of the velocity contour.

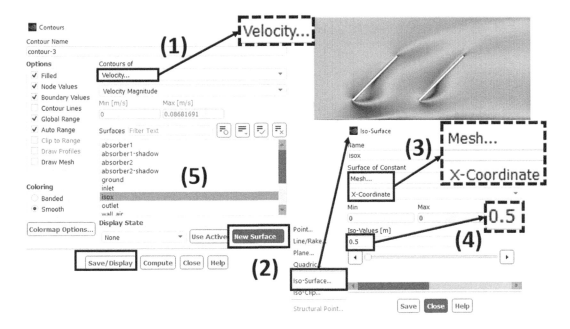

FIGURE 11.44

To visualize the velocity distribution *(1)* in another way generate ***Vectors (Results → Graphics → Vectors)***. Specify the ***Iso-Surface*** as vector location *(2)* and display them (Figure 11.45). It is possible to adjust the scale and number of vectors *(3)* to make the visualization clearer.

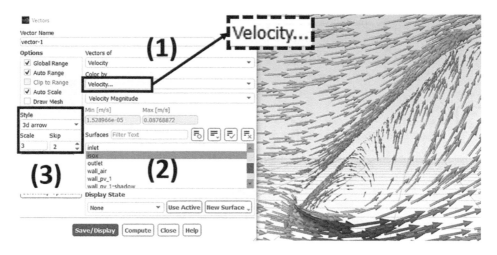

FIGURE 11.45

Another way to visualize the fluid movement is to use pathlines ***Results → Graphics → Pathlines*** (Figure 11.46). Define the display style by ***Velocity Magnitude (1)***. The pathlines should be released from the inlet *(2)*. To save some time during pathlines generation, set ***Steps*** to ***500*** and ***Skip*** parameter to ***50 (3)***. Activate the ***Draw Mesh*** option to display the panels and the ground surface simultaneously with the pathlines *(4)*.

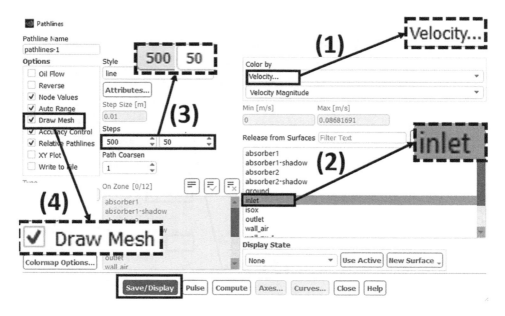

FIGURE 11.46

Tutorial 4 – Photovoltaic Farm

To calculate the average temperature of each absorber and ground, insert a new *Surface Integral (1)*. Define it as *Area-Weighted Average (2)* of *Temperature (3)*. Then choose appropriate surfaces where the average should be calculated *(4)*. Click *Compute* and the results should be displayed in the *Console (5)* as in Figure 11.47.

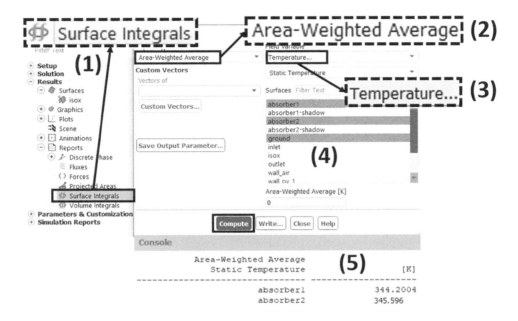

FIGURE 11.47

11.6 ADDITIONAL EXERCISE

In the *Models of Radiation of Solar Calculator* tab change the location of the PV panels to your coordinates (as in Figure 11.27). Generate the *View Factors* again (Figure 11.26), Initialize the solution (Figure 11.40) and start the calculations again (Figure 11.41). Compare the results obtained between two analyzed locations.

For one selected location, change the *Time* parameters in the *Solar Calculator* (as in Figure 11.27) to observe how the shadows from panels are oriented.

Part IV

Wind-Power-Based Technologies

12 Theoretical Background

12.1 DEVELOPMENT OF TECHNOLOGY

Wind formation is possible thanks to the pressure difference that comes from the non-uniform heating of the earth's surface by solar radiation. Due to this discrepancy (the earth is warmed up, especially in equatorial areas), some percent of the solar radiation energy can be converted into air mass movement. An increase in the air temperature results in a decrease in its density, while in case of cooling its density increases. The consequence is a pressure gradient forcing the wind to blow. Another factor influencing the wind is the earth's rotation, generating a Coriolis force, responsible for the air mass acceleration [1]. According to Ref. [2], even ten million MW of power could be generated with the whole available wind energy. However, as in case of other renewable energy sources, its availability strongly depends on geographic location and the time of the year.

Although there are no historical records of the exact time when wind energy was first used by man, however, it is certain that people learned this many centuries B.C. As a result of nature observation, ancient civilizations realized very fast that, if properly streamed, the wind has the potential to move even relatively heavy objects, provided that they are the right shape and size, as well as made of an appropriate material. This led to the invention of sails, already used in ancient Egypt, or vertical-axis drag-force-based windmills constructed in ancient Persia [3].

The mechanism of the latter was not very complicated - the torque was transmitted directly to the mill stone. However, the airstream directing technique was applied here, as can be noted in the "b" part of Figure 12.1.

The wind turbines that we know today were developed at the very beginning of the previous century. James Blyth was one of the pioneers. Apart from many other projects, he proposed the vertical-axis-turbine-based (VAWT) solution that was first used to generate electric power (actually even at the end of the XIX century) [1,4]. Nowadays, the biggest wind turbines are obviously horizontal-rotation-axis-based ones. According to Refs. [2,5,6], intense studies of the airfoils dedicated to horizontal-axis wind turbines (HAWT) started in the 1980s in the National Renewable Energy Laboratory (NREL, USA).

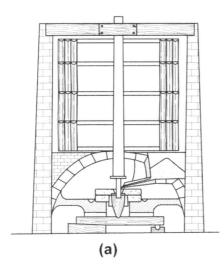
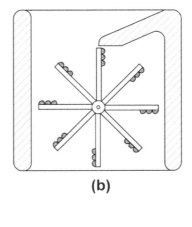

FIGURE 12.1 Drawing of the Persian windmill (a) and top view of the cross section (b) [3].

12.2 STATISTICAL DATA

Figure 12.2 illustrates the installed capacity of wind-power technologies in the top ten countries around the world in 2021. The largest installed capacity is noted for China (329 GW), which has been a leader in this ranking since 2010. Moreover, the five largest onshore wind farms in the world are located in China and this country is still rapidly expanding its wind installations.

The second place belongs to the USA with 133 GW of wind-power installations. These two countries share half of the world's wind-power-based technologies (824 GW). Germany (64 GW) and India (40 GW) take third and fourth places, respectively. All remaining countries have less than 30 GW of installed capacity: Spain 27 GW, the UK 27 GW, Brazil 21 GW, France 19 GW, Canada 14 GW and Sweden 12 GW. It is worth mentioning that the biggest offshore wind farm, with 1.2 GW installed capacity [8], is located in the UK.

Figure 12.3 presents the installed capacity of all the wind-power technologies in the world in the ten-year period, from 2012 to 2021. Generally, in the analyzed time the cumulative installed capacity increased from 267 GW in 2012 to 825 GW in 2021, which represents 26.9% of the renewables installed worldwide. This means that wind power is the leading renewable energy technology just after hydropower [9]. The most significant annual net addition in wind-power installed capacity was observed in 2020 and it was equal to 53 GW.

The onshore technologies significantly dominate the market of wind turbines with a 95% share, whereas offshore facilities have only a 5% share. It is due to technical and economical reasons: in 2020 the total installed cost of onshore wind technology was 1,355 USD/kW, whereas 1 kW of offshore cost was 3,185 USD [7]. Nevertheless, the total installed costs of offshore facilities declined by 33% between 2012 and 2020. It is predicted that the total installed capacity of onshore wind turbines would nearly triple by 2030 (to 1,787 GW) and increase eightfold by 2050, nearing 5,000 GW [9].

Figure 12.4 gives information about energy generation from onshore and offshore wind turbines in ten top countries in 2019. In that year, the global energy production from onshore technologies was 1328 TWh, whereas offshore installations generated 84 TWh. Generally, this disproportion is caused by the different levels of technology maturity. The electricity generation from wind is the highest in China and USA (with 400 TWh and 300 TWh respectively), and the lowest in Italy (about

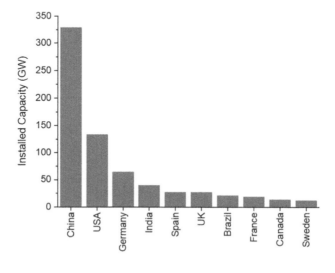

FIGURE 12.2 Top ten countries with the highest installed capacity of wind-power technologies in 2021. (Based on Ref. [7].)

Theoretical Background 211

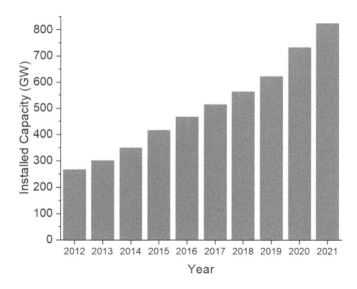

FIGURE 12.3 Cumulative installed capacity of wind-power technologies in the world. (Based on Ref. [7].)

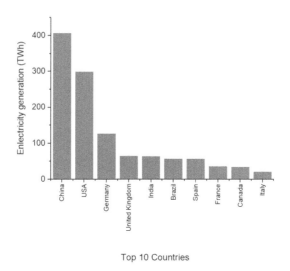

FIGURE 12.4 Top ten countries with the highest electricity generation from wind-power technologies in 2019. (Based on Ref. [7].)

20 TWh). In the United Kingdom, India, Brazil and Spain the energy generation is on a similar level (about 60 TWh). In The United Kingdom, nearly the 50% of the produced energy comes from offshore installations – this is the highest rate among all the presented countries.

12.3 CLASSIFICATIONS AND CHARACTERISTICS

The development of two main types of wind turbines suggests one basis of their division - rotation axis. However, as presented in Figure 12.5, a full classification tree includes another factor – the rotor diameter.

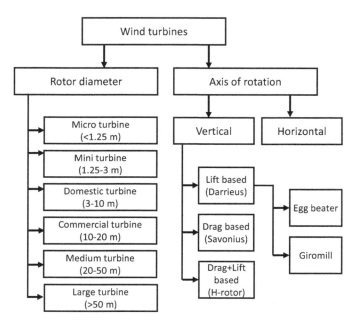

FIGURE 12.5 General classification of wind turbines. (Based on Refs. [10,11].)

As the name goes, in case of the HAWT the direction of windstream that propels the turbine and the rotor axis are generally parallel. To achieve the best possible performance, most of the turbines designed for power generation are equipped with two or three blades. However, when the overall torque is of importance (for example to power directly water pump or other machine), more than three-bladed turbines are applied as well. From the aerodynamics point of view, it has to be noted that both lift and drag forces have great importance here. The presence of the first one results in blade movement, whereas the latter one withstands it. Due to this fact, it is crucial to keep a high lift-to-drag ratio to achieve satisfying efficiency of energy conversion.

Performance issues and technical ability to design and construct really big turbines make HAWT much more popular than VAWT, when it comes to big-scale commercial systems [11,12]. The share changes in case of urban applications, including building-integrated turbines [13,14].

VAWTs can be classified taking into account specifics of the aerodynamic force, acting on turbine blades. The principle of certain device operation can be based on the lift force (*i.e.* Darrieus turbine) or drag force (Savonius turbine), or on both of these forces (like in case of H-rotor turbine). Aerodynamic force in lift-based wind turbines (LBWT) is perpendicular to the air stream direction. The construction of LBWT is relatively advanced, compared to drag-based wind turbines (DBWT). The latter is characterized by the aerodynamic force acting in the direction of the airstream.

The characteristics of the construction and operation of HAWTs and VAWTs are covered in the next subsection.

The third classification method of wind turbines is obviously by the rotor diameter. Technical limitations of turbine sizes are decreasing, which accelerates the increase in wind energy power generation. On the other hand, big farms pose serious issues, among others related to the area usage, selecting appropriate locations, excluding populated regions (among others due to the flickering-shade effect and significant noise generation) and, of course, environmental impact. Discussion of some important challenges facing further wind turbines development, including social acceptance issues, can be found in literature, like [12,13].

A possible solution to the limitations of wind-power generation development is to install many small turbines in built-up, urban areas, wherever it is impossible to apply big devices.

Theoretical Background 213

FIGURE 12.6 Building-integrated wind turbines at the AGH University (Krakow, Poland): spatial geometry of the turbines (a), location on the building elevation (b) and general view of the Energy Center (c).

This solution is based on another classification, applicable in urban conditions, dividing wind turbines into: building-mounted wind turbines (BMWT), building-integrated wind turbines (BIWT) and building-augmented wind turbines (BAWT).

In the case of BMWTs, the building performs the function of a supporting construction that allows locating the turbine in desirable wind conditions. This solution requires taking into account relevant noise and vibration constraints. The most common location of this kind of turbines are roofs.

In contrast to the above discussed turbine class, BIWTs have to be somehow integrated with the building design. This can be done in a few different ways, considering constructional and/or architectural points of view. What is important, also turbine energy system integrated with the building makes it a BIWT. A good example of two (HAWT and VAWT) building-integrated wind turbines is given in Figure 12.6, which presents devices installed in the Energy Center of the AGH University of Science and Technology (Krakow, Poland). Apart from the architectural and constructional aspects the installation is integrated with the university energy network.

In the case of BAWT, the building has to be designed in a way allowing to direct and even augment the air stream flowing through the turbine area, compared with the free standing turbine. Of course, such an approach requires more effort related to the building's aerodynamics and has to be considered in one of the first stages of architectural design.

12.3.1 HORIZONTAL-AXIS WIND TURBINES

The rotor consisting of blades, a nacelle and a hub is the most important part of the whole HAWT construction (Figure 12.7). It transfers the wind's linear motion into shaft rotational energy [10]. The rotor is constantly pointed upwind by the yaw system. The wind flows around the turbine blades and

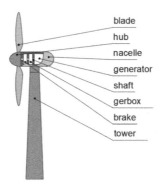

FIGURE 12.7 Construction of a horizontal-axis wind turbine.

generates the lift and drag forces. To increase the lift-to-drag ratio, the airfoil shape varies along the blade length and the blade itself may be also twisted [15]. Usually, HAWTs are equipped with three blades made of epoxy-glass or carbon fiber composites. The blades are mounted to the hub which transfers the rotational energy to the generator. Nevertheless, the generator requires high rotational speed, so a gearbox must be installed. Moreover, the shaft is also equipped with a brakes assembly, which regulates the rotation when the wind is too strong. The shaft, generator, gearbox and brakes are covered by a housing, called nacelle. It protects the machinery from the weather conditions and mechanical damage. The whole rotor is placed on a high tower made of steel and reinforced concrete.

The energy efficiency of HAWTs depends mainly on wind speed and turbulence intensity [16]. In beneficial conditions (strong and steady wind) the efficiency reaches 40%–55% [17]. The nominal power of the turbine increases with the rotor diameter, thus, turbines with 50–80-meter-long blades are constructed [12]. HAVTs are usually used in utility-scale applications, like onshore and offshore wind farms [11].

12.3.2 Vertical-Axis Wind Turbines

Due to the rotation axis, perpendicular to the ground and wind direction, the turbine may work in whatever direction of the wind and does not require the yaw system to point the rotor upwind [10,15]. The blades in VAWTs are located around the central vertical shaft, which transfers the rotational energy to the generator. The gearbox and brake assembly are required, but in contrast to the HAVTs, they are placed near the ground and are easily reached [18]. Due to this, the installation and maintenance costs of VAWTs are significantly lower [17]. They are also preferable in urban areas, because of the low cut-in wind speed, wind-direction independence and small size [10,11]. Operation at low rotational speed also reduces the vibrations and noise generation, so these turbines dominate the already mentioned building-integrated wind systems [11].

The Darrieus (LBWT) turbine is characterized by simple construction and relatively low manufacturing costs. Blades may be straight (H-type/giromill) or curved (egg-beater/phi-rotor) and they are mounted on a vertical framework [17]. In each type, there are usually two or three blades [19] made of epoxy-glass, carbon fiber or aluminum alloy [20].

In an H-type turbine (Figure 12.8), the blades are straight and mounted parallel to the main shaft by support arms. Their cross section usually consists of only one airfoil with constant cord length [11]. In this kind of turbine, the rotor is usually placed on a high tower to provide better operational conditions.

In a phi-rotor turbine (Figure 12.9) blades are curved and directly mounted on the upper and lower hub. This turbine does not require additional blade support, but the whole construction is

Theoretical Background

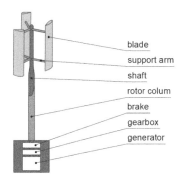

FIGURE 12.8 Construction of an H-type (Darrieus) wind turbine.

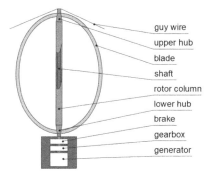

FIGURE 12.9 Construction of a phi-rotor (Darrieus) wind turbine.

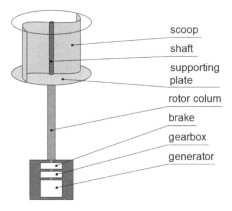

FIGURE 12.10 Construction of the Savonius wind turbine.

prone to vibration, so the guy wires should be installed. They connect the top of the shaft with the ground and provide stiffer turbine operation [15].

In Savonius (DBWT) turbines (Figure 12.10) typical blades are replaced by scoops, which makes their construction the simplest of all wind turbines. There are usually two or three semicircular scoops [17] placed between two circular supporting plates. When wind flows around the scoops, the area of stagnation pressure occurs on their concave side. This generates a positive torque on the advancing blade. At the same time, an opposite torque influences the convex surface of the returning blade. The difference in acting forces makes the scoops spin around the central shaft [21]. Because of the drag-type principle of the operation, these devices harness less wind energy than lift-type turbines of similar scale [18].

To sum up, Darrieus turbines have better aerodynamic performance and higher efficiency than Savonius ones. Nevertheless, the Savonius type provides more reliable operation in unfavorable wind conditions, especially when the wind speed is low [11,22] because they do not need any supportive start system [21].

12.4 FUNDAMENTALS OF ENERGY CONVERSION AND BALANCE

Wind as the movement of a certain mass with a given velocity can be considered as an energy source. Let's consider air moving with velocity v as shown in Figure 12.11.

Assuming, for simplicity sake, that wind passes through a circular surface S, we can see that during an arbitrarily chosen time t, it passes the distance $L = vt$. Therefore, we can conclude that in this way a cylinder with the bottom surface S and height L (Figure 12.11) is created. The volume V of this cylinder is $V = LS$ and the mass of air inside $m = \rho V = \rho LS$.

Kinetic energy E of the air mass inside the cylinder is:

$$E = \frac{mv^2}{2} = \frac{\rho V \cdot v^2}{2} = \frac{\rho LS \cdot v^2}{2} \tag{12.1}$$

This air mass m passes through the bottom during the time $t = \dfrac{L}{v}$, and hence the wind power P is:

$$P = \frac{E}{t} = \frac{1}{2} \cdot \rho S \cdot v^3 \tag{12.2}$$

Unfortunately, the power we can take from the air stream is smaller due to fundamental physical limitations. Let's consider the following situation as presented in Figure 12.12.

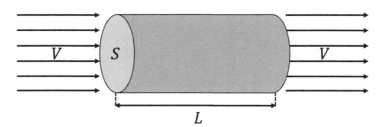

FIGURE 12.11 Scheme for calculation of wind power.

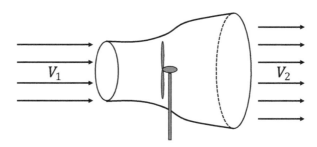

FIGURE 12.12 Wind turbine inside an air stream, the inlet and outlet wind velocity and energy are indicated.

Theoretical Background

Assuming the inlet and outlet wind velocity as v_1 and v_2 respectively, we can calculate the energy loss of the wind energy ΔE which can be taken by a turbine:

$$\Delta E = E_1 - E_2 = \frac{1}{2} \cdot \rho S \cdot \left(v_1^2 - v_2^2\right) \tag{12.3}$$

Since wind velocity varied in the range from v_1 to v_2 and wind velocity near the blades is unknown we assume its value as v.

Let's consider work W done by the wind:

$$W = F \cdot l = Fv\Delta t = \frac{\Delta p}{\Delta t} \cdot v \cdot \Delta t = v\Delta p = mv(v_1 - v_2) \tag{12.4}$$

Here we assume that work is a product of force F and l, a short distance that the wind passes near the turbine blades. Then, according to the second Newton dynamics principle, we expressed force F by the momentum change Δp.

Comparing equations 12.4 and 12.3 because $W = \Delta E$ we obtain:

$$v = \frac{1}{2} \cdot (v_1 + v_2) \tag{12.5}$$

It means that the wind velocity near the turbine blades is simply equal to the arithmetic average of the inlet and outlet wind velocity.

The power of the wind turbine P_T can be expressed as the wind kinetic energy losses rate:

$$P_T = \frac{\Delta E_k}{\Delta t} = \frac{1}{2} \dot{m} \left(v_1^2 - v_2^2\right) \tag{12.6}$$

where $\dot{m} = \frac{m}{\Delta t}$ is the mass stream.

Mass stream $\dot{m}$ can be expressed as:

$$\dot{m} = \rho S \cdot v = \frac{1}{2}(v_1 + v_2) \cdot \rho S \tag{12.7}$$

Putting 12.7 to 12.6 we have:

$$P_T = \frac{1}{4} \rho S (v_1 + v_2)\left(v_1^2 - v_2^2\right) \tag{12.8}$$

Defining power coefficient c_p as a ratio of turbine power to the wind power we can obtain the following formula:

$$c_p = \frac{P_T}{P} = \frac{1}{2} \cdot \frac{(v_1 + v_2)\left(v_1^2 - v_2^2\right)}{v_1^3} \tag{12.9}$$

This formula illustrates that power coefficient depends on outlet wind velocity.

Plot of this formula is presented in Figure 12.13, where parameter b represents $\frac{v_2}{v_1}$.

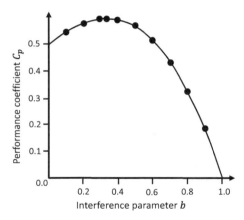

FIGURE 12.13 Power coefficient c_p as a function of interference parameter.

The function c_p reaches its maximum value for $b = 1/3$ (outlet wind velocity $v_2 = 1/3\, v_1$) and is equal $\frac{16}{27} \approx 0.59$. This is the theoretical limit of wind turbine efficiency called Betz limit.

In practice, for real wind turbines efficiencies are definitely smaller and let us consider two cases of (1) horizontal wind turbines (HAVT) and (2) vertical wind turbines (VAWT). Efficiency strongly depends on the so called tip speed ratio (TSR) λ defined as linear speed of the blade tip and the inlet wind speed v_1:

$$\lambda = \omega R / v_1 \qquad (12.10)$$

where ω is angular velocity of the blades and R their length (radius of rotation).

12.4.1 Horizontal-Axis Wind Turbines

When wind passes through the rotor plane and interacts with a moving rotor, a resultant relative velocity at the blade appears, which is the difference between wind velocity and blade tip velocity as illustrated in Figure 12.14.

The relative wind speed w is equal:

$$w = \sqrt{v^2 + u^2} \qquad (12.11)$$

where notation is according Figure 12.14.

The aerodynamics forces on the blade are the lift force F_L and drag F_D and are given:

$$F_L = \frac{1}{2}\rho c W^2 C_L \qquad (12.12)$$

$$F_D = \frac{1}{2}\rho c W^2 C_D \qquad (12.13)$$

where C_L and C_D are, respectively, the lift and drag coefficients, c is the chord length of the blade.

Theoretical Background

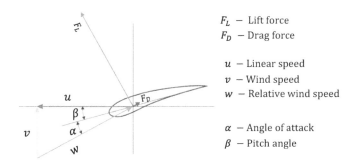

F_L — Lift force
F_D — Drag force

u — Linear speed
v — Wind speed
w — Relative wind speed

α — Angle of attack
β — Pitch angle

FIGURE 12.14 Interaction of wind and blade of HAVT.

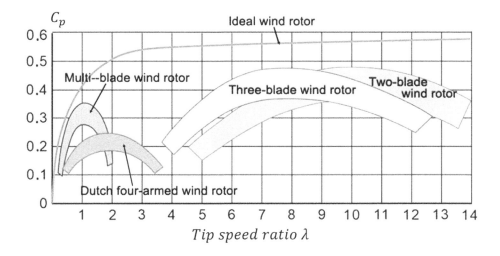

FIGURE 12.15 Power curves c_p for varied HAVT (1,2,3 and many blades).

The coefficient c_p depends on the TSR λ and blades angle β and e.g. the following expression obtained by data approximation can be applied [23]:

$$c_p(\lambda,\beta) = C_1\left(\frac{C_2}{A} - C_3\beta - C_4\right)\exp\left(\frac{-C_5}{A}\right) + C_6\lambda \tag{12.14}$$

where $C_{1,\ldots,6}$ are coefficients obtained from data fitting,

$$\frac{1}{A} = \frac{1}{\lambda + 0.08\beta} - \frac{0.035}{\beta^3 + 1} \tag{12.15}$$

The mechanical torque T can be expressed as:

$$T = c_p(\lambda,\beta)\cdot\frac{\rho S v_1^3}{2\omega} \tag{12.16}$$

The typical c_p curves are illustrated in Figure 12.15.

It is visible that there exists an optimal TSR value λ_{opt} which depends on the number of blades n. In some simplicity optimal TSR can be expressed by the number of blades n:

$$\lambda_{opt} \approx \frac{4\pi}{n} \tag{12.17}$$

12.4.2 Vertical-Axis Wind Turbines

They are composed of two main parts: the blade rotor in a vertical position and a mechanical gear transmission (bevel gear). According to Section 12.3.2, there are many types of VAWT, especially related to blade shapes. Let's consider the Darrieus-type wind turbine with straight blade as a representative [24].

Figure 12.16 shows the main ideas:

Relative wind velocity w can be expressed as:

$$W = \sqrt{V_c^2 + V_n^2} \tag{12.18}$$

$$V_c = \omega R + V_a \cos \Theta \tag{12.19}$$

$$V_n = V_a \sin \Theta \tag{12.20}$$

$$V_a = V(1-a) \tag{12.21}$$

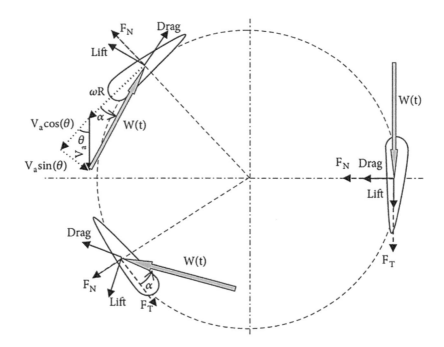

FIGURE 12.16 Scheme of the VAVT, flows and velocities are indicated.

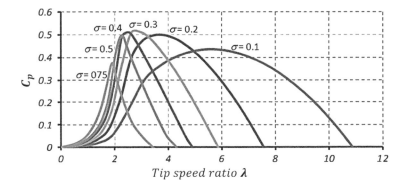

FIGURE 12.17 Power efficiency for a Darrieus turbine in function of TSR for varied values of rotor solidity σ.

The angle of attack is defined as the angle between the resulting air velocity vector and the blade chord. It is expressed as:

$$\alpha(\Theta) = \tan^{-1}\left(\frac{V_n}{V_c}\right) = \frac{(1-a)\sin\Theta}{(1-a)\cos\Theta + \lambda} \qquad (12.22)$$

The c_p coefficient for VAWT (on the example of the Darrieus turbine) can be expressed by the parameter called rotor solidity σ, which is blade area divided per rotor area. Some exemplary curves are presented in Figure 12.17.

REFERENCES

1. T. Corke, R. Nelson, *Wind Energy Design*, CRC Taylor & Francis Group, Boca Raton, FL, 2018.
2. N. Karthikeyan et al., Review of aerodynamic developments on small horizontal axis wind turbine blade, *Renewable and Sustainable Energy Reviews*, 42 (2015), 801–822.
3. https://commons.wikimedia.org/wiki/File:Perzsa_malom.svg (last access: 1.06.2022).
4. Wind turbines - the types & history, https://www.vertogen.eu/wind-turbine-history/ (last access: 14.04.2022).
5. W. A. Timmer, R. P. J. O. M. van Rooij, Summary of the Delft University wind turbine dedicated airfoils, *Journal of Solar Energy Engineering*, 125 (2004), 488–469.
6. J. Radhakrishnan, D. Suri, Design and optimisation of a low Reynolds number airfoil for small horizontal axis wind turbines, *International Conference on Mechanical, Materials and Renewable Energy, IOP Conference Series*, 377 (2018), 1–7.
7. Statistics Data of the International Renewable Energy Agency, www.irena.org (last access: 07.05.2022).
8. World's Largest Offshore Wind Farm Fully Up and Running, available online: https://www.offshorewind.biz/2020/01/30/worlds-largest-offshore-wind-farm-fully-up-and-running/ (last access: 11.06.2023).
9. IRENA (2019), Future of wind: Deployment, investment, technology, grid integration and socio-economic aspects (A Global Energy Transformation paper), International Renewable Energy Agency, 2019, available online: https://www.irena.org/publications/2019/Oct/Future-of-wind (last access: 14.04.2022).
10. Z. Tasneem et al., An analytical review on the evaluation of wind resource and wind turbine for urban application: Prospect and challenges, *Developments in the Built Environment*, 4 (2020), 1–15 (100033).
11. R. Kumar, K. Raahemifarb, A. S. Fung, A critical review of vertical axis wind turbines for urban applications, *Renewable and Sustainable Energy Reviews*, 89 (2018), 281–291.
12. N. A. Ahmed, M. Cameron, The challenges and possible solutions of horizontal axis wind turbines as a clean energy solution for the future, *Renewable and Sustainable Energy Reviews*, 38 (2014), 439–460.
13. K. C. Anup, J. Whale, T. Urmee, Urban wind conditions and small wind turbines in the built environment: A review, *Renewable Energy*, 131 (2019), 268–283.

14. T. Stathopoulos et al., Urban wind energy: Some views on potential and challenges, *Journal of Wind Engineering & Industrial Aerodynamics*, 179 (2018), 146–157.
15. S. Eriksson, H. Bernhoff, M. Leijon, Evaluation of different turbine concepts for wind power, *Renewable and Sustainable Energy Reviews*, 12 (2008), 1419–1434.
16. J. M. Carrasco, E. Galván, R. Portillo, Chapter 4: Wind turbine applications. *Alternative Energy in Power Electronics*, edited by M. H. Rashid. Butterworth-Heinemann, Oxford, 2011, pp. 177–230.
17. A. Tummala et al., A review on small scale wind turbines, *Renewable and Sustainable Energy Reviews*, 56 (2016), 1351–1371.
18. X. Jin et al., Darrieus vertical axis wind turbine: Basic research methods, *Renewable and Sustainable Energy Reviews*, 42 (2015), 212–225.
19. F. Alqurashi, M. H. Mohamed, Aerodynamic forces affecting the H-rotor Darrieus wind turbine, *Modelling and Simulation in Engineering*, 2020 (2020), 1368369.
20. W. Tjiua et al., Darrieus vertical axis wind turbine for power generation I: Assessment of Darrieus VAWT configurations, *Renewable Energy*, 75 (2015), 50–67.
21. N. R. Maldar, C. Y. Ng, E. Oguz, A review of the optimization studies for Savonius turbine considering hydrokinetic applications, *Energy Conversion and Management*, 226 (2020), 113495
22. J. V. Akwaa, H. A. Vielmo, A. P. Petry, A review on the performance of Savonius wind turbines, *Renewable and Sustainable Energy Reviews*, 16 (2012), 3054–3064.
23. N. Derbel, Q. Zhu (editors), *Modelling, Identification and Control Methods in Renewable Energy Systems*, Springer, New York, 2019.
24. B. Zghal et al., Analyses of dynamic behavior of vertical axis wind turbine in transient regime, *Hindawi Advances in Acoustics and Vibration*, 2019 (2019), 7015262.

13 Tutorial 5 – Horizontal-Axis Wind Turbine

13.1 EXERCISE SCOPE

This tutorial presents the procedure of simulating a moving domain including a horizontal-axis wind microturbine *(1)* (Figure 13.1), like the one presented in the classification (Section 12.3), based on the dynamic mesh approach applied in the turbine surroundings *(2)*. In the course of the simulation, it is possible to test whether the wind stream of a certain velocity magnitude is strong enough to cause revolutions of the turbine rotor characterized by the given moment of inertia and mass.

Try to recall from the theoretical background (Section 12.4) the factors influencing turbine ability to rotate and operate efficiently. From the aerodynamics point of view lift and drag forces have great importance here. The presence of the first one results in the blade movement. This exercise allows to assess whether the considered simple blade shape can provide a high enough lift force to move the rotor.

Among others, this exercise can teach you how to:

- use the circular pattern tool to duplicate repeatable objects (like turbine blades) in *SpaceClaim DirectModeler*,
- apply different local sizing functionalities of *Ansys Meshing* to differentiate cell sizes in near-rotor regions and free stream areas,
- create interfaces between stationary and moving mesh areas in *Ansys Fluent*, to allow a simulation of the model with an inconsistent mesh.

Pay special attention to the *Degrees of Freedom (DOF)* settings, including defining of the DOFs number and moment of inertia.

All of the images included in this tutorial use courtesy of ANSYS, Inc.

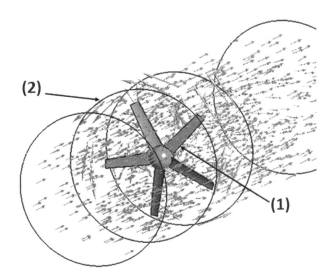

FIGURE 13.1 Cylindrical domain including the turbine rotor considered in the exercise.

13.2 PREPROCESSING – GEOMETRY

Find the *Toolbox* window (LHS) (Figure 13.2). Then drop down the *Analysis System* list, hold *1xLMB* on *Fluid Flow (Fluent) (1)*, drag and drop it in the *Project Schematic* window. Then click *2xLMB ion A2* (*Geometry*) *(2)* cell to launch *Ansys SpaceClaim Direct Modeler*.

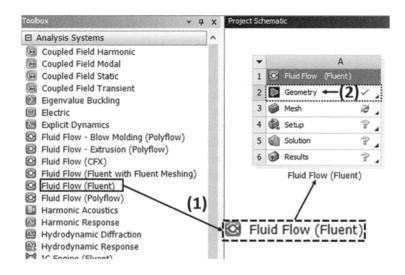

FIGURE 13.2

Find and click the *Sketch (1)* tab on the top bar (as in Figure 13.3). Basic sketching tools are available in the toolbox below. Select *1xLMB* on *Circle (2)*. A set of four buttons appears at the bottom of the workspace window. Click *Select New Sketch Plane (3)* and move the mouse cursor to the Z-axis of the coordinate system *(4)*. If the plane visualization appears, click *1xLMB* on this axis. That's how you can select any sketch plane based on the global coordinate system axes – the created plane is always normal to the clicked axis. Now set *Plan View (5)*. Alternatively, you can set this view by pressing the *V* key from the keyboard.

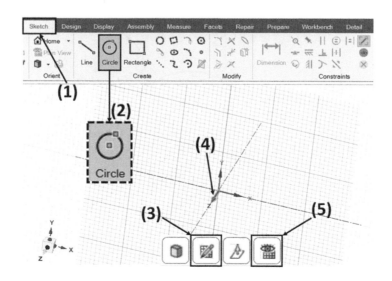

FIGURE 13.3

Tutorial 5 – Horizontal-Axis Wind Turbine   225

To draw a circle, click *1xLMB* on the center of the coordinate system *(1)* (the intersection of the *X* and *Y* axes). Move the mouse cursor anywhere to start drawing, insert a diameter (2) of *120 mm* (keyboard) and press **Enter**. Then click **End Sketch Editing *(3)***. The project switches to the ***Design*** tab (top bar). The ***Pull (4)*** tool is activated automatically (if not – you can click it now). Click *1xLMB* on the surface that has been created from your circle sketch. A small toolbar starting with + and – buttons should appear somewhere around (Figure 13.4). Select the ***Pull Both Sides (6)*** tool (two arrows). Now click and hold *LMB* on the small yellow arrow *(7)* and drag it a bit. The ***Pull*** operation has started and you can insert its dimension *−40 mm (8)*. Press **Enter** to confirm. The cylinder part of the turbine hub is ready. Now it's time to create its front part.

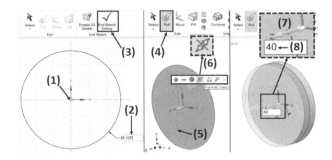

FIGURE 13.4

Go to the ***Sketch (1)*** tab of the top bar and select the ***Circle (2)*** tool from the ***Create*** section. The circle has to be drawn on the front side of the hub cylinder (it's opposite to the *Z-axis* direction) – click *1xLMB* on this surface *(3)*. Have a look at the triad *(4)* – it might help you to mark the appropriate cylinder face. Draw the circle *(5)* (according to Figure 13.5) with a diameter equal to the hub cylinder diameter. Then switch the sketching tool to ***Line (6)***. Create the diameter line *(7)* coincident with the *Y* axis. Now there are two arcs instead of a circle in the sketch. Use the ***Trim Away (8)*** tool from the ***Modify*** section of the top toolbar to remove one arc – just select the above mentioned tool and click *1xLMB* on either of the two arcs *(9)*. Finally, only two components *(10)*: A line and an arc, should remain in the sketch. If so, click ***End Sketch Editing (11)***.

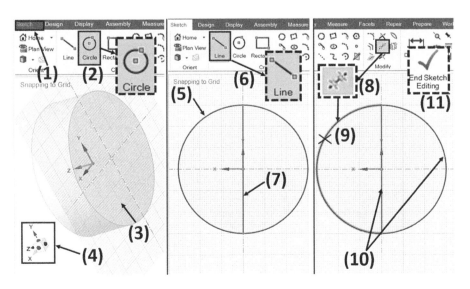

FIGURE 13.5

After **End Sketch Editing**, the project switches automatically to the **Design (1)** tab. Additionally, the **Pull (2)** tool should be activated automatically. If not, you can select it now. Click *1xLMB* on the previously created semicircle *(3)* surface (Figure 13.6). A small yellow arrow appears on it, but this time the surface is not to be pulled in a straight line. Select the **Revolve (4)** function. Then click *1xLMB* on the diameter line *(5)* – from now on, it is the axis of revolution. A small yellow (light gray) rounded arrow appears around the axis of revolution. Click and hold it with **LMB** next. Drag a bit until a small dimension window appears next to the arrow. Insert *180° (6)* and press **Enter** to confirm. The spherical hub front *(7)* is done.

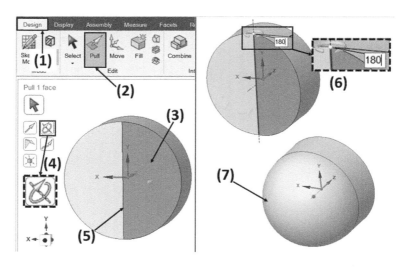

FIGURE 13.6

Stay in the **Design (1)** tab. Have a look at the **Structure** (LHS) window (Figure 13.7) – it should contain one object – **Solid (2)**. It is 3D hub geometry. You can hide and show any object which is in the **Structure** window by clicking the square to the left of this object. Find and click The **Plane (3)** tool on the top toolbar. Then move the mouse cursor to the **Y** coordinate system *axis (4)*. The plane visualization appears *(5)*. Click *1xLMB* on the axis to confirm the plane location. Now this plane is visible in the **Structure** window too.

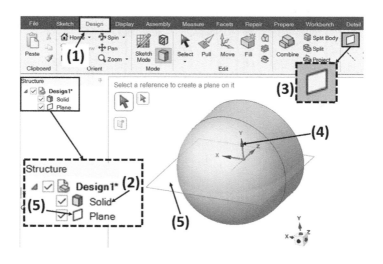

FIGURE 13.7

Tutorial 5 – Horizontal-Axis Wind Turbine

This plane will be used to create the sketch of the blade profile. However, it has to be moved to the appropriate location first. Click the *Move (1)* tool on the top toolbar. Then click *1xLMB* on any edge of the plane (Figure 13.8). An additional triad of the *Move* tool appears in the plane center. Click and hold *LMB* on the vertical *Y* axis *(3)* and drag it up a bit. When the dimension cell appears, insert *61 mm (4)* and confirm the move by *Enter*.

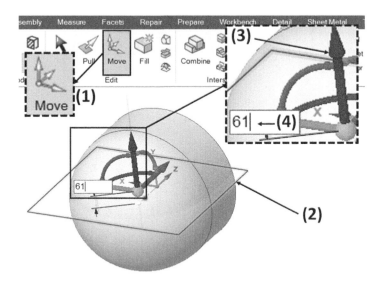

FIGURE 13.8

You can change the view by clicking the *Z*-axis of the global coordinate system triad and using the scroll mouse bottom to see that the created plane is located just above the hub *(1)*, according to Figure 13.9. Now go to the *Sketch (2)* tab (top menu) and select the *Sweep Arc (3)* tool from the *Create* section. Click *1xLMB* on the plane to select it as the sketching plane *(5)*. Then press the *V* key to set the plane view.

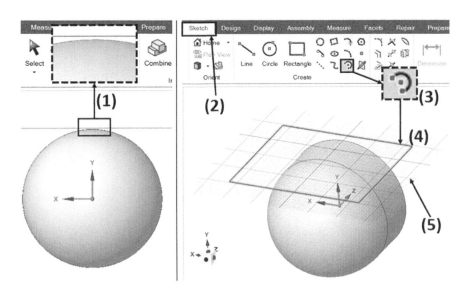

FIGURE 13.9

Create a sketch according to Figure 13.10. Click *1xLMB* somewhere around the location *(1)* to define the arc center point (accurate dimensions will be defined later!). Define the location (*1xLMB*) of the arc start point *(2)* and, analogously, the end point *(3)*. If something goes wrong, you can undo the operations (*Ctrl+Z*). Find the *Dimension (4)* tool in the *Constraints* section of the top toolbar and define all the required dimensions.

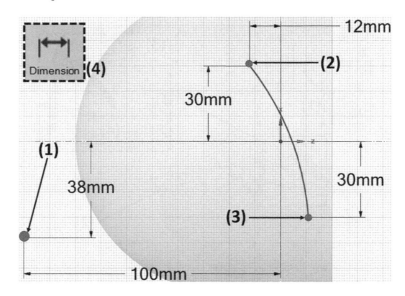

FIGURE 13.10

Now the arc has to be copied. Select the *Move (1)* tool and click *1xLMB* on the already created arc *(2)*, as in Figure 13.11. The triad of the *Move* tool appears. Click and hold its horizontal axis *(3)*, and drag left a bit. When the dimension cell appears, insert *4mm (4)* and confirm by enter. Alternatively, you can just click on the horizontal axis and enter the *4mm* value. To complete the sketch, use the *Line (5)* tool to draw two short sections between the ends of the arcs. Click *End Sketch Editing (6)*.

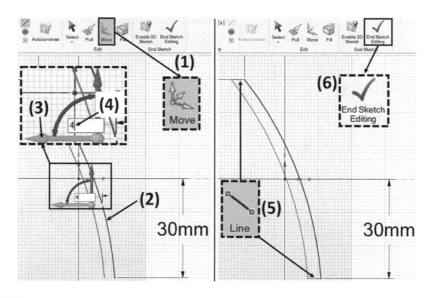

FIGURE 13.11

Tutorial 5 – Horizontal-Axis Wind Turbine

The sketch fills automatically to the surface *(1)*. Deactivate **Solid** in the **Structure** window to make the surface better visible. Set the view *(2)* using the triad (Figure 13.12). Hold **Ctrl** and select all four edges *(3)* of the blade base profile. Copy them by **Ctrl+C** → **Ctrl+V**. The number of objects on the list *(4)* increases: **Solid** *(5)* is the hub, **Surface** *(6)* is made up of the created sketch (it is the blade base profile surface). The last object on the list is a set of **Curves** *(7)* created as a consequence of the copying operation.

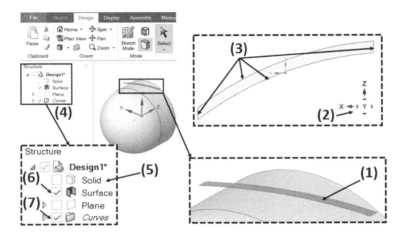

FIGURE 13.12

Have another look at the **Structure** *(1)* window and leave only **Curves active.** *(2)*. It will help you to select all of them. The *Y* axis of the coordinate system (triad) should be facing up *(3)*, as in Figure 13.13. Select the **Move** *(4)* tool again. Click *1xLMB* wherever in the workspace and select all four curves *(5)* by **Ctrl+A**. Then press and hold **Ctrl** and drag up a bit the *Y* axis *(6)* of the **Move** tool triad (using *LMB*). Insert *250 mm* in the dimension cell *(7)*. Confirm by **Enter**.

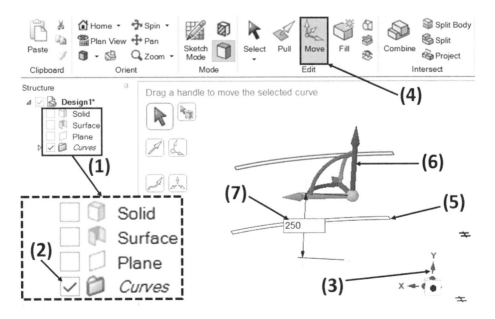

FIGURE 13.13

The created top blade profile (Figure 13.14) has to be scaled. Stay in the *Design (1)* tab. Select all the curves of the top blade profile and click *the pull (2)* tool, according to Figure 13.14. Click *Scale Body (3)* and move the mouse cursor to the left profile arc. Find the center point (4) of the arc (a small triangle appears) and click *1xLMB*. It is the so called scaling anchor point now. Click and hold *LMB* somewhere around the profile and move the mouse (up/down) to start resizing of the profile. The *Scale (5)* cell appears. Start writing the scale value *(6)* before you release the mouse *LMB*. The correct value is *0.7*. Confirm by *Enter*.

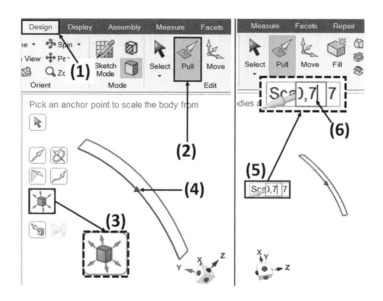

FIGURE 13.14

Click *Select (1)* (Figure 13.15) to make sure that the *Pull* tool is deactivated. Drag the selection box *(2)* in the workspace to select all four curves of the already scaled profile. Create a surface from the curves by clicking the *Fill (3)* tool or the *F* key. The new surface is visible in the workspace *(4)* (move the mouse cursor anywhere between the curves) and the *Structure (5)* window.

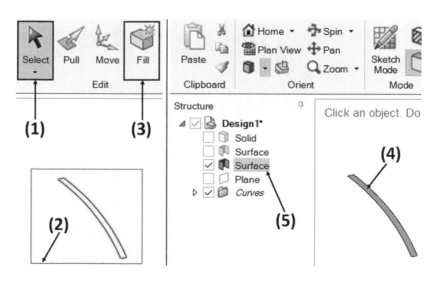

FIGURE 13.15

Tutorial 5 – Horizontal-Axis Wind Turbine

There are two surfaces *(1)* of two blade profiles in the ***Structure*** window now as in Figure 13.16. Remove ***Curves (2)*** from there (***1xRMB*** → ***Delete***). To create a 3D blade, select the ***Blend (3)*** tool from the ***Edit*** section of the top toolbar. Then hold ***Ctrl*** and select subsequently both profile surfaces *(4)*. A projection of the 3D blade geometry between the profiles appears *(5)*. To confirm it, click ***Complete (6)*** or press ***Enter***. A new ***Solid (7)*** object has appeared in the ***Structure*** window. Two surfaces have disappeared.

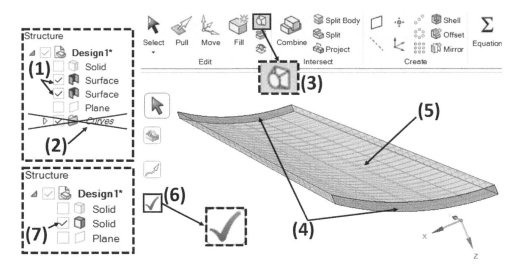

FIGURE 13.16

Now the blade will be fitted to the hub shape, but the objects won't be merged. Activate both ***Solid*** objects *(1)* in the ***Structure*** window, as in Figure 13.17. Select the ***Pull (2)*** tool (top toolbar) and click the ***No Merge (3)*** option in the ***Options – Pull*** window. Click ***1xLMB*** on the bigger bottom surface (close to the hub) of the blade *(4)*. Now click the ***Up To (5)*** function and click ***1xLMB*** on the cylindrical surface of the hub *(6)* (somewhere in its top region). The blade base is fitted to the hub now.

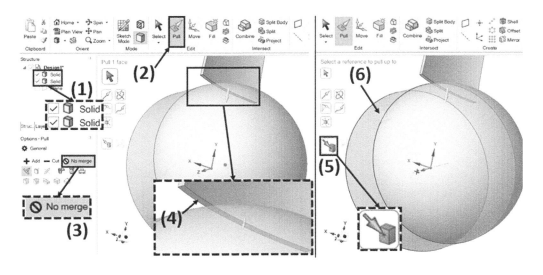

FIGURE 13.17

The remaining blades will be created by pattern as in Figure 13.18. Select the *Circular Pattern* tool from the *Create* section of the top toolbar. Click the *Select Object (2)* function and from the *Structure* window select *Solid (3)* representing the blade. Click the *Direction (4)* function and select the *Z*-axis *(5)* of the global coordinate system. Set *Circular count* to *5* and *Angle* to *360° (6)*. A projection of the blades appears *(7)*. Click *Create Pattern (8)* to confirm.

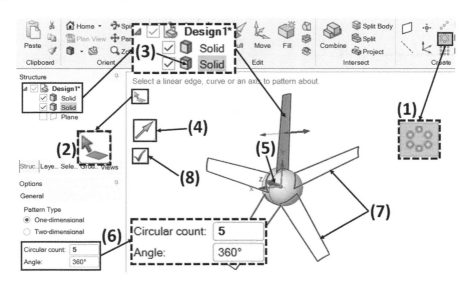

FIGURE 13.18

The Structure window includes a new pattern *(1)*. You can open it to see 5 components representing 5 blades, as in Figure 13.19. Actually, *Plane* is not needed anymore and can be completely removed *(2)*. Click *Select (3)* to make sure that the *Circular Pattern* tool is deactivated. Then select the *Combine (4)* tool to combine all solids in the turbine geometry. Click *1xLMB* on the hub. Click *Select Bodies to Merge (6)* and click *1xLMB* on any blade *(7)*. The blade is merged with the hub now. Hold *Ctrl* and click *1xLMB* on each subsequent blade. Now the *Structure* window includes one *Solid (8)* object (whole turbine) and *Pattern* with one empty *Component1 (9)*. *Delete* this *Pattern*.

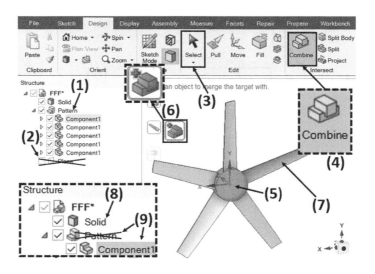

FIGURE 13.19

Tutorial 5 – Horizontal-Axis Wind Turbine   233

The Structure window should include only one *Solid* object – according to *(1)* in Figure 13.20. Go to the *Sketch (2)* tab. Click the *Circle (3)* tool. Select the *X-Y* sketching plane (click *Z*-axis of the triad) and set the plane view (as in Figure 13.20). Create a circle with the center in the coordinate system center *(4)* and with a diameter of *800 mm (5)*. Click *End Sketch Editing (6)* to proceed.

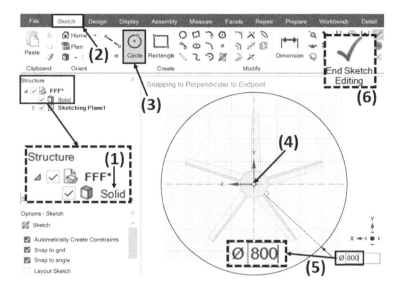

FIGURE 13.20

The Pull (1) tool from the *Design* tab is activated automatically. Before pulling click *No Merge (2)* (below the *Structure* window) to obtain another individual solid in the result. Then click *1xLMB* on the surface *(3)* created from the circle sketch. A small longitudinal toolbar window *(4)* should appear somewhere around, as in Figure 13.21. Click the fourth function from this toolbar – *Pull Both Sides*. Then drag the yellow (light gray) arrow a bit and insert a distance of *500 mm* in the dimension cell *(5)*. Confirm by *Enter*. Second *Solid (6)* (cylinder surrounding turbine rotor) appears in the *Structure* window.

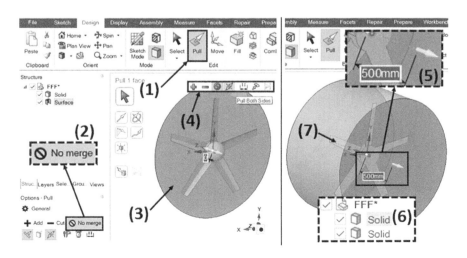

FIGURE 13.21

In a typical fluid flow model, there is no need to consider any solid bodies – if a turbine rotor has to move, it is enough to simulate the movement of its walls (no solid presence inside is required). Thus, the turbine shape will be cut in the cylinder body now. Stay in the *Design (1)* tab. Click the *Combine (2)* tool from the top toolbar, as in Figure 13.22. Click the *Select Target (3)* function and select the cylinder body *(4)*. Now switch to the *Select Cutter (5)* mode and click *1xLMB* on the solid representing the turbine rotor *(6)* in the *Structure* window. It is easy to recognize which object represents the turbine – just activate or deactivate subsequent solids in the *Structure* window and check when the turbine *(7)* appears.

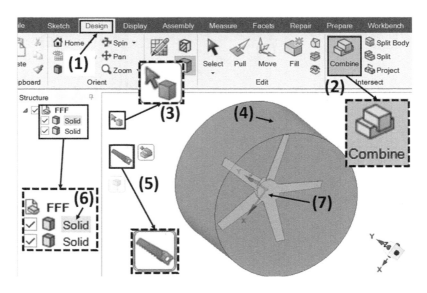

FIGURE 13.22

After cutting, there are three *Solid* objects in the *Structure* window, according to *(1)* in Figure 13.23. Remove completely (*1xRMB → Delete*) two unnecessary geometry parts – only the cylinder with the cut turbine – shape void *(2)* should remain in the project. To make sure which solid should remain, deactivate all solids in the *Structure* window and move the mouse cursor subsequently to each solid. They will be visualized in the workspace window. Only one will be shown

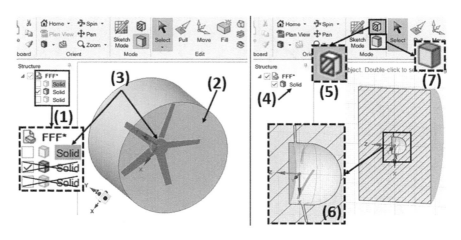

FIGURE 13.23

Tutorial 5 – Horizontal-Axis Wind Turbine   235

with the turbine shape inside *(3)*. This one is appropriate to stay in the project – it should be the only *Solid* in the *Structure* window *(4)*. You can also check whether the turbine-shape void exists in this cylinder using the *Section Mode (5)* tool (X keyboard shortcut) – just select a section plane (analogously as sketching plane), for example, the *Y-Z* one to see the cylinder interior *(6)*. Click the *3D Mode (7)* tool (D keyboard shortcut) to deactivate the section.

Click the *Select* button from the top toolbar and select one of the two circular cylinder surfaces (the one opposite to the *Z* direction), according to *(1)* in Figure 13.24. Copy this surface by *Ctrl+V* → *Ctrl+Z*. Individual *Surface (2)* appears in the *Structure* window. Select the *Pull (3)* tool and activate the *No Merge* function *(4)*. Then pull out the surface to *700 mm (5)* to obtain the second cylinder – the air way before the rotor area (*Z* is the air stream direction).

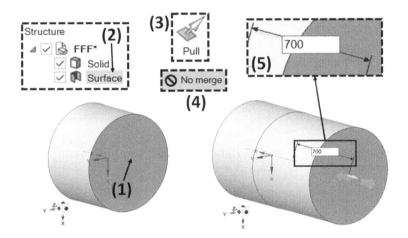

FIGURE 13.24

Turn the geometry according to *(1)* in Figure 13.25 and use *Pull (2)* with *No Merge (3)* to create another cylinder behind the rotor section – this time *1500 mm (4)* long. Now the three *Solid* objects *(5)* are available in the *Structure* window. Note that the 'outlet' cylinder is longer than the 'inlet' one- it is recommended to create a domain that surrounds an object in this way. In this exercise, the size of the domain is restricted by the mesh limits (you work using student license software). Under other conditions, the 'inlet' cylinder should have the length of at least five turbine diameters and the 'outlet' cylinder – of 10 diameters.

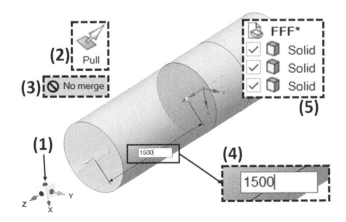

FIGURE 13.25

Spatial geometry of the turbine rotor surroundings is ready. Three cylindrical **Solid** objects *(1)* create a complete domain. Click *1xLMB* heading *(2)* – just above these objects (it is **FFF*** in Figure 13.26). Go down to the **Properties** table and make sure that the **Share Topology** option is set to **None** *(4)* – it is important from the point of view of the methodology applied in the further part of the exercise. Shut down **SpaceClaim DirectModeler** (just click the cross in the left top corner).

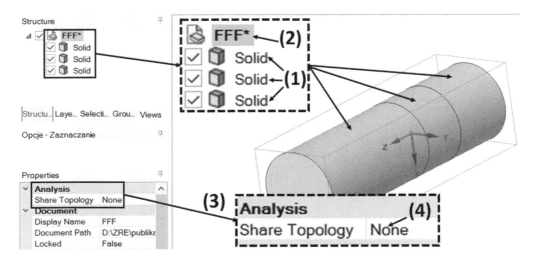

FIGURE 13.26

13.3 PREPROCESSING – MESHING

Save *(1)* the project from the main project window level, as in Figure 13.27. Then click *2xLMB* on the **Mesh (A3)** cell *(2)* to launch the **Ansys Meshing** module.

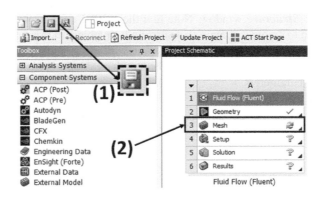

FIGURE 13.27

The **Outline** window (Figure 13.28) includes a project tree with different categories. Of course the most important tab is **Mesh** *(1)*. Note that there is a lightning thumbnail next to it. It means that a mesh has to be generated. Before it is done, certain global and local settings have to be defined. Global settings (applying in the whole domain) are available after clicking *1xLMB* on the **Mesh** tab in the **Details of "Mesh"** window below the project tree. Due to the **Analysis Systems** mode of the project (**Fluid Flow (Fluent)** – see Figure 13.2), **Physics Preference** is automatically set to

Tutorial 5 – Horizontal-Axis Wind Turbine

CFD and *Solver Preference* – to *Fluent (2)*. Find the *Element Size* cell below and change the value to **2.5e-002 m** *(3)* (obviously you can just type *0.025*). The geometry imported to *Ansys Meshing* includes two contact regions *(4)* between three cylinders. Due to that, the project tree includes the *Connections (5)* tab which will be discussed later.

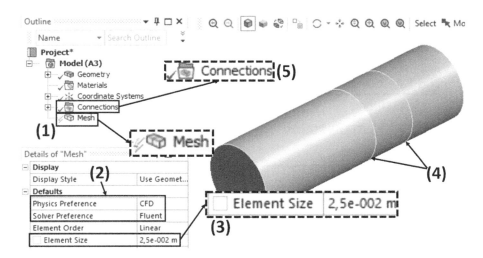

FIGURE 13.28

Open (+) the *Connections* tab (Figure 13.29). You'll see two contact regions. Click the first one *(1)* to check the location of the contact surfaces belonging to the neighboring cylinders. The source surface and the target surface create a contact region. By rotating the geometry, you can see which cylinder is *Contact Body* and which is *Target Body (2)*. A detailed view of the *Contact Body* with the source surface *(3)* and *Target Body* with the target surface *(4)* is given in two additional windows – these bodies are shown as opaque objects. Note that *Contact Body* is the rotor – it includes the source surface. This information will be important a bit later.

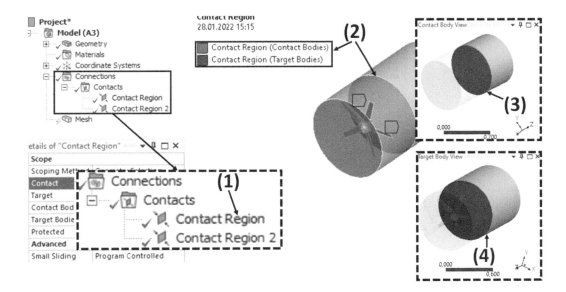

FIGURE 13.29

Now check the location of the source surface and the target surface of the *Contact Region 2 (1)* as in Figure 13.30. Note that again *Contact Body* (with the source surface) is the rotor domain *(2)*, while *Target Body* (with the target surface) is the outlet cylinder domain *(3)*.

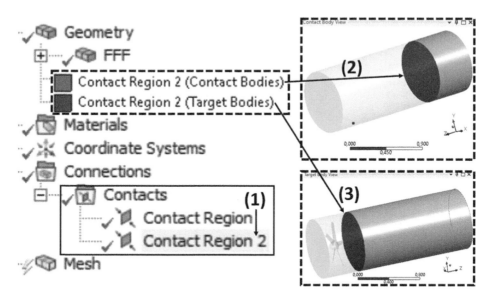

FIGURE 13.30

Selection names allow to easily define mesh local settings, as well as the boundary conditions and other settings in the solver preprocessor. Let's define selection names for our model now. Turn the model in a way allowing you to see the inlet face (on the shorter cylinder), as in Figure 13.31. The triad *(1)* of the global coordinate system can be useful. Find the selection mode bar *(2)* and click the *Face* selection mode (or *Ctrl+F*). Click *1xLMB* on the inlet surface and press the *N* key to display the *Selection Name* window. Enter the *inlet (4)* name for this surface and confirm by *OK (5)*.

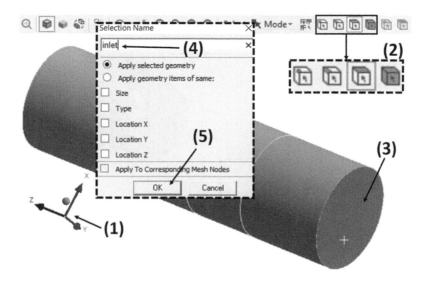

FIGURE 13.31

Tutorial 5 – Horizontal-Axis Wind Turbine   239

Turn the model to see the outlet face. Pay attention to the **Z**-axis direction *(1)* (Figure 13.32). Use the ***Face* (2)** selection mode again, to create (analogously to the inlet case) selection names: ***outlet* (3)** for the outlet face and ***wall-ext* (4)** – together for the three external cylinder surfaces. Then Switch the selection mode to ***Body* (5)** (you can also press ***Ctrl+B***) and create three individual selection names for the three cylinders: ***fluid-stat-1* (6)**, ***fluid-rotor* (7)** and ***fluid-stat-2* (8)**. All of the created selection names are available now in the ***Named Selections* (9)** tab of the project tree. Note that the selection names referring to bodies have a bit different thumbnails than the ones referring to faces.

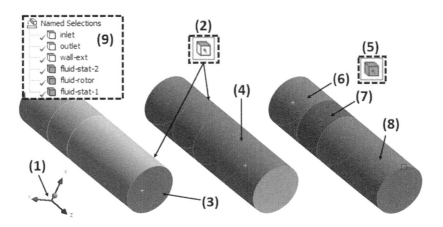

FIGURE 13.32

A series of local settings have to be applied in the case of the rotor area. To make it easier, select the ***Body* (1)** selection mode and select both external cylinders *(2)*, as in Figure 13.33. Click ***1xRMB*** wherever in the workspace and, from the drop-down list *(3)*, select ***Hide Body* (4)**. Only the rotor cylinder *(5)* remains in the workspace now.

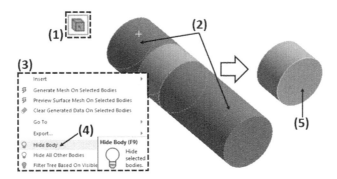

FIGURE 13.33

Switch to the ***Face* (1)** selection mode (Figure 13.34) or just press ***Ctrl+F*** and click any external face of the cylinder to select it. Then press ***Ctrl+A***. All the faces of the visible rotor cylinder are selected now. Note that apart from the three external faces, there are also a number of internal faces due to the turbine separation from the cylinder. Press and hold ***Ctrl*** to deselect three external faces (***LMB***). Now only turbine faces are selected. The contours of the internal faces are visible *(3)*. Press the ***N*** key and add the blade's *(4)* selection name. Check whether it is available on the ***Named Selections*** list *(5)*.

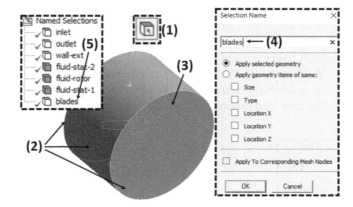

FIGURE 13.34

Use the *Face (1)* selection mode and the *Ctrl* key to select together two side faces *(2)* of the cylinder, as in Figure 13.35. Click *1xRMB → Insert (3) → Sizing (4)*. Go to the *Details of "Face Sizing"* window below the project tree and find the *Defaults* tab. This time the default *Element Size (5)* is fine, because the role of the defined face sizing is to keep the cell faces on the rotor domain cylinder as large as in the two external stationary cylinders (*0.025 m* is set there). Note that a default element size in any local sizing is equal to the defined global cell sizing. A new object – *Face Sizing (6)* is available below the *Mesh* tab of the project tree.

Let's add the next local sizing in a different (alternative) way than previously. Go to the *Mesh*

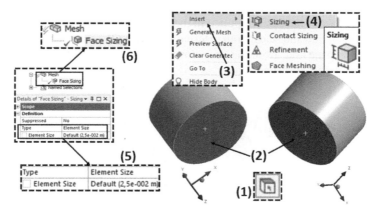

FIGURE 13.35

(1) tab (Figure 13.36) on the top menu bar. Click the *Body (2)* selection mode and select the cylinder. Click *Sizing (4)* on the top toolbar. Go to the *Details of "Body Sizing"* window. One body is selected for this sizing *(5)*. Go to the Defaults tab and change the Element size to *0.02 m* (or *2.e-002 m*) *(6)*. A new object – *Body Sizing (7)* is available below the *Mesh* tab of the project tree.

Tutorial 5 – Horizontal-Axis Wind Turbine

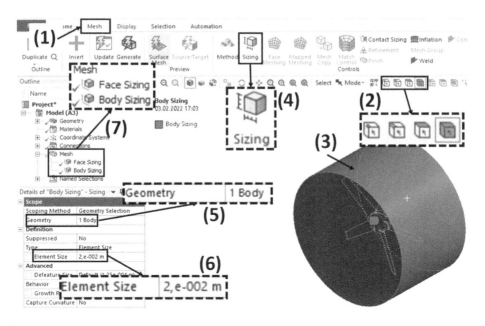

FIGURE 13.36

Another method of defining local sizing uses selection names to determine an appropriate geometry element(s)m as in Figure 13.37. Let's test it. Click *Sizing (1)* on the top toolbar. Go straight to the details window and open the (+) *Scope* tab. Change *Scoping Method* from *Geometry Selection* to *Named Selection (2)*. Select *blades (3)* from the list below. As you can see this method is convenient in the case of hard-to-reach model elements. Change *Element Size* to *0.01 m (or 1.e-002 m)* *(4)*. The third local sizing object appears in the Outline window (5).

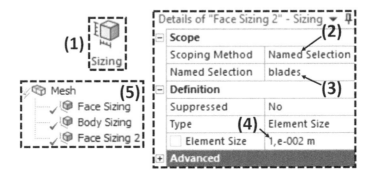

FIGURE 13.37

The Inflation tool allows to get a finer mesh in the near wall regions. This can be helpful in the case of the blade's face. To add inflation, first, select the *Body (1)* selection mode (Figure 13.38) and click *1xLMB* on the cylinder *(2)*. Then find the *Inflation (3)* tool on the top toolbar. Go to the *Details of "inflation"* window below the project tree. Note that the cell to the right of *Geometry* (in the *Scope* tab) already includes *1 Body (4)* – the one previously selected (cylinder). Go down to the *Definition* tab and change *Boundary Scoping Method* to *Named Selections (5)*. Click the cell to the right from *Boundary* to drop down the list and select *blades (6)* (confirm selection by pressing *Enter*). Below, insert required inflation settings *(7)*: *Inflation Option* → *Total Thickness*, *Number of Layers* → *5*, *Growth Rate* → *1.2*, *Maximum Thickness* → *0.01 m*.

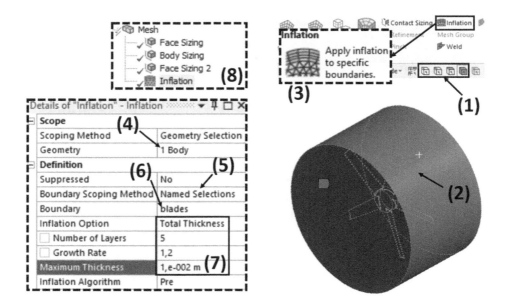

FIGURE 13.38

Click *1xRMB* anywhere in the project workspace → *Show All Bodies (1)*. Now three cylinders are visible again, as in Figure 13.39. It is time to generate mesh. In general, it can be done in several ways: By clicking *1xRMB* on the *Mesh (2)* tab → *Generate Mesh* or by clicking the *Generate (3)* button on the top toolbar. However, these methods create the whole mesh in one step. In this case, the mesh has to be generated gradually. Hold the *Ctrl* key and using *LMB select* both external cylinders (*Body* selection mode) → *1xRMB* → *Generate Mesh On Selected Bodies (4)*.

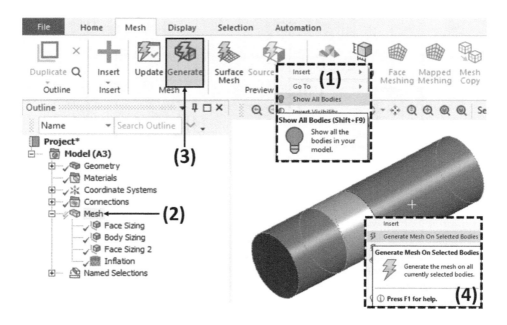

FIGURE 13.39

Tutorial 5 – Horizontal-Axis Wind Turbine

Repeat the above described procedure to create a mesh in the rotor cylinder region *(1)*. After that the thumbnail next to the *Mesh (2)* tab should change from the lightning to the check mark (Figure 13.40). Note that due to the regular shape (and lack of shared topology of the three subsequent domain parts) of the external cylinders, the mesh in these regions *(3)* is different (hexahedral) than in the rotor cylinder (tetrahedral) *(4)*. Mesh inconsistency is visible in the contact area – the external cylinder nodes are not coincident with the ones in the rotor cylinder *(5)*.

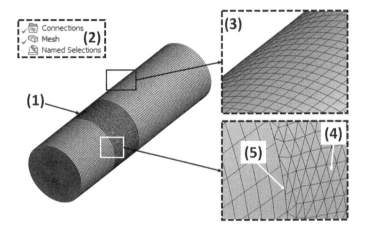

FIGURE 13.40

The Details of "Mesh" window (Figure 13.41) includes, among others, two tabs allowing to assess the mesh features. Open the (+) *Quality* tab and find the *Mesh Metric (1)* cell. There is a drop-down list with *None* next to it. Select *Skewness (2)* from this list. *Skewness* informs how much deformed or degenerated mesh cells are. The lower skewness, the better cell. *Average* skewness *(3)* is around *0.2*, which is a really good result (it is recommended to keep it below 0.4~0.6). It has to be noted that *Standard Deviation* is quite high (cell skewness values differ significantly in the mesh). The highest skewness is *0.89*, which is still below the critical level of *0.95* (it is recommended to

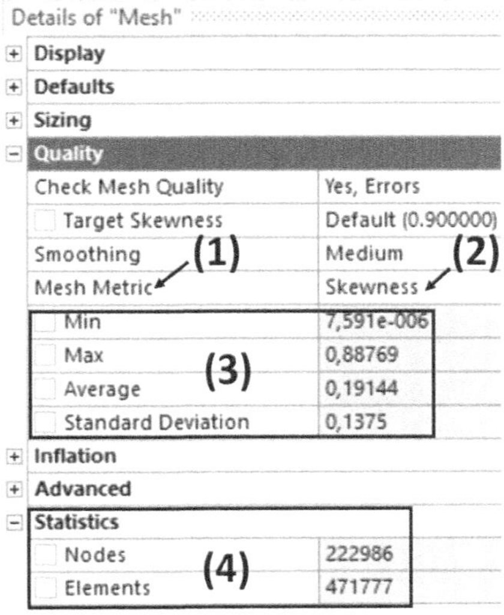

FIGURE 13.41

avoid higher values!). Go down to the ***Statistics (4)*** tab which provides information about the number of nodes and elements. Remember that in the case of student license software, the maximum number of nodes and cells has to be below ***512,000***.

Activation of skewness analysis (see Figure 13.41) causes display of the histogram (Figure 13.42). Of course, you can resize the histogram window and relocate it in the ***Ansys Meshing*** GUI, to make the analysis easier. Bars correspond to cell groups characterized by a certain quality. Note that according to the legend above the histogram, there are tetrahedral *(1)* cells (***Tet4***- four corner nodes), hexahedral *(2)* cells (***Hex8***- eight corner nodes) and wedge – prismatic *(3)* ones (***Wed6***- six nodes). You can click a certain bar or bars (with ***Ctrl*** key) to see cell locations in the grid. To see all cells again – just click ***1xLMB*** anywhere on the histogram white background. After mesh assessment, close the ***Ansys Meshing*** module to proceed.

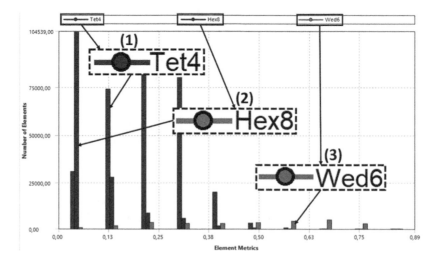

FIGURE 13.42

13.4 PREPROCESSOR – SOLVER SETTINGS

Save (1) the project again in the ***Ansys Workbench*** main project window (Figure 13.43). The created mesh has to be manually updated now. Click ***1xRMB*** on the ***A3 Mesh (2)*** cell and select ***Update (3)***. Wait until the thumbnail in ***A3*** cell changes to the check mark. Then click ***2xLMB*** on ***A4 Setup (4)*** cell to launch ***Fluent Launcher***.

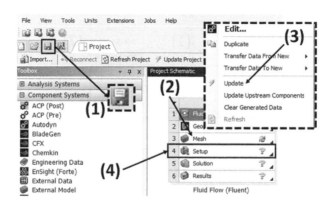

FIGURE 13.43

Tutorial 5 – Horizontal-Axis Wind Turbine

In the *Fluent Launcher* window (Figure 13.44), dimensions are automatically set to *3D (1)*. Due to working with the student software version, you can carry out parallel solutions at most four *Solver Processes (2)* (to speed up the computations). Click *Start (3)* to launch *Ansys Fluent*.

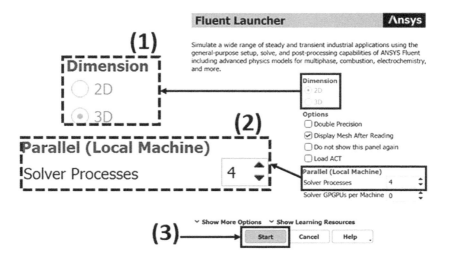

FIGURE 13.44

The Outline View (1) window (Figure 13.45) includes a project tree – like the one in the *Ansys Meshing* module. Subsequent tabs of the project tree allow to define, among others, model physics and boundary conditions, solution parameters and certain postprocessing functions. Usually, there is no need to use all the tabs (certain default settings are fine). First, the *General (2)* tab is active. The *Check (3)* button allows displaying certain geometry information in the console (below the workspace) *Display... (4)* includes tools to display selected elements of the model (faces, bodies, *etc.*). *Report Quality (5)* displays a short quality report in the console. The *Time* frame below allows to switch between the steady-state and transient model. Set *Transient (6)* – this exercise concerns the movement of the turbine rotor.

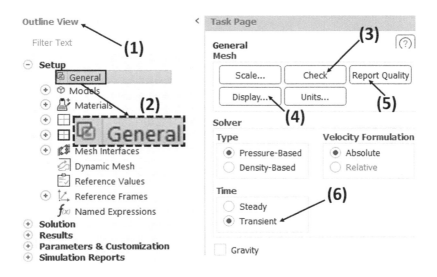

FIGURE 13.45

Click **2xLMB** the **Models (1)** tab. Go to the **Models (Task Page)** window to the right of the **Outline View**, according to Figure 13.46. The third tab in this window is **Viscous... (2)** and concerns turbulence modeling – click it **2xLMB**. In the **Viscous Model** window, switch **Model** from **k-omega** to **k-epsilon (2 eqn) (3)**, which is a much better selection in the case of a relatively coarse mesh. Change the **k-epsilon Model** mode to **Realizable (4)**. In general, the **Near Wall Treatment** section concerns wall functions. However, in this exercise the choice of a more advanced approach – **Enhanced Wall Treatment (5)** is required. Click **OK (6)** to confirm settings.

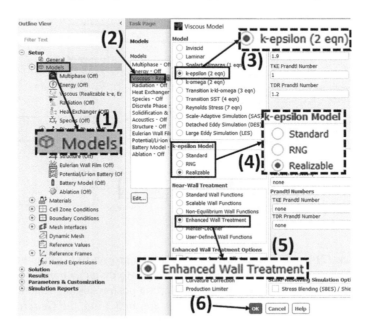

FIGURE 13.46

Click **2xLMB** in the **Materials (1)** tab. Materials are divided into two groups **(2)**, as in Figure 13.47. **Air** and **Aluminum (3)** are default – they are fine in this exercise. If required, the **Create/Edit... (4)** window can be opened to modify the material properties. Furthermore, new materials can be created or added from the library there.

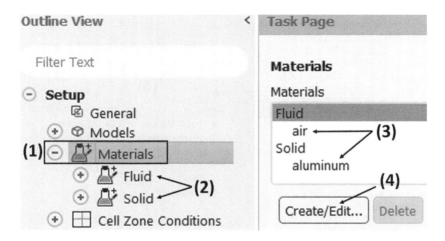

FIGURE 13.47

Tutorial 5 – Horizontal-Axis Wind Turbine

Click *2xLMB* on the *Cell Zone Conditions (1)* tab. The subtab *(2)* below is named *Fluid* (Figure 13.48) and there are no other subtabs (like *Solid*) – it means that all cell zones (subdomains) in the model are defined as fluids. This subtab includes three cell zones *(3)* with names coming from selection names defined in *Ansys Meshing*. A list of cell zones is also available in the *Cell Zone Conditions* (*Task Page*) window *(4)*. If you click *1xLMB on* any zone, you can see its *Type (5)* below. If required, the type can be switched from fluid to solid there, but not in this exercise – the domain material here is the air. Additional domain properties (not required), the frame or mesh motion, sources or porous regions can be defined after clicking *the dit… (6)* button.

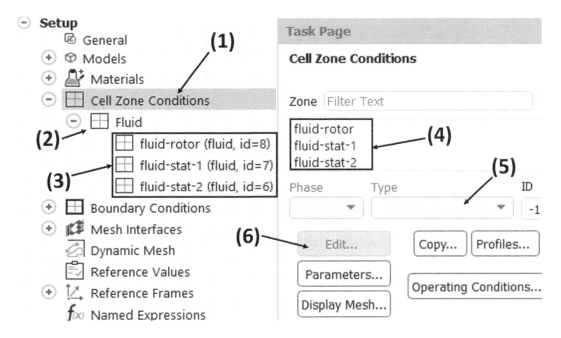

FIGURE 13.48

Defining boundary conditions (BCs) is one of the solver setup key steps. Click *2xLMB* on the *Boundary Conditions (1)* tab in the *Outline View* window (Figure 13.49). You can see that boundaries have been preliminarily classified and they exist in different subtabs *(2)* – *inlet*, *interface*, etc. All BCs are also listed in the *Boundary Conditions* (*Task Page*) window. Click *1xLMB* on any BC to check its *Type* on the list below. For example, in the case of *outlet* it should be a *pressure outlet* while in the case of *blades* it should be *wall*. Now find and click *2xLMB* on *inlet (3)*. In the *Velocity Inlet* window the only one available tab is *Momentum*, due to the lack of any models (like heat transfer or species transport) apart from the isothermal fluid flow. Go to the *Velocity Magnitude (5)* cell. The value has to increase gradually in subsequent time steps until it reaches a certain stable level (*2.5 m/s*). To insert an expression into the cell *(7)*, click the small triangle next to the *fx (6)* symbol and select the *expression* from the drop-down list. Then insert the expression as follows:

2.5 [m s^-1]*(1-exp(-Time/0.5[s]))

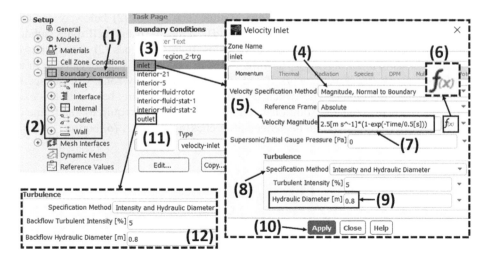

FIGURE 13.49

Additionally, **Specification Method** *(8)* in the **Turbulence** frame has to be changed to **Intensity and Hydraulic Diameter** – this diameter is known (cylinder diameter) so it is recommended to apply this information. Do not change the **Turbulence Intensity** value, but insert a **Hydraulic diameter [m]** of *0.8 (9)*. **Apply** *(10)* the BC definition to proceed. Now click **2xLMB** on the outlet *(11)* and, analogously as in the **inlet** case, change the **Specification Method** to **Intensity and Hydraulic Diameter** and set the diameter to *0.8 [m] (12)*.

Open (*2xLMB*) the **Mesh Interfaces** *(1)* tab. The mesh interface is a specific kind of boundary condition allowing to conserve consistency of the variable fields in a domain with an inconsistent mesh. In such a case the interface is created between two contact faces. As spotted during the mesh analysis (see Figures 13.29 and 13.30), there are two contact regions *(2)* in the model (Figure 13.50). The **Mesh Interfaces** window includes the **Boundary Zones** list with four boundaries *(3)* described as sources (*src*) or targets (*trg*). Each couple of *src* and *trg* boundaries create one interface – contact region. Two created interfaces are listed on the **Mesh Interfaces** list *(4)*. As you can see, the interfaces were created automatically. They do not require further editions, but it is possible thanks to the **Edit…** *(5)* option. If required (for example in case of wrong automatic creation), the **Manual Create…** *(6)* option can be used to manage the creation of the interface. Close the **Mesh Interfaces** window to proceed.

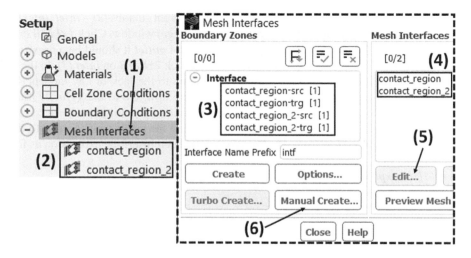

FIGURE 13.50

Tutorial 5 – Horizontal-Axis Wind Turbine

Open the ***Dynamic Mesh (1)*** tab (***Outline View*** window) to activate submodels allowing the turbine movement related to the wind impact. ***Activate Dynamic Mesh (2)*** as in Figure 13.51. Activate ***Six DOF (3)*** (Degrees of Freedom) on the ***Options*** list and click ***Settings… (4)*** below to proceed.

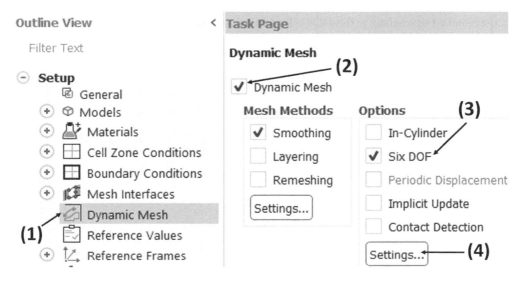

FIGURE 13.51

In the ***Options*** window (Figure 13.52), click ***Create/Edit… (1)***. In the ***Six DOF Properties*** window, insert ***Name*** – for example, *rotor (2)*. Activate ***One DOF Rotation (3)***. Insert ***Mass [kg]*** of ***1.3538 (4)*** and put *1* in the **Z**-axis cell ***(5)*** of the ***Axis*** table below. Insert ***Moment of Inertia [kg m²]*** of ***0.01086 (6)*** and click ***Create (7)***. Then click ***OK (8)*** to close the ***Options*** window.

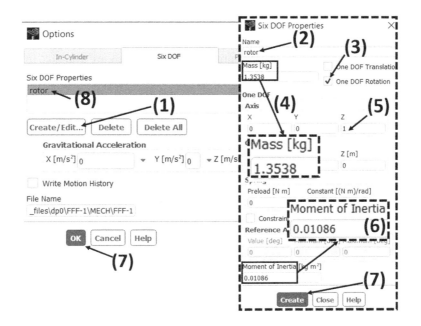

FIGURE 13.52

Click **Create/Edit... (1)** in the **Dynamic Mesh** (**Task Page**) window to define **Dynamic Mesh Zones** (appropriate window appears). Select **blades (2)** from the **Zone Names** drop-down list. Select **Rigid Body (3)** from the **Type** list below and click **Create (4)** at the bottom. The first **Dynamic Mesh Zone** has been defined **(5)**. Now use exactly the same procedure to define: *contact_region_src*, *contact_region_2_src*, *fluid_rotor* and *wall-ext-fluid-rotor*. All these zones will be able to move. Note that source faces belong to the rotor cylinder domain (see Figures 13.29 and 13.30) – that's the reason why they're defined as able to move (**Rigid Body**). Close the **Dynamic Mesh Zones** window. The list **(6)** includes 5 **Dynamic Mesh Zones** now (Figure 13.53).

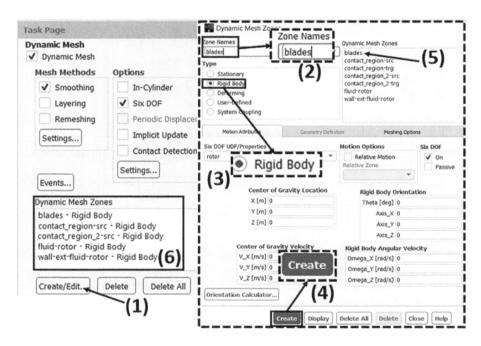

FIGURE 13.53

Further solver setup concerns the solution process – these settings are available in the tabs below the **Solution (1)** main (bold) tab (Figure 13.54). Click *2xLMB* on the **Methods (2)** tab. In this example, the segregated **SIMPLE (3)** algorithm is used for **Pressure-Velocity Coupling**. The **Spatial Discretization** scheme can be set individually for each equation available on the list **(4)**. In simple terms, we can say that different schemes provide a different approach to an approximation of the transported variable on the wall of the adjacent cells. Generally, higher order schemes have greater accuracy, but if you are not experienced, leave the default settings. Go to the **Controls (5)** tab now. The **Solution Controls** table **(6)** includes **Under-Relaxation Factors** (**URFs**) for calculating different fluid flow parameters. Increasing certain **URF** (at most to *1*) can reduce the solution time in the case of a stable solution, but can also cause convergence problems. Reducing **URFs** stabilizes the solution, but can significantly extend the solution time. URFs don't need to be modified in this exercise.

Tutorial 5 – Horizontal-Axis Wind Turbine

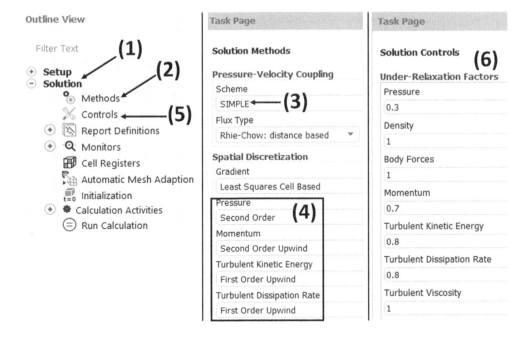

FIGURE 13.54

Open the *Monitors (1)* tab and click *2xLMB* on *Residuals (2)*. The *Residual Monitors* window (Figure 13.55) allows to edit settings of residuals. The *Options (3)* frame settings allow you to decide whether residues have to be displayed in the console, and as curves on the chart during the solution process. Thanks to the *Iterations to Plot (4)* option you can also decide how many iterations can be displayed on the chart. In the case of transient computations, which usually have a significant number of iterations, it is reasonable to set a lower value of *Iterations to Plot*. It will make the analysis of residuals more clear. Go to the *Equations* table. The first column title is *Residual (5)*. All transport equations are listed below. Find the *continuity (6)* row. Now go to the *Absolute Criteria (7)* column and change the value for continuity to *0.0001 (8)*. It increases the required convergence level of this equation. Other residual levels don't have to be modified. Click *OK (9)* to proceed.

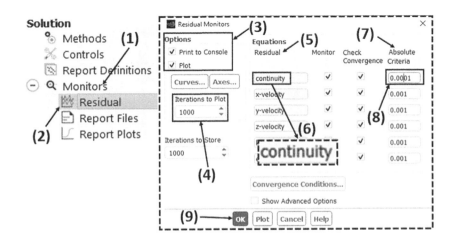

FIGURE 13.55

To display the selected variables (like moment of force – torque) on charts during the solution process, appropriate *Report Files* have to be defined. Click *1xRMB* on the *Report Files (1)* tab → *New... (2)*. In the *New Report File* window (Figure 13.56) click *New (3)* → *Force Report (4)* → *Moment... (5)*. The *Moment Report Definition* window appears.

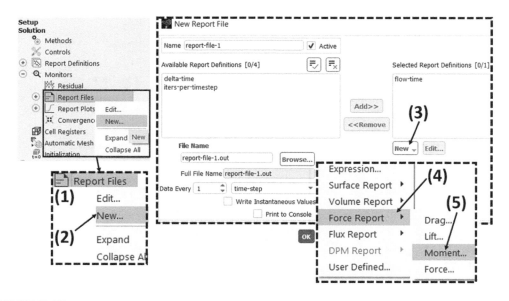

FIGURE 13.56

Insert the name of the monitor – for example, *mom (1)*. Change *Report Output Type* to *Moment (2)*. From the *Zone* list select *blades (3)* as in Figure 13.57. In the *Create* section, it is enough to select *Report Plot* (displaying the value on the chart during the solution). The report file above allows to save the txt file with the monitor value for each time step. Click *OK (5)* to proceed.

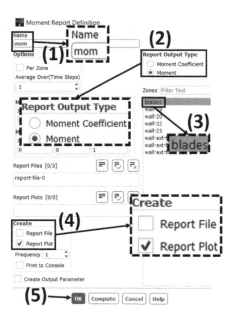

FIGURE 13.57

Tutorial 5 – Horizontal-Axis Wind Turbine

New monitor *(1)* appears on the *Available Report Definitions* list, as in Figure 13.58. Another monitor will be the average air velocity at the *inlet* to the domain – it will allow you to check whether the expression describing velocity as a function of a time step (see Figure 13.49) works correctly. Click *New (2)* again. From the drop-down list select *Surface Report* (*inlet* zone is surface) -> *Mass-Weighted Average… (4)*.

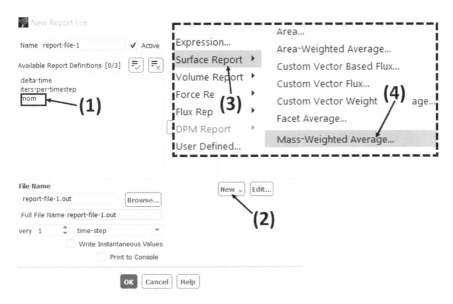

FIGURE 13.58

Insert *Name* (Figure 13.59) – for example *vel-in (1)*. Selected *Report Type* should be *Mass-Weighted Average (2)*. *Field Variable* settings are respectively *Velocity…* and *Velocity Magnitude (3)*. Then select *inlet (4)* from the *Surfaces* list. Select *Report Plot* (optionally *Report File*) in the *Create (5)* section and confirm settings by *OK (6)*.

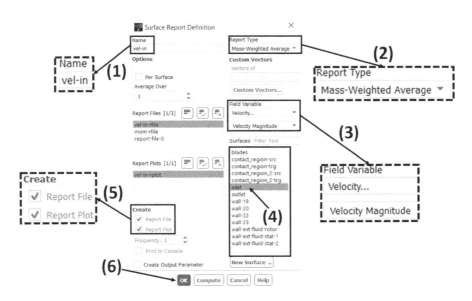

FIGURE 13.59

Initial model conditions have to be set before the solution process starts. Click **2xLMB** on the **Initialization** tab (Figure 13.60). Go to **Task Page** and switch **Initialization Methods** to **Standard Initialization (2)**. In the **Z Velocity [m/s]** (which is aligned with the mean air flow direction) cell you can insert **0.001**, which helps to converge the solution in the first time step. Turbulence scalars **(4)** can stay as default because at the beginning of the simulation turbulence production is insignificant. Click **Initialize (5)** to proceed.

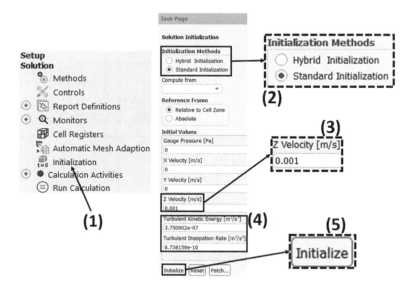

FIGURE 13.60

In the **Calculation Activities** tab (**Outline View** window) set **Auto Save every 5 time steps** (otherwise only data from the last time step of the simulation will be saved!). Then go to the **Run Calculation (1)** tab (Figure 13.61). The **Check Case (2)** button (**Task Page**) allows to display

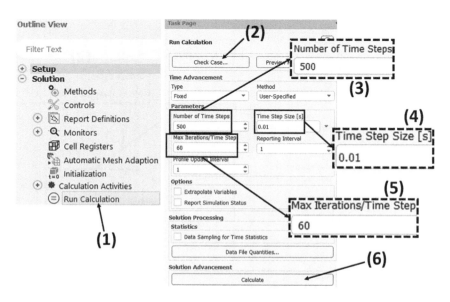

FIGURE 13.61

Tutorial 5 – Horizontal-Axis Wind Turbine

recommendations regarding the defined project. Usually, if no technical mistakes were made, the solver only suggests checking mesh (if it was not done at the beginning, it can be done optionally now and improving certain advection schemes (see Figure 13.54) – it can be ignored in this exercise (higher order advection schemes improve solution accuracy but obstruct its convergence). Set *Number of Time Steps* to *500 (3)* and *Time Step Size [s]* to *0.01 (4)*. These two values together mean that the total simulation time is *5 s*. Of course, if you want, you can simulate longer turbine operation as well. Set *Max Iterations/Time Step* to *60* and click *Calculate (6)* to launch simulation.

During the simulation, look at the legend – *Residuals (1)*. It includes a description of the residual curves that are plotted in chart *(2)* as in Figure 13.62. Each residual is related to an imbalance in one transport equation. Thus, the number of curves corresponds to the number of transport equations. It is desirable to obtain sawtooth shape curves during transient simulations. The peaks correspond to the transition to the next time step. Charts representing monitors (moment of force of turbine blades and inlet wind velocity) can be displayed by switching top tabs *(3)*. Accurate residual values are displayed in each iteration of each time step in the *console (4)*. The simulation progress is visualized by the progress bar *(5)* below the console. If required, the solution can be stopped at the end of each iteration or time step *(6)*. Although an individual iteration does not take a long time, the whole simulation can be relatively time consuming, due to a great number of time steps.

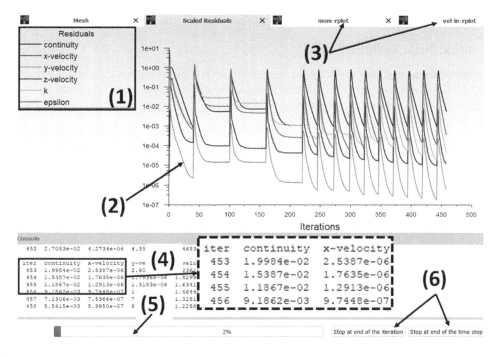

FIGURE 13.62

The end of the solution process is signaled by a communicate in the pop-up window. When it appears, klick *OK* and have a look at the charts that have been created (Figure 13.63). The velocity at the inlet was increasing in each time step, according to the applied expression (boundary condition definition) and after about 2.5 s *(2)* it reached the highest magnitude. The moment force curve *(3)* is opposite the velocity one. Negative values come from the relation between the forces in the system and its location in the reference frame (axis directions). Of course, stabilization of the second parameter also occurs after about 2.5 s *(4)* of the simulation.

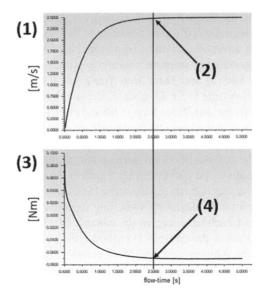

FIGURE 13.63

13.5 POSTPROCESSING

Close *Ansys Fluent* and save the whole project in the *Ansys Workbench* main project window (see Figure 13.43). Look at the *A4* (*Setup*) and *A5* (*Solution*) cells (Figure 13.64) – in both cases check mark thumbnails to inform that these stages of the project are completed. Click *2xLMB* on *A6* (*Results*) *(2)* cell to launch the postprocessing module.

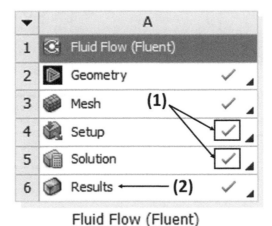

FIGURE 13.64

The Outline (1) tab in the left top corner includes a project tree with setting tabs. Later, newly added post-processor tools (visualizations) will also appear there. Before you add any visualization, only wireframe of the geometry *(2)* is visible as in Figure 13.65. Find on the project tree and click *2xLMB* the *Wireframe (3)* tab to edit it. Properties can be modified in the *Details of Wireframe* window below. Among others, you can change *Color Mode* and *Line Width*. Change one or both properties and *Apply (5)* to test these functionalities.

Tutorial 5 – Horizontal-Axis Wind Turbine

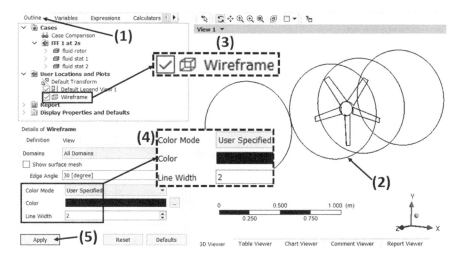

FIGURE 13.65

The Project tree includes all geometry objects (Figure 13.66). They are grouped into three sub-domain tabs: *fluid rotor*, *fluid stat 1* and *fluid stat 2*. Any object can be displayed – activate (mark) *blades (2)* below the *fluid rotor* tab. Then click it *2xLMB* and go to the *Details of blades* window below. *Mode and Color (2)* can be defined. First, change the *Color and Apply (3)*. Color change concerns only the selected object – *blades (4)*.

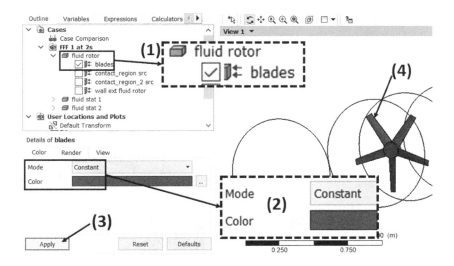

FIGURE 13.66

Now change the *Mode* from *Constant* to *Variable (1)*, as in Figure 13.67. Any selected surface can be used to display the contour of the selected variable field. In the case of wind turbine blades, it is worth checking pressure distribution – set *Variable* to *Pressure* and *Range* to *Local (2)*. Local range mode results in adjusting the maximum and minimum on the legend to the extremes of the displayed field. A number of visualization options like *Color Scale*, *Color Map* and *Contour* type are available below *(3)* – for the first time they can stay as default. Click *Apply (4)*. The range of pressure changes *(5)* is displayed above. The contour *(6)* and the legend *(7)* are displayed in the workspace.

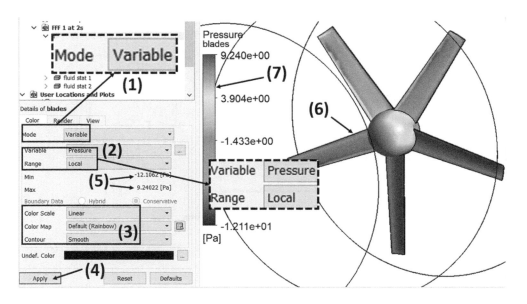

FIGURE 13.67

Set *Color Mode* to *Constant* again and go to the *Render (1)* tab of the *Details of blades* window, as in Figure 13.68. *Transparency (1)* of blade coloring can be activated and adjusted (*0* to *1*) here. Test this setting. Apart from other functionalities, the interesting one is displaying mesh. Activate the *Show Mesh Lines (3)* option and increase *Line Width (4)* – for example to *2*, to make the grid better visible. Then click *Apply (5)*. Grid lines should be displayed on the blade surfaces. This option can be useful when creating a presentation or checking if a potentially spotted variable anomaly in any simulation results might come from deformed cells.

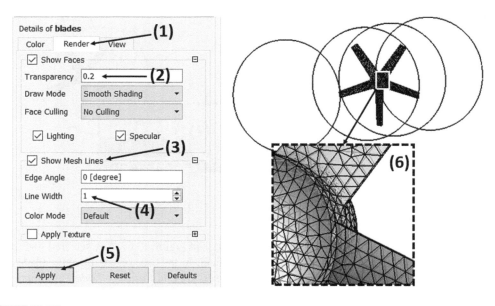

FIGURE 13.68

Tutorial 5 – Horizontal-Axis Wind Turbine

The top toolbar (Figure 13.69) includes result visualization tools *(1)*. Select **Contour**. Click **OK** in the **Insert...** *(2)* window (optionally of course you can insert any name with no special characters). Then go down to the **Details of Contour 1** window. Have a look at the **Geometry** *(3)* tab. Although functionalities in this tab are similar to the ones available in the previously discussed case (see Figures 13.67 and 13.68), **Contour** allows to carry out more detailed analyses. Set **blades** as **Locations** for visualization, **Variable** to **Pressure** and **Range** to **Local** (see Figure 13.67) and go **to the Render** *(4)* tab. Activate the **Show Contour Lines** *(5)* option. Click **Apply** *(6)*. **Contour** appears as the new object in the project tree *(7)*. Contour lines *(8)* are visible at the turbine surface. This function helps to show clearly changes in the variable field.

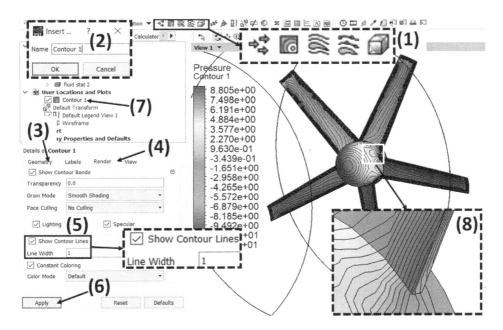

FIGURE 13.69

The vector tool allows to display velocity vector fields on planes, surfaces and in 3D domains. Deactivate **Contour** in the project tree and go to the visualization tools *(1)* on the top toolbar, according to Figure 13.70. This time select **Vector** and click **OK** in the **Insert...** *(2)* window. Go to the **Details of Vector 1** window. In the **Geometry** tab find the **Locations** drop-down list. To select three subdomains (three cylinders) together, click three dots to the right *(3)*. Go to the displayed **Location Selector** window. Hold the **Ctrl** key and click subsequently *(LMB)* three headings – subdomains: *fluid rotor, fluid stat 1* and *fluid stat 2* *(4)*. Confirm by **OK**. Now insert *200* in the **Factor** *(5)* cell – it reduces the number of the displayed vectors which clarifies the visualization. Go to the **Symbol** *(6)* tab, change **Symbol** to **Arrow 3D** and **Symbol Size** to *2* *(7)*. Click **Apply** *(8)* to display visualization.

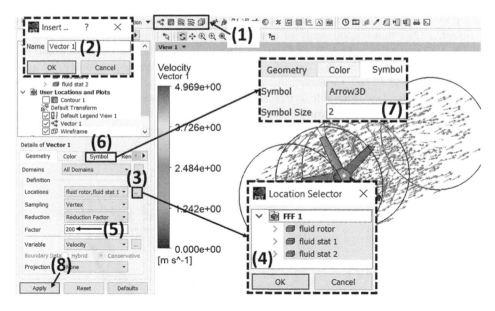

FIGURE 13.70

You can analyze velocity vectors in different areas of the domain (Figure 13.71). Note that the vectors close to the inlet are homogeneously distributed and have equal length *(1)*. In the near-turbine regions, velocity distribution is dynamic and strongly depends on the specific location. Air hits the blades *(2)* and changes direction rapidly. Sparse, long vectors representing swirling air (due to turbine revolutions) are visible around the rotor *(3)*.

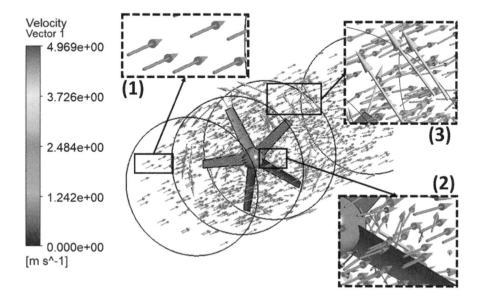

FIGURE 13.71

Tutorial 5 – Horizontal-Axis Wind Turbine

Of course thanks to setting the auto saving in all 5 steps, it is possible now to analyze the already created visualizations in each of them. To switch between the time steps, find the ***Timestep Selector*** tool on the top toolbar – it is the clock thumbnail *(1)*. All the available time steps are listed in the ***Timestep Selector (2)*** window – just click *2xLMB (3)* any of the rows and after loading the timestep you will see the movement of the rotor and a change in the variable fields if any of them are active. Then select the first time step and click the small thumbnail of a film reel (RHS of the ***Timestep Selector*** window). The ***Animation (4)*** window allows to run an animation (play button) and change timesteps using the slide bar *(5)*. ***Repeat (6)*** allows to set the number of simulation runs in series. The Infinity symbols activate running simulation in a loop. It has to be switched off, if you want to use the ***Save Movie (8)*** option. ***Close (9)*** the Animation window when you finish testing its functionalities (Figure 13.72).

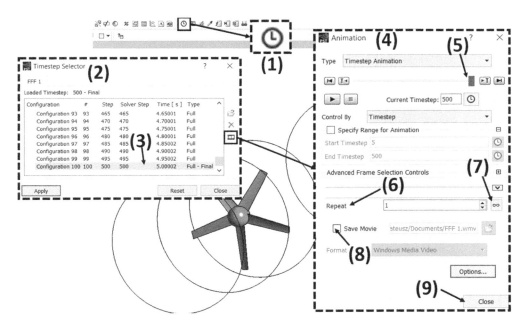

FIGURE 13.72

14 Tutorial 6 – Vertical-Axis Wind Turbine

14.1 EXERCISE SCOPE

This tutorial is devoted to the comparison of two models representing vertical-axis wind turbines, as in Figure 14.1. Both cases have the same rotor construction, with two blades *(1)*. The first turbine is a free-standing one, while the other one is equipped with additional stationary blades *(2)* – the yaw system directing the wind stream toward the rotor *(3)* and surrounding the rotor blades. Simulations allow to compare the torque and the theoretical power obtained in the conditions of constant rotor axial velocity.

According to the classification in the theoretical part (see Section 12.3, Figure 12.5), this exercise considers the drag based Savonius micro turbine. According to the already discussed energy conversion fundamentals, yawing can increase the air stream passing through the rotor area and wind harnessing efficiency. However, the aerodynamic drag generated by the additional stationary construction elements has to be taken into consideration.

This exercise helps to get familiar with specific, useful software functionalities, for example:

- creating surfaces from sketches and duplicating repeatable geometry elements applying the patterns in ***Ansys Design Modeler***,
- applying inflation to get fine mesh in the vicinity of the rotor blades and also to generate regular cells in the transition mesh regions,
- defining moving mesh in a selected cell zone to obtain rotor movement in a dynamic (transient) simulation.

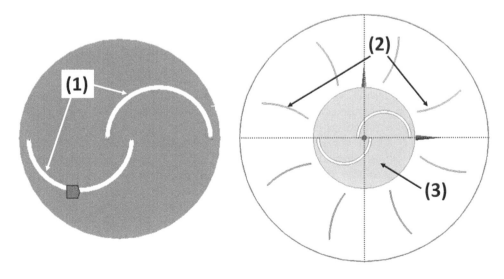

FIGURE 14.1 Top view of the turbine rotor and the yaw system considered in the exercise.

Note that VAWT model can be easily reduced to the 2D case in the preliminary study stage. Pay attention to the possibility of duplicating whole project stages (modules) in the Ansys Workbench main project window.

All of the images included in this tutorial use courtesy of ANSYS, Inc.

14.2 PREPROCESSING – GEOMETRY – PART 1

Find the *Geometry* module in the LHS toolbox, as in Figure 14.2. Then drag (*LMB*) and drop it in the project workspace. The *Geometry (1)* module should appear in the *Project Schematic* window. Spatial geometry for this project will be prepared using the *Ansys DesignModeler (DM)* tool. To open *DM*, click *1xRMB* on the *A2* cell and select the *New DesignModeler Geometry...* option. Then wait a moment for the module launch

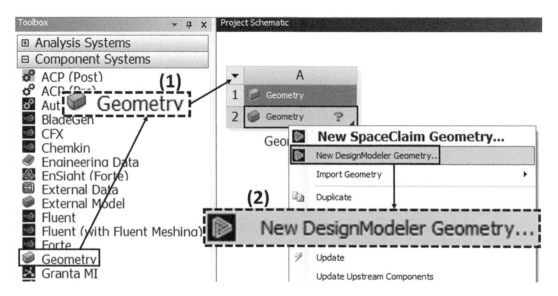

FIGURE 14.2

The Tree Outline window (Figure 14.3) includes the project tree *(1)*. Selecting any project tree component activates its *Detail View (2)* table below. Three axes of the global coordinate system *(3)* allow to recognize the current location and position of any designed object. Although you can use mouse buttons and the scroll wheel to resize, move and turn the object, it is also possible to apply navigation options *(4)*: *Rotate* option – hold the scroll wheel and move the mouse anywhere, *Pan* option – hold *Ctrl* from the keyboard together with the scroll wheel and again move the mouse, *Zoom* – just scroll. *F7* is *Zoom to Fit* key. Additionally, you can change the object position by clicking an appropriate triad *(5)* axis. Creation of an object has to be preceded by a sketch. Select *XY Plane (6)* from the project tree (*1xLPM*) and click the *New Sketch (7)* function. To set the sketching plane view, use the *Look At Face/Plane/Sketch (8)* function. If required, you can change the project units using the *Units* tab from the top toolbar. Now go to the next step (Figure 14.4).

Tutorial 6 – Vertical-Axis Wind Turbine

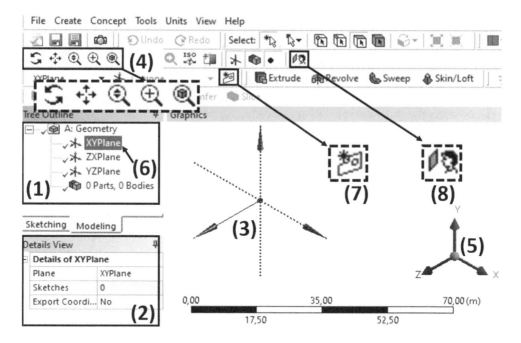

FIGURE 14.3

Click the ***Sketching (1)*** tab to activate the sketching mode, as in Figure 14.4. Remember that at any moment of designing the sketch, you can return to the reference view using the already mentioned ***Look At Face/Plane/Sketch (2)*** option. ***Ruler (3)*** shows the current size of the sketching workspace. Have a look at the ***Sketching Toolboxes*** window tabs *(4)*. They include different groups of sketching operations that can be browsed using arrows *(5)*.

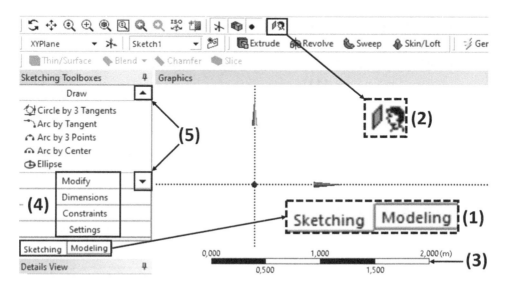

FIGURE 14.4

Time to create the first element of the geometry – turbine blades. Go to the ***Draw (1)*** tab of the ***Sketching Toolboxes*** window. Find and click the (***1xLMB***) ***Arc by Center*** tool. Before drawing the first sketch element, it's always good to adjust the sketching field area to the created object. Use the mouse scroll to resize the ***ruler (3)*** to around ***0.400 (m)***. Then click ***1xLMB*** on the horizontal axis ***(4)***, as in Figure 14.5. You should see ***the*** letter ***C***, which means ***Coincident*** with the axis. Then click ***1xLMB*** again somewhere on the ***LHS*** of the vertical axis ***(5)*** – it is the arc start point. Then analogously select the end point ***(6)***. The arc is ready.

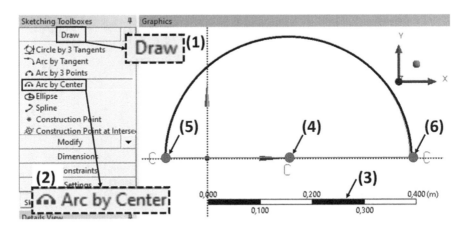

FIGURE 14.5

Go to the ***Dimensions (1)*** tab. Click (***1xLMB***) ***General (2)*** which is a kind of universal dimensioning tool – you can use it to define length, a diameter, or a radius. Click the arc edge anywhere, and click ***1xLMB*** again, somewhere out of this edge, to add the first dimension – as ***R1*** in Figure 14.6. Distances between objects (like points) have to be defined using other tools. Select ***Horizontal (3)*** to add the distance between the coordinate system center and the arc start point – as ***H2***. Values of the defined dimensions can be set in the ***Details View*** window – set ***H2 to 0.04 m*** and ***R1*** to ***0.2***. If the length unit in your project is mm, you have to set ***40*** and ***200 mm***, respectively!

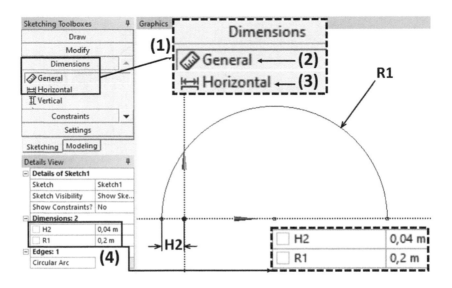

FIGURE 14.6

Tutorial 6 – Vertical-Axis Wind Turbine

As previously described, add another, a bit bigger arc, as in Figure 14.7, with **R2** radius of **0.22 m** **(1)**. Two arcs have to be connected by straight sections – use the **Line (2)** tool available in the **Draw** tab to add two lines closing the turbine blade sketch. To add the line click **1xLMB** at its start point and again at the end point. During the creation of the line, the letter **P** (which means **Point**) should appear next to the mouse cursor (as **C** previously) when the cursor is exactly on the arc start or end point.

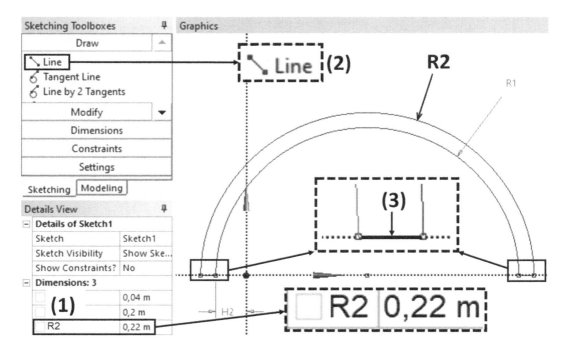

FIGURE 14.7

Now it is time to sketch the second blade profile **(1)**, as in Figure 14.8. Make sure that the smaller arc has a radius (like **R3**) of **0.2 m** and the bigger one (like **R4**) -**0.22 m**. Another required sketch element is the circle that limits the rotor area. Use the **Circle (2)** tool from the **Draw** tab to create it (diameter **D=0.82 m**). You can use any constrain tool from the **Constrains** tab of the **Sketching Toolboxes** window to correct your sketch. For example, you can use the **Horizontal (3)** tool to make sure that short, straight lines between arcs are horizontal (you just have to click **1xLMB** line when the tool is active). Furthermore, you can use the **Coincident (4)** tool to make the circle center coincident with the vertical and the horizontal axis of the coordinate system (to locate the circle in the system center). Press **Ctrl+Z** if information about over constrained sketch appears.

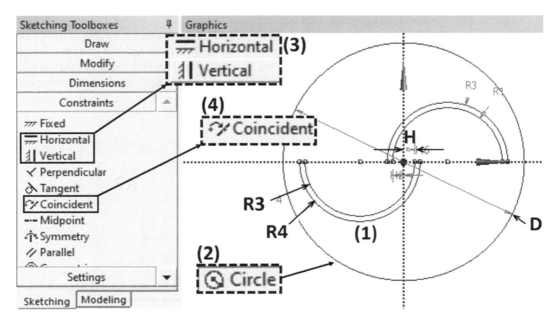

FIGURE 14.8

Go back to the *Modeling (1)* mode to create the first surface as shown in Figure 14.9. Select the *Surface From Sketches (2)* tool from the *Concept* tab (top toolbar). Go to the *Details View* window and click *1xLMB* in the *1 Sketch* cell, next to the *Base Object* one. *Apply/Cancel* buttons should appear. It means that now you can decide from which sketch the surface should be created. Click *Sketch1 (4)* below *XYPlane* in the *Tree Outline* window. Now go back *to the Details View* window and change *Operation (5)* to *Add Frozen* (it results in creating an independent face or body treated as a fluid). Click *1xRMB* on the *SurfaceSk1 (6)* thumbnail in the *Tree Outline* window and select *Generate* to create the surface.

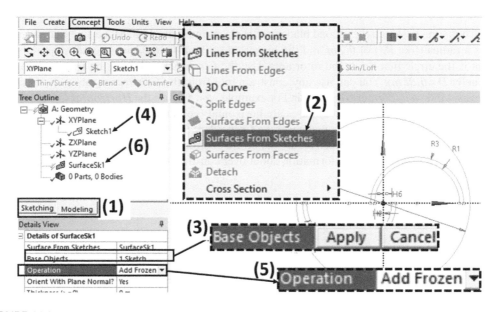

FIGURE 14.9

Tutorial 6 – Vertical-Axis Wind Turbine

After generating the surface, the space between the rotor sketch edges is filled *(1)*, as in Figure 14.10. **Sketch 1** determines the result of the **SurfaceSk1** operation; thus, this sketch is available below the operation *(2)* in the **Tree Outline** window. A new physical object – **Surface Body** has appeared below *1Part, 1 Body* tab *(3)*. You can click it *1xLMB* to see its details *(4)* in the **Details View** window.

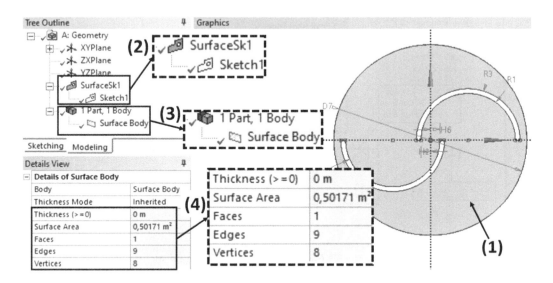

FIGURE 14.10

Add another sketch *(1)* on the **XYPlane** (you can find a detailed procedure in Figure 14.3). Add **NewSketch** *(2)* and go to the **Sketching** *(3)* mode to activate **Sketching Toolboxes**. Create two other circles *(4)* with center points in the center of the global coordinate system, as in Figure 14.11. The smaller circle has the same diameter as the one from the previous sketch (Figure 14.8). To set this diameter you can add a new dimension or use the **Coincident** constrain tool. Larger circle has a diameter $D = 2\,m$ *(5)*.

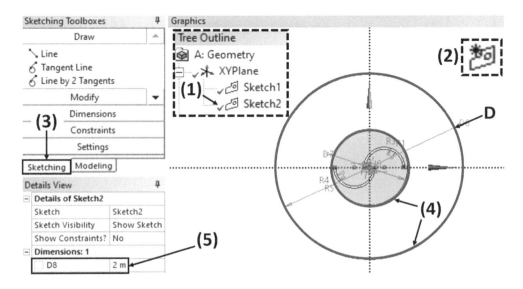

FIGURE 14.11

Go to the **Modeling (1)** mode and create another surface using the **Surface From Sketches (2)** tool from the **Concept** tab. Filling of the ring surrounding the rotor (3), and the second **Surface Body (4)** object should appear, as in Figure 14.12.

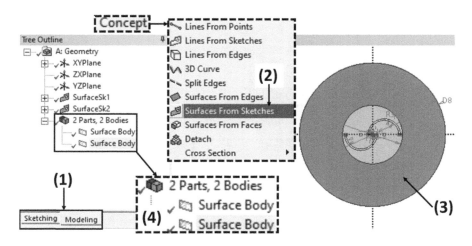

FIGURE 14.12

Add the last required sketch on the **XYPlane (1)**. It represents distant surroundings of the turbine – the rectangle with a circle inside (2), as in Figure 14.13. It is common to calculate the enclosure dimensions based on the characteristic length of a turbine. This is done to minimize the influence of the domain boundaries on the flow characteristics. The diameter of the circle should be the same as the larger circle in the previous sketch (**Sketch2**, Figure 14.11). To determine the dimensions of the rectangle, you can subsequently use the **General (3)**, **Horizontal (4)** and **Vertical (5)** tools. Create the third surface from **Sketch3**.

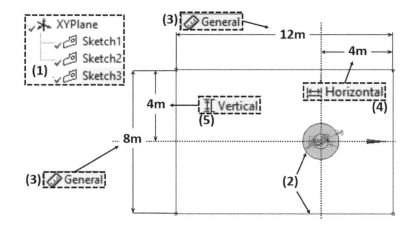

FIGURE 14.13

Let's summarize. You have sketched three parts of the spatial geometry and generated three surface bodies that are visible in **3 Parts 3 Bodies** tab of the project tree now, as in Figure 14.14: a rotor with blades (1), annular close turbine surrounding (2) and rectangular distant turbine surrounding? (3).

Tutorial 6 – Vertical-Axis Wind Turbine

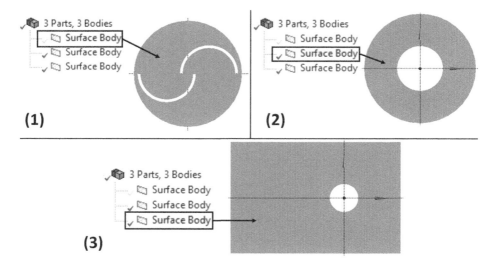

FIGURE 14.14

Surrounding parts have to be topologically consistent (together they create stationary domain). Select both surrounding surface bodies (***Ctrl+1xLMB***) from *3 Parts 3 Bodies* tab *(1)*, as in Figure 14.15. Then click *1xRMB* and select ***Form New Part (2)***. Now two surface bodies are grouped under ***Part (3)*** and they share topology. The geometry is ready, so you can close DesignModeler.

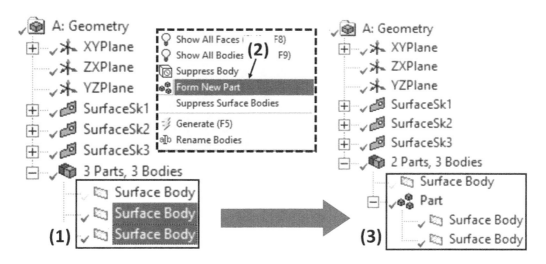

FIGURE 14.15

14.3 PREPROCESSING – MESHING – PART 1

Find the ***Mesh (1)*** thumbnail in the ***Component Systems*** list of the ***Toolbox*** window. Drag and drop the ***Mesh*** module *(2)* next to the already existing ***Geometry*** module, as in Figure 14.16. Click and hold *1xLMB* on *A2* cell (***Geometry***), drag the mouse cursor to *B2* cell and drop the link *(3)*. Now you can launch the ***Ansys Meshing*** module – click *2xLMB* on *B3 Mesh (4)* cell.

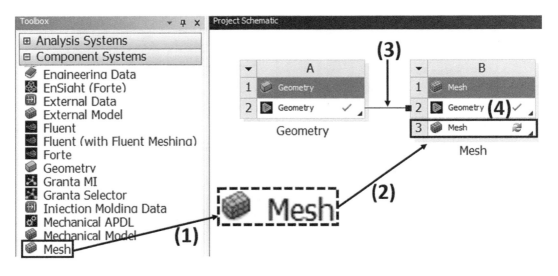

FIGURE 14.16

The *Outline* window includes the project tree. Find the *Mesh (1)* thumbnail at the bottom, as in Figure 14.17. The lightning symbol means that the mesh with current settings is not generated yet. Click *1xLMB* on *Mesh (1)* and go to the *Details of "Mesh"* window to change the selected global settings in the *Defaults (2)* tab: *Physics Preference -> CFD*, *Solver Preference -> Fluent*, *Element Order -> Linear*. Now go down to the *Sizing* tab. Check if the default settings are: *Use Adaptive Sizing: No (3)*, *Capture Curvature: Yes (4)* and *Capture Proximity: No (5)*. *Capture…* functions allow to refine mesh cells in specified domain regions (curvatures/proximities).

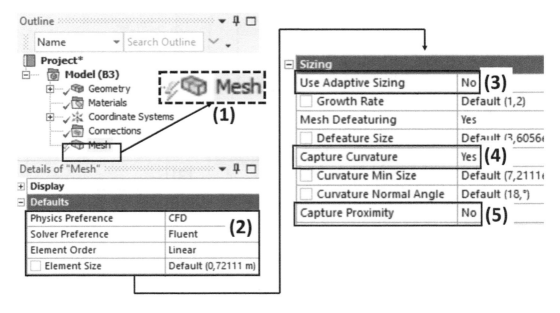

FIGURE 14.17

Tutorial 6 – Vertical-Axis Wind Turbine

Click *1xLMB Mesh (1)* thumbnail. Find selection filter – four cube symbols *(2)*, above the project workspace, as in Figure 14.18. Click *Face (Ctrl+F)* and click *1xLMB* rotor face *(3)* in the project workspace (its color should change to green). Click *1xRMB*, and select *Insert (4)* -> *Sizing (5)*. Now the local cell size can be added to the rotor area. Go to the cell to the right of the *Element Size (6)* one and set *0.005 m*. You can change project units using the *Units (7)* button if required. New *Face Sizing* has been added to the project tree *(8)*, below the *Mesh* tab.

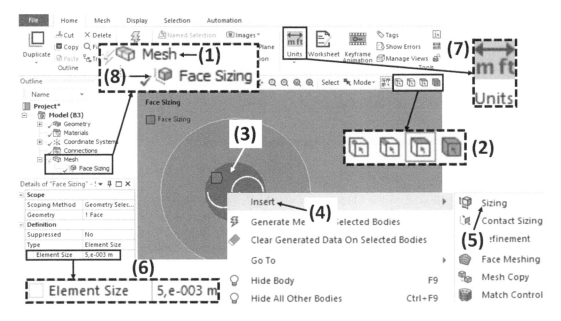

FIGURE 14.18

Analogously, (Figure 14.18) add *Face Sizing* to close (ring) the turbine surroundings *(1)*, *0.01 m* and distant turbine surroundings *(2)*, *0.05 m*. Note that whenever you apply local sizing, the global option *Capture Curvature* is deactivated (see Figure 14.17). Three *Face Sizing* objects *(3)* should appear below the *Mesh* tab, as in Figure 14.19.

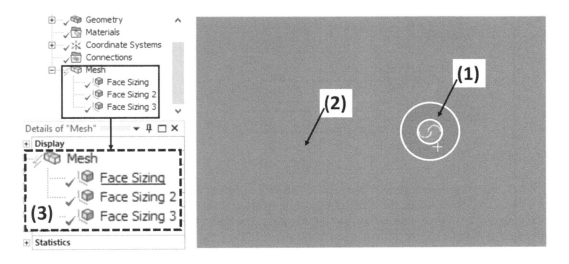

FIGURE 14.19

It is easier to apply further local functions when two outer model parts are hidden. Use again selection modes – select **Face (1)**. Now hold **Ctrl** and select both close and distant rotor surrounding parts by clicking **1xLMB** on each of them *(2)*, as in Figure 14.20. Click **1xRMB -> Hide body (3)**.

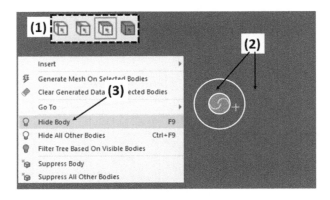

FIGURE 14.20

This time the cell size on the blade edges will be defined (Figure 14.21). Chose the **Edge (1)** selection mode. Click **1xLMB** anywhere in the project workspace window and **Ctrl+A** from the keyboard. All edges of the rotor domain have been selected. Hold the **Ctrl** key and just deselect the external circle edge. Now only eight blade edges *(2)* are selected. Of course, you can also select edges with the **Ctrl** key one by one. As previously, add sizing (**1xRMB -> Insert -> Sizing**). Go to the **Details of "Edge Sizing"** window. You can check whether 8 edges are selected in the **Geometry (3)** cell. Set **Element Size (4)** of **0.0005 m**. **Edge Sizing** should be visible below three **Face Sizing** objects *(5)* on the project tree.

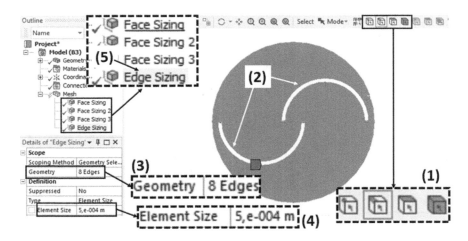

FIGURE 14.21

Click anywhere in the project workspace **1xRMB -> Invert Visibility**. Now two outer geometry parts are visible. Click the **Edge (1)** selection mode and select the smaller circle edge *(2)*, as in Figure 14.22. Add another sizing. Set the **Element Size** to **0.005 m**. When the **Element Size** is already set, you can see the visualization of the mesh cells on the circumference *(3)*. Furthermore, another **Edge Sizing** object should appear in the project tree *(4)*.

Tutorial 6 – Vertical-Axis Wind Turbine

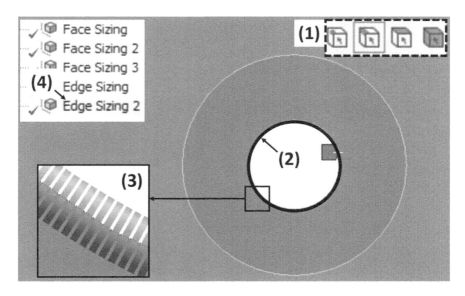

FIGURE 14.22

The Inflation tool allows to refine mesh in the boundary layer region. Use again the *Inverse Visibility* option to show only the rotor region. Go to the *Mesh (1)* tab of the top bar and select the *Inflation (2)* tool. In the *Details of Inflation* window click *1xLPM* in the cell *(3)*, to the right of the *Geometry* cell. Use the *Face* selection mode to select the rotor face *(4)*, as in Figure 14.23. Apply your selection. Click *1xLPM* in cell *(5)* to the right of the *Boundary* cell. Select all of the blade edges *(6)* using the *Edge* selection mode. 8 edges in total should be selected. *Apply* selection. Set the required inflation properties *(7)*: *Inflation Option – First Layer Thickness*, *First Layer Height -0.0001 m*, *Maximum Layers -9*, *Growth Rate – 1.2*. The new inflation object is available in the *Outline* tree *(8)*.

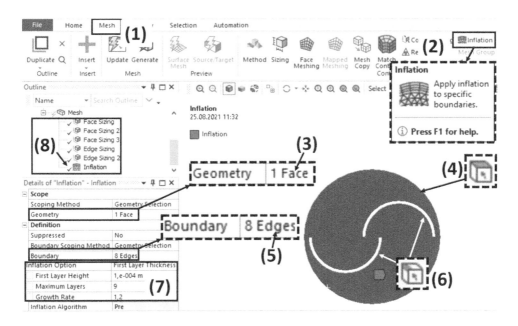

FIGURE 14.23

Add analogously *Inflation* in the rectangular region *(1)*, on the round edge *(2)*. You can hide any other objects (Figure 14.24). Set *Inflation* properties *(3)*: *Inflation Option – First Layer Thickness*, *First Layer Height -0.01 m*, *Maximum Layers -50*, *Growth Rate -1.2*. *Inflation 2 (4)* object appears in the *Outline* window.

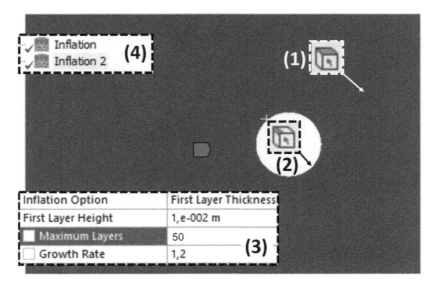

FIGURE 14.24

Before the mesh generation, set *Named Selections* which allow the solver to automatically recognize the boundary and the object types. First, click *1xLMB* on the *Mesh (1)* tab of the project tree. Then use the *Face* selection mode to select the rotor face *(2)*. Press the *N* button (keyboard). In the displayed *Selection Name* window enter the *fluid-rotor (3)* name and click *OK (4)*. The first object in the *Named Selection (5)* tab appears, as in Figure 14.25.

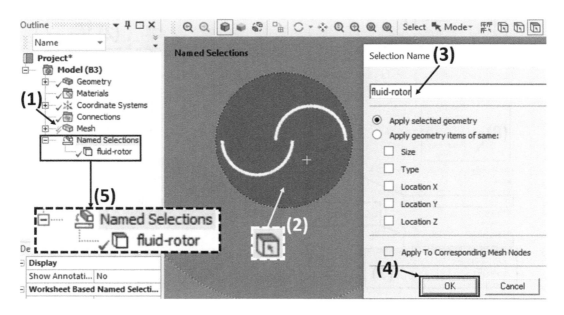

FIGURE 14.25

Tutorial 6 – Vertical-Axis Wind Turbine

Now set analogously other *Selection Names*, according to Figure 14.26: ***fluid-stat (1)*** (***Face***), ***inlet (2)*** (***Edge***), ***outlet (3)*** (***Edge***), ***wall-ext (4)*** (***Edge***). Then you can have a look at the list below the *Named Selections (5)* to check whether all the necessary objects are there.

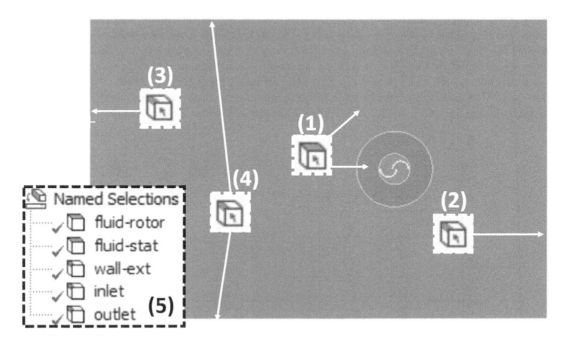

FIGURE 14.26

Add three remaining *Selection Names*: ***wall-blades***, for 8 blade edges *(1)*, ***interface-1*** for the external edge of the rotor region *(2)*, and for the internal ring edge *(3)*. Check on the *Named Selection* list whether the three new objects *(4)* have been added there, as in Figure 14.27.

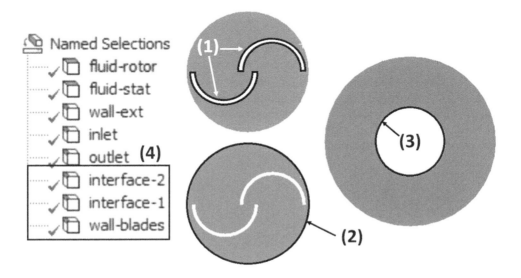

FIGURE 14.27

It's finally time to generate the mesh. First, click the ***Mesh (1)*** tab of the project tree, according to Figure 14.28, to activate any required meshing options. In case of relatively simple projects, the whole mesh can be generated in this one step – then the ***Generate (2)*** button can be used. However, mesh generation order matters in many projects which include several different meshing regions, as in this case. Thus, the mesh should be generated individually on subsequent subdomains. Use the ***Face*** selection mode to select the rotor region *(3)*, click ***1xRMB -> Generate Mesh On Selected Bodies (4)***. You can clear mesh from any region in exactly the same way, by the ***Clear Generated Data On Selected Bodies (5)*** option. Meshing progress is displayed in the left bottom corner *(6)*. Click the progress bar to see details *(7)*. To see the generated mesh click the ***Mesh*** tab of the project tree. Now, analogously generate mesh on the remaining faces – first, on the ring region of the close turbine surrounding *(8)*, second, on the rectangular, distant turbine surrounding *(9)*.

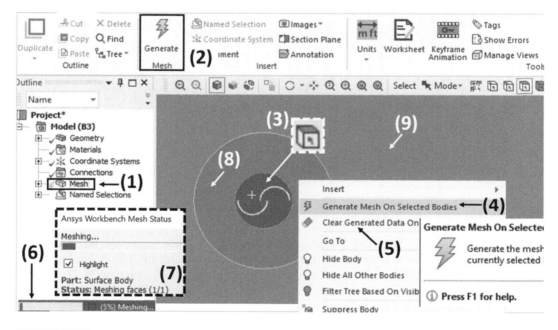

FIGURE 14.28

Figure 14.29 presents the complete mesh for the project's computational domain. Detailed views of the mesh in certain regions are given in additional windows, below the project's overall view: dense mesh in the boundary layer – close to the blade edges *(1)*, inconsistent mesh on the connection edge between the moving rotor and its stationary close surrounding *(2)*, consistent mesh between two stationary surrounding regions *(3)*, relatively coarse mesh far away from the turbine blades *(4)*.

Tutorial 6 – Vertical-Axis Wind Turbine

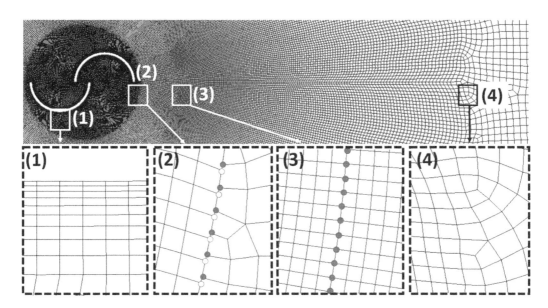

FIGURE 14.29

14.4 PREPROCESSING – SOLVER SETTINGS – PART 1

Find the ***Fluent*** module in the ***Component Systems (1)*** tab. Drag and drop it into the ***Project Schematic*** window, to the left of the previously added modules (2), as in Figure 14.30. Create a link *(3)* between *B3* and *C2* cells. The lightning sign in *B3* cell means that the mesh has to be updated. Click *1xRMB* on *B3* cell -> ***Update (5)***. During this process, the Fluent input file containing all the information about the previously generated mesh is created. Wait until the mesh is updated and the lightning mark turns into the check mark (the same as in *B2* and other cells). Now click *2xLMB* on *C2* cell to launch the solver.

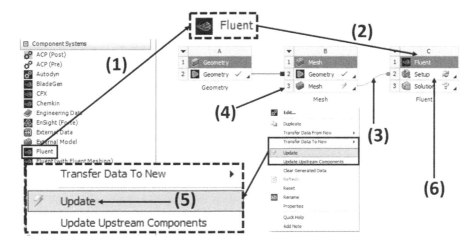

FIGURE 14.30

The *Fluent Launcher* window (Figure 14.31) includes preliminary solver settings. The number of dimensions *(1)* is predefined. Student *AWB* license allows to split the solution process into 4 parallel parts. Set *Solver Processes (2)* to *4*. Now click *Start (3)* and wait for the solver to launch.

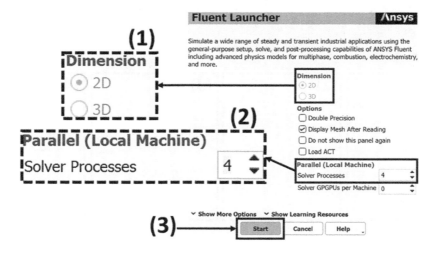

FIGURE 14.31

Like in the case of *Ansys Meshing*, the *Outline View* window in *Ansys Fluent* includes a project tree *(1)* with subsequent steps of the solver setup (Figure 14.32). The *General* tab of this tree is open as default just after the solver launch. *Check (2)* allows to display in the console overall information about the model and plot errors/warnings (such as unmatched interfaces). *Report Quality (3)* is used to display mesh quality reports (orthogonal quality and aspect ratio). *Display…(4)* allows to activate or deactivate the view of any named object in the model. After selecting *Display…*, the available objects can be found on the *Surfaces* list of the *Mesh Display (5)* window. Select any set of surfaces from the list and click *Display* below to try this functionality. Close the *Mesh Display* window and set the *Time* mode to *Transient (6)* – the simulation of a moving object (rotor) is time dependent.

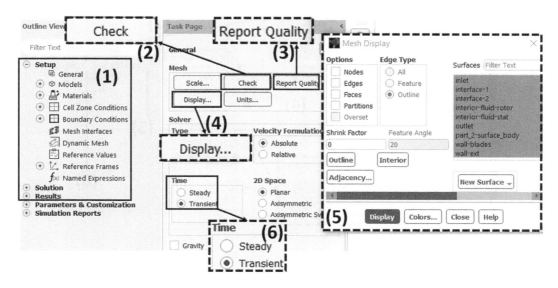

FIGURE 14.32

Tutorial 6 – Vertical-Axis Wind Turbine   281

Click *2xLMB* on the *Models* tab in the *Outline View* window. Find and click *2xLMB Viscous (SST k-omega)* tab *(1)*, which includes turbulence modeling main settings. In the *Viscous Model* window, you can find the available turbulence modeling approaches (*Model* list, as in Figure 14.33). In this case, the *k-omega (2) SST (3)* model is appropriate due to a fine mesh in the boundary layer (close to the turbine blades). Click *OK (4)* to close the window.

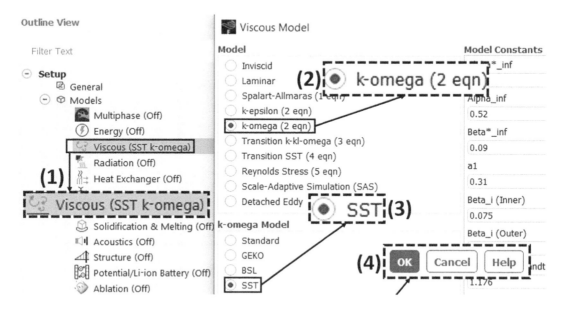

FIGURE 14.33

Open the *Materials (1)* tab, which includes materials available in this project. They are displayed in the *Task Page* window and divided into *Fluid* and *Solid* sections *(2)*, as in Figure 14.34. *air* and *aluminum* are always available as default. Air is fine for this project, so there is no need to create any new material. However, it can be done using the *Create/Edit* option *(3)*.

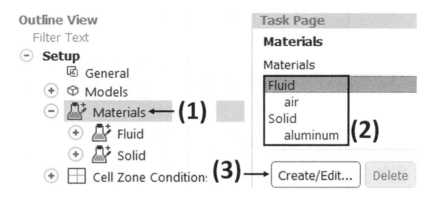

FIGURE 14.34

Open the *Cell Zone* conditions tab. Two zones are available there, as in Figure 14.35. Select the *fluid-rotor* zone by *1xLMB*. Check the *Type* list below the zone list – it should be *fluid* domain. Click *Edit… (3)*. The *Fluid* window appears. *Air* is the default zone material *(4)* (*Material Name*). The Rotor has to be the moving zone, thus *Mesh Motion* option has to be activated *(5)*. The rotor center is located in the center of the global coordinate system, so no changes are required in the *Rotation-Axis Origin (6)* section. Set *Rotational Velocity* to *12.56 [rad/s] (7)*. Click *Copy To Mesh Motion (8)* and check in the *Mesh Motion (9)* tab whether all settings are copied there. *Apply (10)* settings.

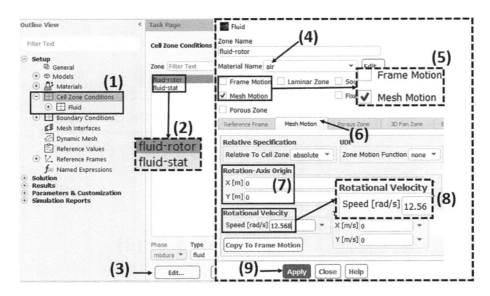

FIGURE 14.35

Click *2xLPM* on the *Boundary Conditions* (BCs) tab. The list of the existing BCs (Figure 14.36) appears below *(1)*. Furthermore, all the BCs are available in the *Boundary Conditions* window (*Task Page*) on the right. Click *1xLMB* in the *inlet (2)* to check the automatically assigned BC *Type (3)*. Due to the previously created *Selection Name* it is *velocity-inlet*. Click *Edit (4)* below to open the *Velocity Inlet* BC window. The model is isothermal with no species or multiphase transport, thus, only the *Momentum (5)* tab is active. Set *Velocity Magnitude* to *2 [m/s] (6)* and *Apply (7)*. Close this window and check the outlet BC type – it should be a *pressure-outlet*.

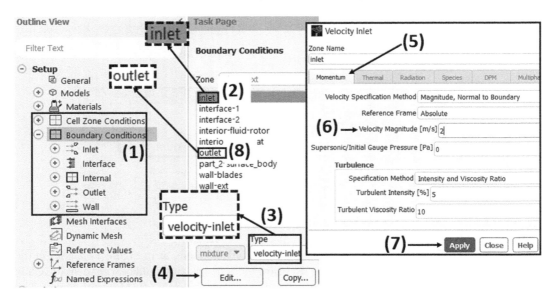

FIGURE 14.36

Tutorial 6 – Vertical-Axis Wind Turbine

Inlet and *outlet* are graphically defined (Figure 14.37) by arrows *(1)*. Check the remaining BC types: two interface sides *(2)*, interiors *(3)* (filling of two subdomains) and walls *(4)*. An additional object is *part-2-surface-body (5)*. It is the edge connecting two parts of the stationary domains. Its *Type* has to be set to *interior (6)*.

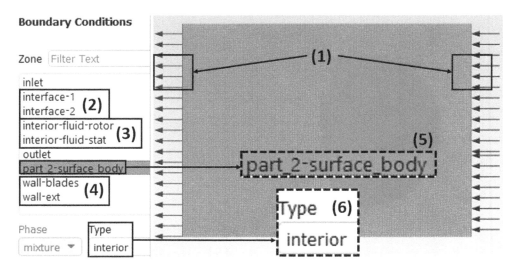

FIGURE 14.37

To avoid shear stress on the external model boundaries (which are not physical walls), open the *wall-ext (1)* BC window (*2xLMB* on *wall-ext*) and in the *Wall (2)* window that appears (Figure 14.38) switch *Shear Condition (3)* to *Specified Shear*. If the *Shear Stress (4)* components are equal *0*, there is no shear stress. *Apply (5)* settings.

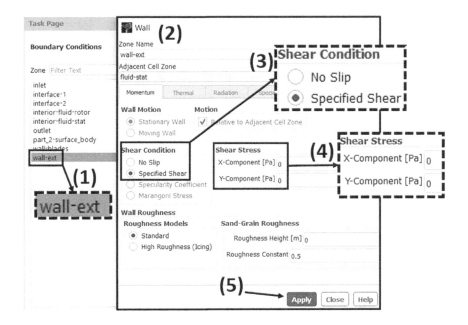

FIGURE 14.38

Click **2xLMB** on the **Mesh Interfaces (1)** tab. On the left side of the **Mesh Interfaces** window, mark (*1xLMB*) subsequently *interface-1* and *interface-2* objects *(2)*. Click **Create (3)** below. If a new interface *(4)* appears on the right-side list as in Figure 14.39, close *(5)* the window.

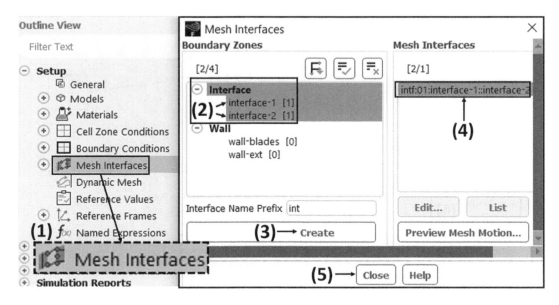

FIGURE 14.39

Open the **Reference Values (1)** tab. Select *inlet (2)* from the **Compute From** list, according to Figure 14.40. It means that the required reference values will be set based on this BC. Although the model is reduced to 2D case, certain dimensions allowing to compute forces acting on the turbine have to be defined. Set **Depth (3)** to *1.5 [m]*. This parameter corresponds to the turbine height. Set **Length (4)** to *0.76 [m]*. It corresponds to the turbine diameter (it's a kind of characteristic dimension).

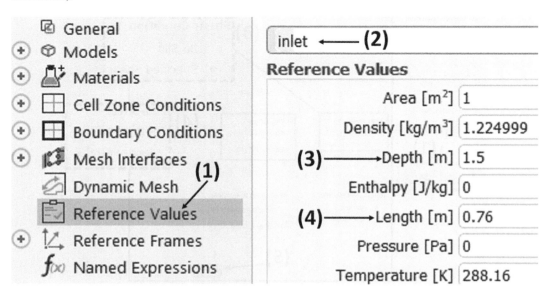

FIGURE 14.40

Tutorial 6 – Vertical-Axis Wind Turbine

Open the **Methods (1)** tab. Appropriate **Pressure-Velocity Coupling Scheme** is **Coupled (2)** (Figure 14.41). Contrary to the **SIMPLE** or **SIMPLEC** schemes, the **Coupled** scheme does not use a segregated (serial) momentum solution or continuity equations. Activate **Auto Select (3)** of the **Flux Type**. **Spatial Discretization** of the **Momentum**, **Turbulent Kinetic Energy** and **Specific Dissipation Rate** should be set to **Second Order Upwind (4)**.

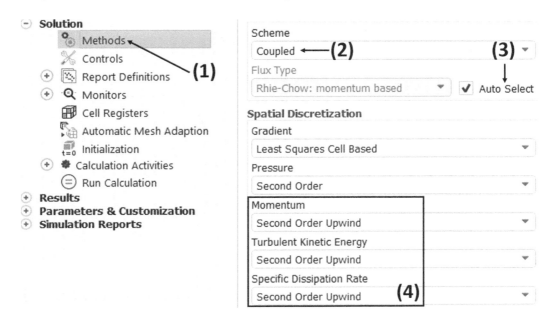

FIGURE 14.41

Click **1xRMB** on **Report Definition (1)** (**Outline View** window) → **New** → **Force Report** → **Moment (2)**. The **Moment Report Definition** window appears (Figure 14.42). Insert any report **Name**, for example, **moment-1 (3)**. Switch **Report Output Type** to **Moment (4)** and select **wall-blades (5)** from the **Zones** list. Make sure that the **Report File** and **Report Plot** options in the **Create** section and **Print to Console** are switched on and click **OK (6)** to close the window.

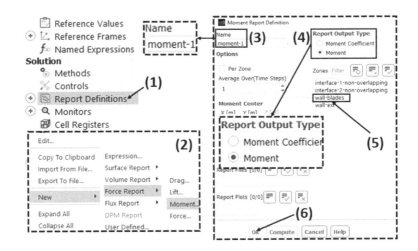

FIGURE 14.42

The first computation process always has to be initialized, to provide the initial data required to run the first solution iteration. Open the *Initialization (1)* tab and set the *Standard Initialization (2)* in the *Initialization Methods* section, according to Figure 14.43. Select *inlet (3)* from the *Compute from* the list. It means that for the initialization purpose, flow parameters are assumed to be the same as on the *inlet* BC. Thus, for example, *X velocity (4)* is automatically set to *-2 [m/s]* (according to the reference frame). Click *Initialize (5)*. Done!

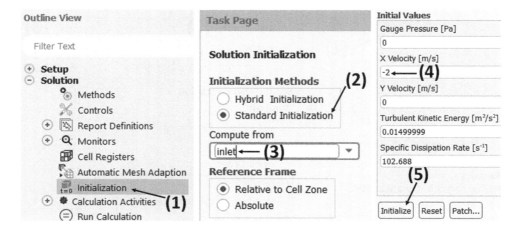

FIGURE 14.43

Click *2xLMB* on the *Calculation Activities (1)* tab and set *Autosave Every 10 Time Steps (2)*. Now you can check whether the simulation has been initialized. Go to the *Graphics* tab below the *Results* tab, as in Figure 14.44, and click *2xLMB* on *Contours (3)*. If the *Contours* tab is active (available), it means that the case has been initialized. Select *Velocity...(4)* from the *Contours of* the list. Select *Velocity Magnitude (5)* from the list below. Then subsequently click *1xLMB* on *interior-fluid-rotor* and *interior-fluid-stator (6)* in the *Surfaces* window. If you want to obtain a continuous contour, do not select any surface. Click *Save/Display (7)* to generate the velocity contour – it appears in the workspace window. New *contour-1* object *(8)* is now available below the *Contours* tab.

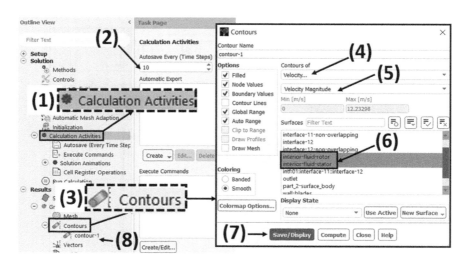

FIGURE 14.44

Tutorial 6 – Vertical-Axis Wind Turbine

Functionalities of the *Solution Animations* (Figure 14.45) tab allow to track results of any simulation in progress. Open this tab (*2xLMB*) and set *Record after every 10 time-step (1)*. Switch *Storage Type* to *In Memory (2)* (displayed data will not be available in any folder after simulation). If you want to save the frames, select another *Storage Type* option. Select the already created *contour-1 (3)* from the *Animation Object* list (velocity contour will be displayed in every 10 timesteps). Click *OK (4)*.

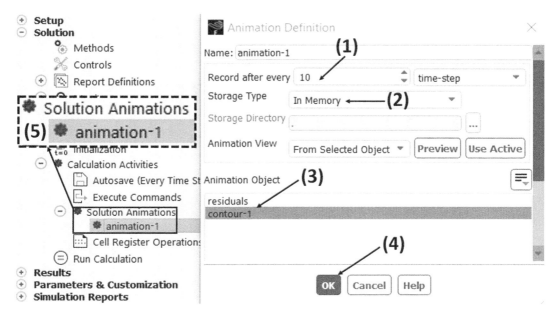

FIGURE 14.45

Open the *Run Calculation (1)* tab (Figure 14.46). Find and set the following parameters: *Number of Time Steps (2) -1000*, *Time Step Size [s] (3) -0.005*, *Max Iterations/Time Step (4) -20*. Timestep was calculated based on the rotational speed and it depends on the angular speed. The 1000 Time Steps (TS) times 0.005 [s/TS] gives 5 seconds. This is the total simulation time. Click *Calculate (5)* to launch the solution process.

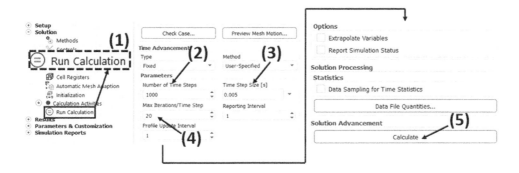

FIGURE 14.46

The chart that appears in the workspace window (Figure 14.47) includes a set of residual curves corresponding to the imbalance in transport equations that are being solved. The legend *(1)* (*Residuals*) allows to identify the curves. The repetitive peaks *(2)* come from the model solving in subsequent timesteps – turbine rotation changes the model conditions, influencing a temporary solution divergence (peak). After each 10 timesteps, the current velocity contour is displayed in *Contours of Velocity Magnitude [m/s]* tab *(3)*.

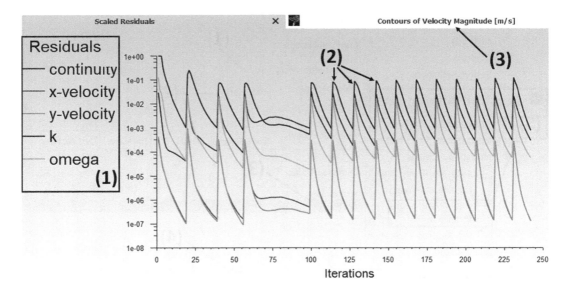

FIGURE 14.47

Thanks to the defined report (see Figure 14.42), the chart representing the dynamics of the moment on the turbine blades is created throughout the simulation, as in Figure 14.48 (this chart is exemplary and can differ from the one that will be obtained in the course of your simulation). To see it, go to the appropriate tab *(1)* above the chart window. Although in the first stage of the simulation, the moment is decreasing (due to initial conditions), it stabilizes after a certain number of timesteps *(2)*. Fluctuations come from the rotor zone rotations, changing temporary conditions in the system. The moment computed in each time step can be read in the console *(3)*. After simulation you can use the report file to process the data, for example, to create the chart in any tool (like Origin or MS Excel). You can find the report file in: *your-model-folder_files → dp0 → FLU → Fluent*. It is *rfile* *(4)* type. It is worth noting that the results from selected timesteps (see Figure 14.45) are also saved there *(5)*. Close the solver and save the whole project.

Tutorial 6 – Vertical-Axis Wind Turbine

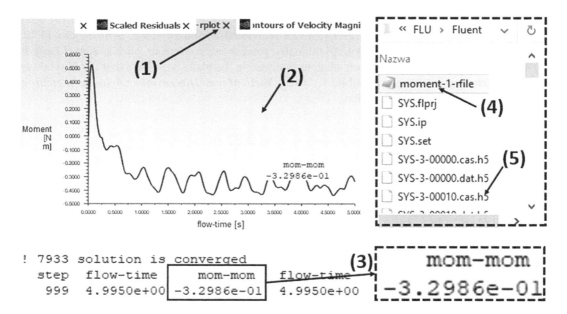

FIGURE 14.48

14.5 PREPROCESSING – GEOMETRY – PART 2

Further part of the exercise covers an analysis of the alternative turbine design (with stationary blades). Most of it is analogous to the first part. Thus, the first project workflow can be duplicated. Hold **LMB** to select *A*, *B* and *C* modules using the box selection mode, as in Figure 14.49. Then click *1xRMB* on *C1* (*Fluent*) cell *(2)* and select **Duplicate** *(3)* from the drop-down list. Copies of modules *(4)* appear. Click *2xLMB* on *D2* (*Geometry*) cell *(5)* to open **Design Modeler** and add the required construction parts.

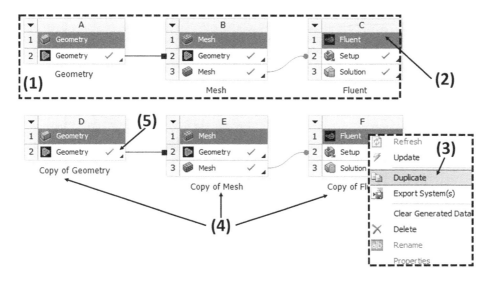

FIGURE 14.49

Remember the procedure of creating geometric objects (for example Figures 14.3–14.10). Add *New Sketch (1)* on the XY Plane *(2)*. Create an individual stationary blade, as in Figure 14.50. Define required dimensions (*R*, *H*, *V*). Additionally, you can use constraints, such as *Equal Length (3)* and *Vertical (4)* to define the short sections connecting the blade arcs. Note that the *H* symbols refer to the *Horizontal* dimension, while *V* – to the *Vertical* one (*Dimensions* tab in the *Sketching Toolboxes*). *G* is a *General* dimension.

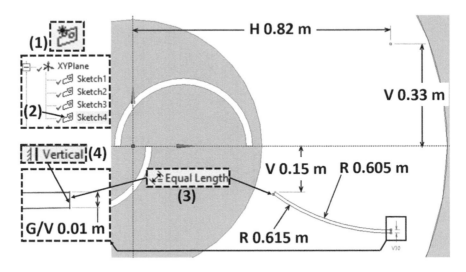

FIGURE 14.50

Switch to the *Modeling (1)* mode and open the *Concept (2)* drop-down list (Figure 14.51). Select the *Surface From Sketches* tool to add the stationary blade filling. In the *Details View (4)* window click *1xLMB* in the *Apply/Cancel* cell (5), to the right of the *Base Object* one. Select the right sketch *(6)* from the outline tree (*Sketch 4*) and click *Apply*. Switch *Operation* to the *Add Frozen (7)* mode. Then find the edited *SurfaceSk4 (8)* operation in the *Tree Outline* window, *1xRMB* → *Generate*.

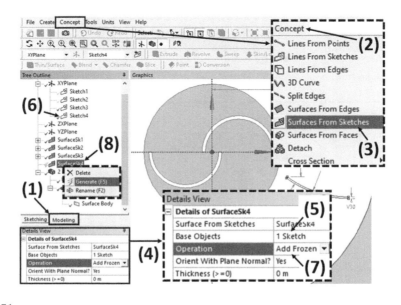

FIGURE 14.51

Tutorial 6 – Vertical-Axis Wind Turbine

New *Surface Body (1)* appears below the *...Parts,...Bodies* tab, as in Figure 14.52. Drop down the *Create (2)* list and select *Pattern (3)* to duplicate the stationary blade. In the *Details View (4)* window change *Pattern Type* to *Circular (5)*. Then click *1xLMB* in the cell to the right of the *Geometry (6)* cell and select the appropriate object – you can click *1xLMB* on the *Surface Body* on the list in the *Outline Tree* window or just select it directly in the workspace window. Remember to *Apply* your selection in the cell *(6)*. Now click *1xLMB* in the cell *(7)* below and select *XYPlane (8)* from the project tree. *Apply* selection and set *7 Copies (9)* of the duplicated object. Find a *Pattern* with the lightning sign in the *Outline Tree* window, *1xRMB* → *Generate*.

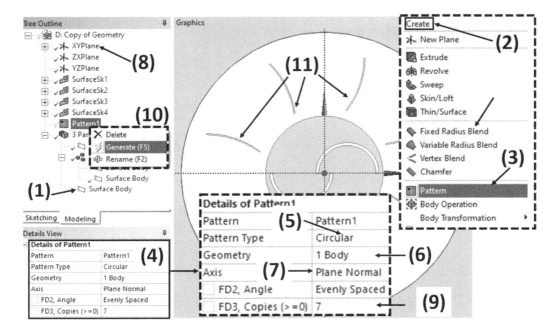

FIGURE 14.52

The created stationary blades have to be subtracted from the fluid domain. Drop down the *Create* list again and select *Boolean (2)*, according to Figure 14.53. Define settings in the *Details View (3)* window: *Operation (4)* → *Subtract*, *Target Bodies (5)* → ring stationary domain (*1xLMB* on *(6)*, *Apply*), *Tool Bodies (8)* → all *Surface Body* objects *(9)* representing stationary blades, *Preserve Tool Bodies? (10)* → *No*. *Generate* the *Boolean (11)* operation.

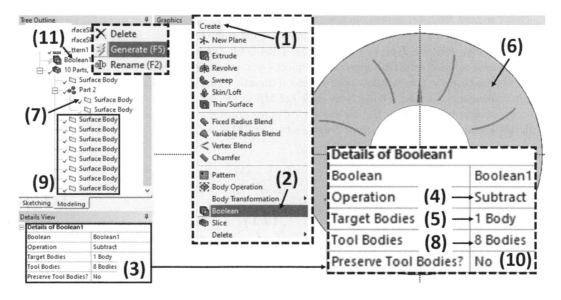

FIGURE 14.53

Let's summarize. **SurfaceSk4 (1)** operation allowed to add one stationary blade. It was duplicated by **Pattern**. Blades were subtracted from the domain by **Boolean1 (2)**. After that, there are only three Surface body objects again (3). Stationary blades are represented by empty spaces (4) in the domain (Figure 14.54). Close Design Modeler module.

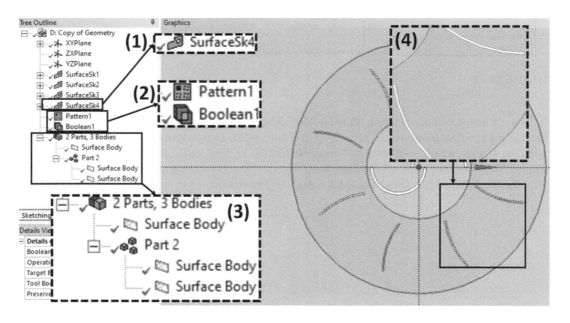

FIGURE 14.54

Tutorial 6 – Vertical-Axis Wind Turbine

14.6 PREPROCESSING – MESHING – PART 2

Click *2xLMB E3* (*Mesh*) cell *(1)* to launch *Ansys Meshing* module. Confirm that upstream data should be read *(2)* as in Figure 14.55.

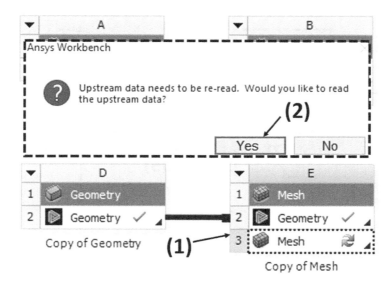

FIGURE 14.55

All the previously defined sizing functions *(1)* apply (Figure 14.56). Add new sizing to define the cell size on the stationary blade edges. Use the *Edge (2)* selection mode to select the edges of each stationary blade *(3)*. You can hide geometry parts except the ring one, to make selection faster (then use *Ctrl+A* and just deselect two circles). Click *1xRMB* → *Insert (4)* → *Sizing (5)*. *Edge Sizing 3* appears in the *Outline Tree* window. Check in the *Details of Edge Sizing 3* widow whether the number of selected edges (it should be *32*) is correct *(7)*. Set *Element Size (8)* to *0.0025 m*.

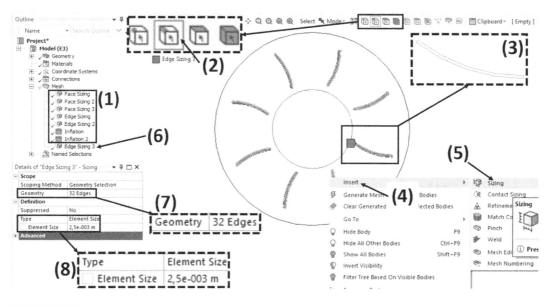

FIGURE 14.56

Again use the ***Edge (1)*** selection mode to select stationary blade edges *(2)* (Figure 14.57). Press the *N* key and define the Selection Name ***wall-stator (3)***. It appears at the bottom of the ***Named Selections*** list *(4)*. Generate mesh for subsequent domain parts in order *(5)*, *(6)*, *(7)*. Use the ***Generate Mesh On Selected Bodies (8)*** option (*1xRMB*).

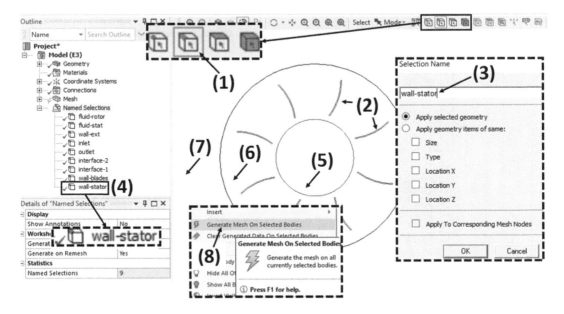

FIGURE 14.57

Find the ***Connections (1)*** tab in the ***Outline Tree*** window (Figure 14.58). Open it and click *1xRMB* on ***Contacts (2)***. Select ***Delete (3)*** from the list. Contacts describe the edges common for two different domain parts. They were not used in the first solver preprocessor setup, thus, they have to be removed here too.

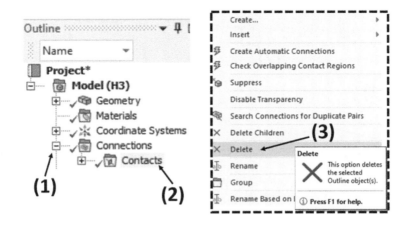

FIGURE 14.58

Tutorial 6 – Vertical-Axis Wind Turbine

14.7 PREPROCESSING – SOLVER SETTINGS – PART 2

It is a good idea to save the whole *Ansys Workbench* project before the next step. Then click *1xRMB* on *E3* (*Mesh*) cell *(1)* → *Update* *(2)*, according to Figure 14.59. Wait until the lightning mark turns into a check mark. Then click *2xLMB* on *F2* (*Setup*) cell *(3)* and wait for the *Fluent Launcher* window (see Figure 14.30). Remember to set *4 Solver Processes* there before you click *Start*.

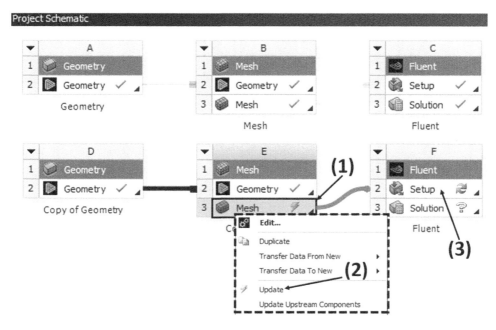

FIGURE 14.59

Loading data during launching the solver is not performed in exactly the same way as in the case of the first turbine model. First of all, *Information (1)* about new zones appears, as in Figure 14.60. Click *OK* in this window. Additionally, the communication about the new zone is displayed in the console. According to *Warning (2)*, the new zone that has been found is *wall-stator*, so it is correct (it has been just added as the *Selection Name* in the *Meshing* module). To make sure that this zone has been correctly implemented in the solver setup, select the *File* drop-down list on the top task bar → *Recorded Mesh Operations...* *(3)*. The *Recorded Mesh Operations and Incoming Zones* window appears. Click *Reload (4)*, which should work as confirmation that the new incoming zone is valid. Confirm (*Yes*) in the *Question (5)* window that you want to proceed. The solver setup is already done (it has been duplicated with settings). Model physics is valid also in this case, so after *Initialization* (as in the first turbine case, see Figure 14.43) you can *Run Calculation* (see Figure 14.46).

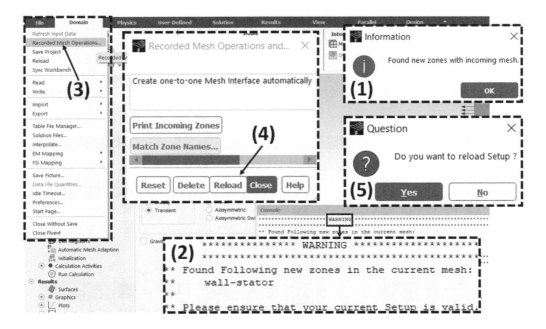

FIGURE 14.60

After the solution process any required field variable contours can be created as previously (see Figure 14.44) in the **Contours** tab. Create a chart (in any software) including two moment curves from the first *(1)* and the second *(2)* turbine cases, as in Figure 14.61 (absolute values presented) to compare the moments in two simulated cases. Furthermore, you can calculate average values and compare theoretical mechanical power on the turbine shafts. This power is the product of angular velocity (known – defined in solver) and moment.

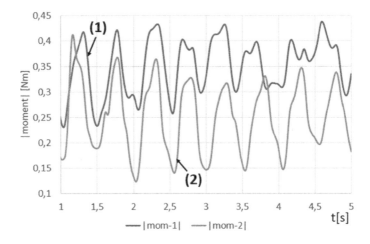

FIGURE 14.61

The solver application can be closed and the whole project is saved now. After that find the **Results** module on the **Component Systems** list, drag and drop it in the **Project Schematic** window. Create two links *(1)* connecting the solver module cells with **Results** *(2)*, according to Figure 14.62: **C3** (**Solution**) with **D2** (**Results**) and **G3** (**Solution**) with **D2** (**Results**). Then click **2xLMB D2** (Results) cell.

Tutorial 6 – Vertical-Axis Wind Turbine

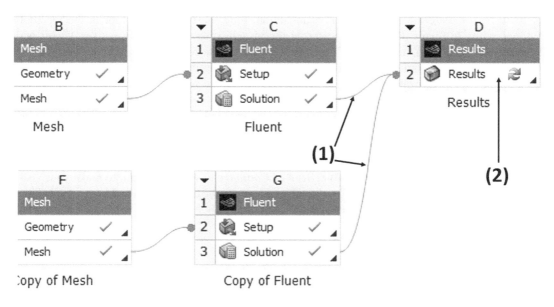

FIGURE 14.62

14.8 POSTPROCESSING

Results module launch can take some time due to a large number of the data sets that have to be transferred. Both turbine cases are available for postprocessing now, as in Figure 14.63. The turbine frame can be moved using the ***Pan (1)*** tool. Switch the display mode to two vertical windows *(2)*. Then you can decide which case has to be displayed in a particular window. The list of cases *(3)* is available just below the navigation tool bar. Case order (***Case 1***, ***Case 2***) depends on the order of creating links between the modules (previous step). Click ***2xLMB*** the ***Case Comparison (4)*** mode. Select ***Case Comparison Active (5)***. Then ***Synchronize camera in active views (7)*** and ***Apply (7)*** settings.

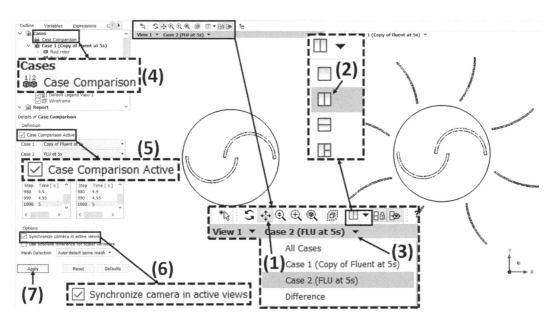

FIGURE 14.63

It's time to create the first **Contour** visualization. Find the **Contour (1)** tool in the top toolbar, according to Figure 14.64. It is located in the visualization section, between the **Vector** and **Streamline** tools. Go to the **Details of Contour 1** window **(2)** to set up the required properties. The **All Domains** option should be selected from the **Domains** drop-down list. Now the location where the variable field visualization is going to be displayed has to be selected. When the individual area (for example only rotor area) has to be displayed it can be done using the **Location** drop-down list. In this case, a multiple area selection is required. Find and click the three dots (**"..."**) button to the right of the **Location** drop-down list. The **Location Selector (3)** window appears. Hold the **Ctrl** key and select two areas in each of the two cases (four areas in total) **(4)**: *fluid rotor symmetry 1*, *fluid stat symmetry 1*. Click **OK** and go back to the **Details of Contour 1** window. Displayed **Variable** should be **Pressure**. **Global Range** is fine. Set **# of Contours** to **45**. If you want to make your visualization clearer, you can pass to the **Render** tab and activate the **Show Contour Lines** option. Then click **Apply** to display **Contour**. It clearly shows how the stationary blades influence the pressure field. Can you associate it somehow with the average moments?

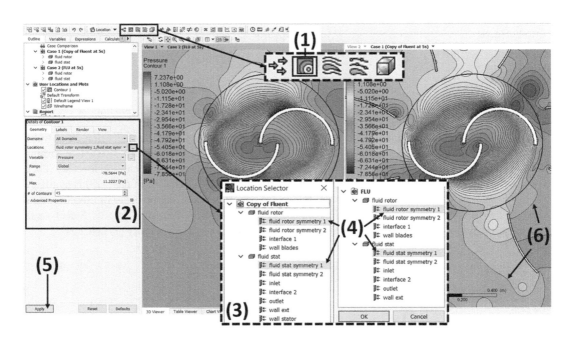

FIGURE 14.64

Legend view *(1)* can be customized (Figure 14.65). Click **2xLMB Default Legend View 1** tab. The **Definition (3)** tab and the **Appearance (4)** tab are available in the **Details of Default Legend View 1** window. In the second tab, **Text Size** and **Aspect (5)** can be modified (set respectively **0.8** and **0.1**). The format of numbers and their **Precision** can be changed too. You can set the **Fixed (6)** format. Furthermore, the t**ext Height** setting **(7)** is available there. You can set **0.035**.

Tutorial 6 – Vertical-Axis Wind Turbine

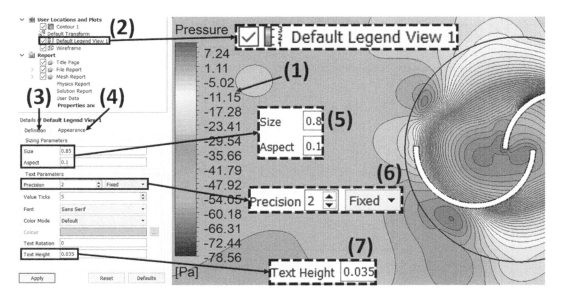

FIGURE 14.65

Because the simulation was transient, all the data for the selected time sets are available (according to the solver settings – see Figure 14.44). You can browse these data in the ***Results*** module. Find the ***Timestep Selector*** option – it is the clock thumbnail *(1)* on the top toolbar, as in Figure 14.66. The ***Timestep Selector*** window *(2)* with the list of all available timesteps appears. Click *1xLMB 4.15 s (3)* timestep. The ***Sync Cases*** (***By Time Step***) *(4)* and ***Match*** (***Same Step***) *(5)* options should be checked. Click ***Apply*** to see the turbine frame movement. Mark and click *2xLMB Contour 1 (6)* in the ***Outline*** window to see the ***Details of Contour 1*** window again. ***Variable*** (7) can be changed to another one – for example, ***Velocity***) ***Color Scale*** and ***Color Map (8)*** modes can be customized. Remember to click ***Apply (9)*** to display the new velocity contour.

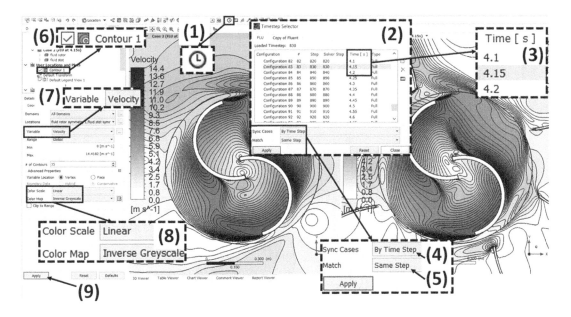

FIGURE 14.66

Zoom out the turbines as in Figure 14.67. Pay attention to the scale of the flow disturbances and air velocities in the case of the reference turbine (1) and the one equipped with the stationary blades (2). In the second case, you can consider the possible impact of the flow disturbance scale on the possibility of installing several turbines positioned closely together.

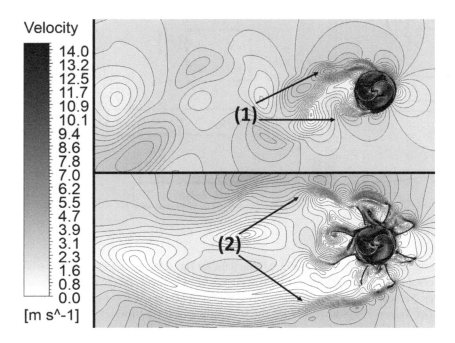

FIGURE 14.67

Another analysis option which might be interesting is the variable field difference generator. Change the view option in one of the two workspace windows (*Case 1* or *Case 2* drop-down list *(1)*) to *Difference (2)*, according to Figure 14.68. Then you can switch the view mode *(3)* to an individual window *(4)*. The contour presents now the result of the subtraction of the velocity fields in the two studied cases.

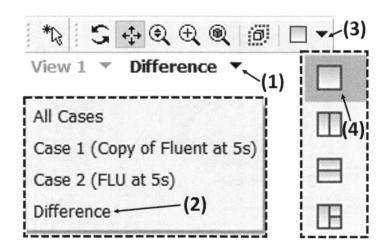

FIGURE 14.68

Tutorial 6 – Vertical-Axis Wind Turbine   301

Sometimes, in case of the velocity distribution analysis, the *Vector* visualization is clearer than the *Contour* one: Let's apply this visualization for the analysis of the phenomena occurring close to the stationary blades in the second turbine design. Deactivate *Contour* in the *Outline* window. First of all, deactivate the *Difference* mode. Then find and click the *Vector (1)* tool (thumbnail of three arrows to the left of the *Contour*, as in Figure 14.69). Go to the *Details of Vector 1* window. Set *Location* to the *fluid stat symmetry (2)*. The *Factor* parameter in frames of the *Reduction Factor* approach allows for to reduce the number of displayed vectors. It makes it easier to analyze the visualization. Set Factor to *20 (3)*. It means that only every twentieth vector will be displayed. Value *Range* and *Color Scale* can be customized in the *Color (4)* tab. The *Symbol (5)* tab allows to scale the vectors. The *Vector* visualization appears in the *Outline* window *(6)* – you can activate and deactivate the visualization there (as *Contour 1* previously). Pay attention to the impact of the stationary blades on the velocity vectors. Curvatures are directing the air stream into the rotor, as in case of *(7)* and *(8)*. Furthermore, circulation regions (eddies) as *(9)* or *(10)* can be observed.

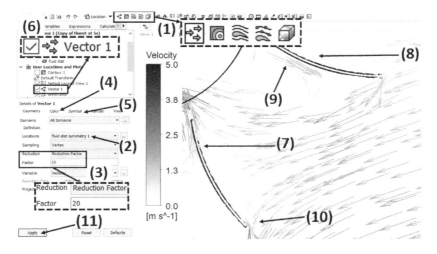

FIGURE 14.69

Part V

Biomass-Based Small-Scale Energy Applications

15 Theoretical Background

15.1 DEVELOPMENT OF THE TECHNOLOGY

Before the popularization of fossil fuels, in the pre-industrial age, biomass was the essential energy carrier for many, then developing industry branches. However, biomass gained a significant role in human history much earlier, actually at the beginning of our civilization, when man learned how to use it to prepare meals, warm up at campfires, and light his dwellings.

During the period of the industrial revolution (18th and 19th centuries), different forms of biomass were gradually replaced by fuels characterized by much higher energy density. Coal, natural gas, and petroleum became easily available. Despite the rapid decrease of interest in biofuels in developed countries, the popularity of woody biomass or agricultural wastes as the energy source remained unchanged.

In many third world countries, especially in rural areas, primitive devices consisting of three stones aligned appropriately, performing the function of primitive field stoves are still very common. According to Ref. [1], due to the low efficiency of such a solution, a pot placed on the stone stove can take only around 5% of the fuel combustion heat. Currently, (that is, India), a very popular solution in Asian countries is the so-called "chulha" stove. It is a very simple device allowing one to put one dish in the vault above the combustion chamber. Hot exhaust gas from the chamber flows around the pot during cooking, causing at the same time significant air pollution in the room. Its harmful effects on human health, as well as various attempts to remove the device from common use, have been widely discussed in the literature, including Refs. [2,3].

Due to the necessity for ensuring energy security, and significant growth of the awareness of environmental issues (including global warming), at the end of the 20th century, biomass was no longer associated with primitive applications like the ones mentioned above. Furthermore, in the last decade, the popularity of biomass has been growing, and it is now considered to be one of the most reasonable renewable energy sources (RESs) for electric power generation. According to the literature [4], biomass is the third world's largest renewable energy source, after solar and wind energy.

In the time of the 1970s energy crisis, searching for efficient substitutes for fossil fuels initiated a modern approach to the generation of "green power," based on the combustion of solid fuels in power plants and heat plants. Scandinavian countries can be treated as pioneers of heat and power generation from biomass. Furthermore, many advanced devices for small-scale applications (house heating) have been developed in Germany and Austria and are current leaders of the European woody-biomass-based small-scale energy devices market (especially technology of stoves and fireplaces) [4].

The first big step toward big-scale biomass energy utilization was its co-combustion in industrial coal boilers. However, due to a growing concern about climate change, the search for other, more efficient, and environmentally friendly techniques of biomass thermochemical conversion (based on dedicated devices) has been intensified [5].

15.2 STATISTICAL DATA

Figure 15.1 shows the installed capacity of biomass-based energy technologies (biogas, liquid and solid biofuels, municipal waste) in the top ten countries around the world in 2021. The largest installed capacity was noted for China (30 GW), Brazil (16 GW), and the USA (14 GW). It is worth mentioning that China advanced in this ranking from the fifth place in 2010 to the leading position

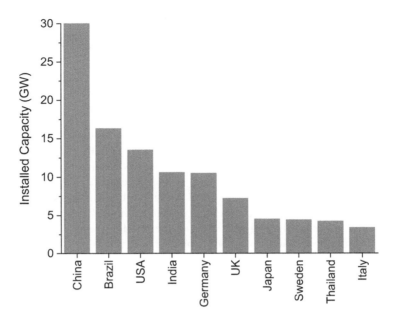

FIGURE 15.1 Top ten countries with the highest installed capacity of biomass-based power technologies in 2021. (Based on Ref. [6]).

in 2021. India took the fourth place with 11 GW of biomass installations. Germany took the fifth place with 10 GW of biomass installations All the remaining countries had less than 10 GW of installed capacity: The United Kingdom 7 GW, Japan 5 GW, Sweden 4 GW, Thailand 4 GW, and Italy 3 GW. The first five countries shared more than half of the world's installed biomass capacity in 2021 (143 GW).

Figure 15.2 compares the installed capacity of biomass-based technologies in the world between 2012 and 2021. In this period, the total installed capacity increased nearly twofold: From 77 GW in 2012 to 143 GW in 2021. The biomass-based technologies were dominated by the conversion of solid biofuels with 70% world's market share in 2021. The second most popular technology was based on biogas (15% share), followed by the utilization of renewable municipal waste (14% share). The annual net additions in the installed capacity of biomass-based technologies significantly fluctuated over the analyzed time. The global-average total costs of installations were also unstable in that period. The lowest prices were noted in 2012: 1513 USD/kW and the highest in 2013: 3028 USD/kW [6].

Figure 15.3 shows the electricity generation from biomass technologies in ten top countries in 2019. In this rank, the first place belongs to China (82 TWh), the second to the USA (63 TWh), and the third to Brazil (55 TWh). In the analyzed year, these three countries generated about 36% of the global electricity production from biomass, with the highest share of solid biofuels (nearly 70%). Not far behind is Germany with 50 TWh of produced energy. Each of the remaining countries generated less than 40 TWh of electricity from biofuels. Generally, in China, Germany, The United Kingdom, and Japan, bioenergy is becoming less and less competitive with solar and wind technologies, regarding the cost per unit of energy due to the policy transition.

Theoretical Background

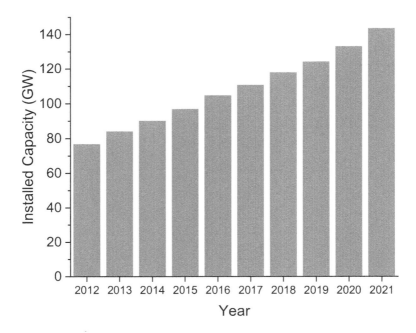

FIGURE 15.2 Cumulative installed capacity of biomass-based power technologies in the world. (Based on Ref. [6].)

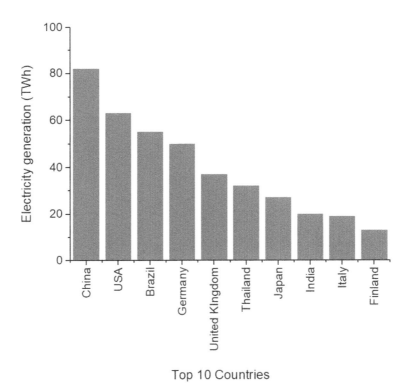

FIGURE 15.3 Top ten countries with the highest electricity generation from biomass-based power technologies in 2019. (Based on Ref. [6].)

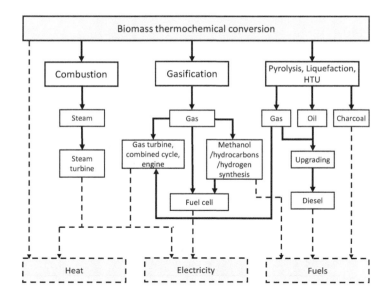

FIGURE 15.4 Biomass thermochemical treatment ways. (Based on Ref. [4].)

15.3 CLASSIFICATIONS AND CHARACTERISTICS

Several basic energy biomass types can be listed, including woody biomass, grassy biomass, industrial wastes, aquatic biomass, or manure (fertilizers). Many different, more extended, classifications can be found in the literature, see Refs. [5,7]. Detailed characteristics of different energy biomass types have been presented, for example, in Refs. [7,8]. Of course, not all biomass waste can should be used as an energy source. In some cases, it would be uneconomical, that is, when the raw material is overly dispersed and low energy density makes harvesting and transportation unprofitable. It is often more efficient to use biomass as a production material (for example, forage, paper, or construction industry) [9,10].

Three basic ways to convert biomass into heat, electricity, or processed biofuels are [4]

- thermochemical treatment, including direct combustion, gasification, and pyrolysis,
- biochemical treatment, like anaerobic digestion and fermentation,
- physicochemical treatment, including esterification.

Although nowadays all of the above-mentioned methods have industrial significance, in this book only thermochemical conversion is considered in detail. This is, among others, due to the characteristics of the practical CFD simulation examples that are discussed further. Figure 15.4 presents a simplified biomass thermochemical conversion scheme. When it comes to small-scale energy applications, devices based on direct combustion (especially!) and gasification have the biggest market significance.

15.3.1 DIRECT-COMBUSTION-BASED TECHNOLOGIES

Direct combustion shares 97% of the world's biomass energy production [11]. Biofuel is burned in the presence of oxygen to produce heat and electricity (in Combined Heat and Power (CHP) units) [12]. Due to a wide variety of biomass types, the construction of a boiler must be adapted to the specified fuel [13]. There are two main technologies of boilers applied in small-scale biomass combustion systems: Grate firing (Figure 15.5) and fluidized bed (Figure 15.6) [14].

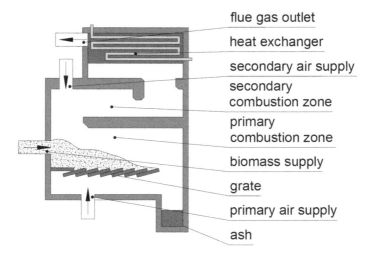

FIGURE 15.5 General construction of a biomass boiler with fixed bed.

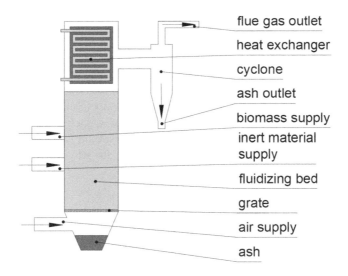

FIGURE 15.6 Construction of a grate biomass boiler with fluidized bed.

In biomass boilers that use grate firing, fuel is loaded and burned on a fixed (in wood-log boilers) or moving grate [13]. This element is made of stainless or refractory steel. The fuel can be stored in an external silo, which automatically supplies the combustion chamber. Primary air is supplied under the grate, whereas secondary air is supplied above the grate [15]. The flue gas flows around the heat exchanger and heats up the working medium. Due to its simplicity, grate firing is the most popular technology of biomass combustion [14]. The temperature of combustion varies from 850°C to 1400°C [15]. The remaining ash may be removed from the combustion chamber manually or automatically.

The second technology uses a fluidized bed, that is, mixed particles of biomass and heat transfer medium (typically sand), which are in constant suspension. The movement of the bed is provided by fans, which blow the air through a perforated bottom plate [15]. This method is characterized by lower operational temperatures (700°C–1000°C) than grate firing, so emissions of NO_x and SO_x are reduced [14]. Moreover, combustion in a fluidized bed requires a relatively low excess air coefficient [15].

Small heating biomass-fired heating units that are used in households can be classified according to several different criteria. The first one is the heat transfer method. In the simplest case (that is, fireplaces and stoves), heat is released through radiation and convection directly into the room in which the device is installed. The other way is heat transportation using an operating medium, which can be air (for example, in the so-called "hot air distribution" systems and air heat exchangers) or water (in the case of classic boilers equipped with a water jacket or in other devices with water heat exchangers). Another classification of heating units is based on the way of solid fuel feeding – it can be manual or automatic. The type of solid fuel (it can be wood logs, wood chips, pellets, straw bales, and others…) is the basis of the third classification method.

Another classification concerning fixed bed combustion takes into account the organization of the combustion process, that is, how the exhaust gas from combustion reaches the combustion chamber outlet and what is the relation between the combustion air stream (or streams) and the gaseous combustion product stream. Countercurrent and co-current combustion-based devices (Figure 15.7) have to be listed here. Furthermore, the two above-mentioned types of devices are characterized, consequently, by combustion in the whole volume of the fuel load (countercurrent) or just in some of its part (co-current).

When combustion air is fed from the bottom of the combustion chamber, it promotes a rapid increase in the region of chemical reactions, which, in practice, means that the whole fuel load burns (Figure 15.7a). Additionally, if fuel is loaded at the top of its stack (for example, fresh wood logs), hot exhaust gas from the bottom of the fuel stack flows through its top layers, increasing fuel heating and drying rate. Rapid ignition of a big stack area results in problems with complete combustion in the gaseous phase and, consequently, in high pollutant emissions. The reason is a relatively low temperature, which in the ignition phase can be even lower than 500°C, and oxygen deficiency. This method is relatively inefficient and typical of the simplest devices, such as classic fireplaces (cast or iron inserts). It should be noted here that the combustion air stream flow and fresh fuel loading occur in opposite directions – that is the simplest explanation for why this technology is called countercurrent combustion.

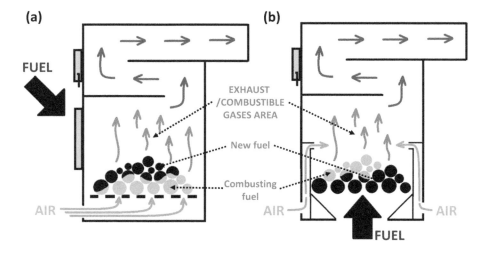

FIGURE 15.7 Two ways of combustion organization: countercurrent (a) and co-current (b) combustion in fixed bed heating unit.

Theoretical Background

Fuel feeding from the bottom of the fuel bed can be realized in the case of automatic devices, like pellet or wood chips boilers. Here, the air is provided above the solid fuel, and it can be said that both fuel and air streams have the same direction. Thus, it is co-current combustion. This method allows to keep efficient control over the combustion process because it takes place only in a limited fuel bed volume. It results in a low environmental impact on the device operation.

15.3.2 Technologies for the Reduction of Environmental Impact of Particulate Matter (PM) Emissions

Combustion of biomass in small-scale boilers is connected with larger emissions of particulate matter (PM) than in the utility of medium-scale facilities [16]. It is due to two reasons: Firstly, large devices have higher combustion efficiency. Secondly, they are usually equipped with advanced installations for flue gas cleaning [12]. PMs released into the atmosphere negatively affect air quality and human health. The problem of PM emissions from the domestic heating sector, especially during winter season, is urgent in many countries [17].

In the case of a complete conversion of solid organic matter, PM consists only of mineral substances. However, in practice such a situation is unlikely. Due to this fact, the emission of PM results in a loss of energy coming from the fuel carried by solid particles to the chimney. The inorganic part can lead to increased emissions of gaseous pollutants as well, especially when it consists of sodium, potassium, zinc, or sulfur. Besides primary methods that improve the burnout of the solid particles (its organic matter) moving upward the combustion chamber (appropriate time of particle stay, temperature, and turbulences in the reaction zone), there are secondary methods, presented in Figure 15.8.

Technologies of flue gas cleaning for small-scale biomass boilers are still under development. Certain innovative approaches to the matter that are currently being studied are not marked on the diagram in Figure 15.8. One such method assumes adding cheap additives, such as limestone, to the feedstock [16]. Another uses catalytic filters that are made of metal mesh, covered with catalysts

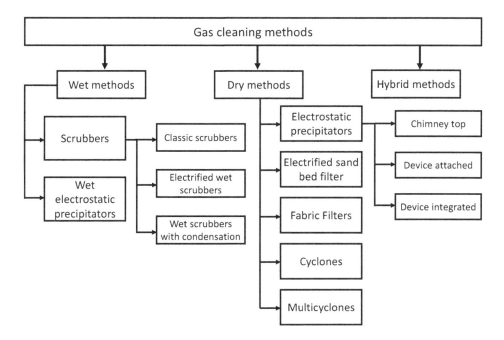

FIGURE 15.8 Methods of cleaning flue gas from biomass combustion. (Based on Refs. [17,18].)

(such as pricey platinum), and placed inside a home stack [16]. The next method is derived from the automobile industry and is based on a miniature heat exchanger, which consists of a pipe bundle. Hot flue gas flows through the pipes and gradually cools. Due to the temperature gradient, the particulates move to the internal surfaces of the tubes and thereby the flue gas is cleaned [16].

Below only two most common secondary technologies of gas cleaning are briefly described. However, a detailed description of these and other technologies can be found in the literature (for example, Refs. [17,18]).

Cyclones are the most commonly used technology, allowing separation of particulate matter. They use centrifugal force to capture particles from the swirling gas stream. Thanks to the above-mentioned force, particles move toward the external walls of the device and strike it, which results in a rapid velocity reduction. Then, due to gravity, the separated fraction sinks to the bottom of the container or is redirected again to the reactor. To capture fine particles, it is required to apply small-radius cyclones. The whole system can include a series of cyclones to improve gas cleaning efficiency. When one device consists of several cyclones (operating in parallel), it is called a multicyclone. Multicyclones can also operate parallel in frames of one system. Although cyclones are relatively cheap, small, and have a simple design, they are inefficient in the case of very fine particles (less than 20 μm). Moreover, there is a problem with erosive wearing and high flow resistance which complicate the operation of the biomass-combustion-based system. For example, the application of the multicyclone presented in Figure 15.9a, connected to the 100 kW straw-fired batch boiler (AGH University, Faculty of Energy and Fuels) makes it necessary to apply a fan generating additional underpressure on the outlet from the multicyclone to the duct connecting it with the chimney.

Electrostatic precipitators (ESPs) are commonly used to clean exhaust gas from energy boilers. This method of particle capturing is characterized by high efficiency, reaching even 80% [16]. The operation principle is based here on the electrostatic field effect on the particulate matter. Electrodes are powered by a constant current with voltage reaching even around 100 kV (depending on the boiler power and the exhaust gas stream parameters). The function of the two types of electrically insulated electrodes is the generation of a strongly heterogeneous electric field, interfering with the solid particle (electrically charged in the field) flow trajectory. Efficiency of the precipitator operation depends on the specific resistance of the particle, its size, chemical composition, and temperature, as well as the moisture content in the gas-carrying particulate matter.

Unlike cyclones, precipitators do not generate high flow resistance. Furthermore, contrary to the common intuitive feeling, the devices are not so energy-intensive. However, the investment cost is significantly higher and the operation requires fulfilling certain safety rules (due to the risk of combustible dust explosion). Additionally, precipitators are sensitive to changes in the physical and chemical properties of the cleaned gas and dust.

FIGURE 15.9 (a) Multicyclone connected to the flue of the straw-fired boiler, and (b) electrostatic precipitator dedicated to small biomass-fired devices, like stoves and fireplaces.

Theoretical Background   313

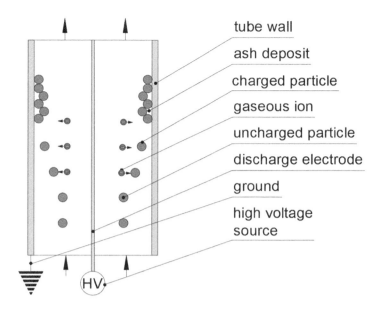

FIGURE 15.10 Construction of an electrostatic precipitator.

Although ESPs are usually used for big-scale applications, it is also possible to design small units, suitable to be mounted on small biomass-based devices. Figure 15.9b shows the original student's project (AGH University, Faculty of Energy and Fuels) dedicated to reducing particulate matter emission from the accumulation stove-fireplace.

General design of the ESPs for domestic-scale applications is shown in Figure 15.10. This one is made of a single-wire electrode placed inside the stainless steel in a stack. The wire is connected to the high-voltage supply (5–20 kV) and ionizes the flue gas as it flows through the stack. The charged particles are driven by the electric field and deposited on the internal surface of the grounded pipe, whereas the clean flue gas leaves the stack [17].

15.3.3 GASIFICATION-BASED TECHNOLOGIES

Differences between the thermochemical processes taking place in combustion and gasification are described in Section 15.3. Now, different ways of classifying gasification technologies will be discussed. One of the important issues influencing the gasification characteristics is what kind of gasifying agent is used. Several different gases and their mixtures can be applied, like air, oxygen, water steam, or even carbon dioxide. The first one from the list above is the most popular and the only one available in the case of the simplest small devices, due to its widespread availability and relatively simple construction of the feeding system. However, because of a high nitrogen content, efficiency of the air-based gasification is usually relatively low, compared with other agents. Of course, the type of agent applied in gasification decides the produced gas composition and heating value. Figure 15.11 presents an illustrative comparison of the main component contents in gas coming from the process run with different agents.

Detailed descriptions of the chemistry and kinetics characteristic of the process undergoing with different agents can be found in numerous literature sources, for example, Refs. [4,5,19–22].

Another, worth mentioning factor influencing the process characteristics is the biomass particle size. Of course, due to the thermal resistance issue, the smaller the characteristic dimension of a particle, the more efficient the process.

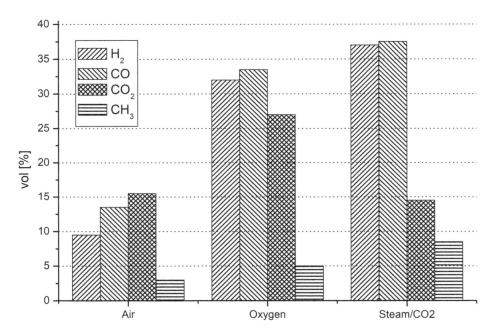

FIGURE 15.11 Syngas composition (average) with different gasifying agents. (Based on Ref. [19])

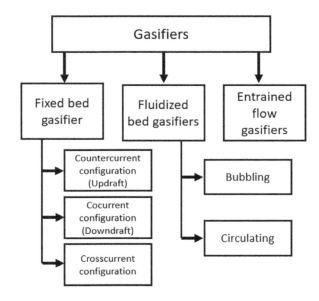

FIGURE 15.12 Simplified classification of small-scale biomass gasifiers. (Based on Refs. [11,19].)

Furthermore, depending on the process energy source, gasification technologies can be classified as autothermal (consumption of heat from the partial fuel combustion) or isothermal (heat supplied from the external source)

The classification based on the device design covers reactors that run the gasification process in various ways. Selected types of biomass gasifier constructions that have practical significance are shown in Figure 15.12. Although three main types of devices are given there, actually only two

Theoretical Background

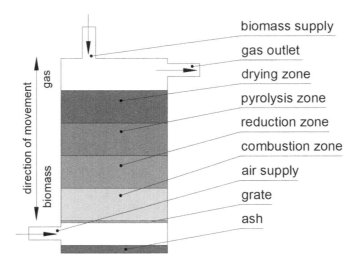

FIGURE 15.13 Construction of updraft (countercurrent) gasifier.

basic groups are usually listed: Fixed bed gasifiers (sometimes also "moving bed gasifiers" – in the case of moving grade application) and fluidized bed gasifiers.

Fixed bed gasifiers, analogously to direct combustion-based devices are divided according to the direction of fuel and agent feeding. In the case of the countercurrent reactor (Figure 15.13), fuel supplied at the top falls down through the gasifier body, while gasifying agent enters it from the bottom. Thus, flow directions are opposite to each other.

Downward-moving biomass is dried and pyrolyzed in the high-temperature area. In addition to the pyrolysis products and steam from the drying zone, CO and H_2 are produced in the reduction zone. Then pyrolysis products are lifted to the relatively low-temperature region (top part of the reactor), where intensive tar production takes place. Due to such organization of the process, the tar content in the produced gas is high, which determines the utilization in burners, (direct firing [23]) which does not require very clean fuel. Many different studies of syngas burners based on different technologies are currently conducted (for example, Refs. [24–27]). Due to the gas outlet located on top of the reactor, the temperature of the produced gas is quite low (around 200°C–400°C [11]), which results in relatively high thermal efficiency of the device. Some reaction heat is used to heat up and dry the fuel. Although, as already mentioned, this type of gasifier has a limited range of applications due to the low quality of the gas, these reactors are less sensitive to the fuel size and its moisture content than the downdraft ones. The residue remaining in the gasifier is ash, collected at the bottom of the reactor.

In the case of downdraft gasifiers (Figure 15.14) fuel is also fed at the top of the reactor, but, unlike in the previously described devices, the direction of the flow of the gasifying medium is co-current with the stream of the biomass particles [15].

The upper part of the biomass bed is dried and moisture is removed. Lower, in the pyrolysis zone, the volatile gases are released. The products of pyrolysis and the output from the drying zone are forced to pass through the oxidation zone, so the combustible substances are burned. This method results in much higher gas quality, thanks to the possibility of tar thermal decomposition (cracking) in the high-temperature (more than 1000°C) area (central and bottom parts) of the reactor. The gas outlet is located at the bottom of the gasifier, so its temperature is much higher than in the updraft devices (even around 1000°C [19]). In contrast to the previously discussed type of gasifiers, the downdraft ones are sensitive to fuel parameters. Particle stay in the reaction zone is relatively short (due to the agent flux direction), so an excessive size of the particle or its too high moisture content

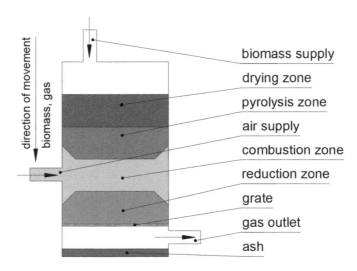

FIGURE 15.14 Construction of a downdraft (co-current) gasifier.

may affect the char conversion degree. The remaining ash is collected under the grate at the bottom of the gasifier. The obtained gas must be cooled and then it is well-suited for direct use in engines or turbines [23]. Scaling of the downdraft reactor makes some problems, due to the inhomogeneity of the conditions of chemical processes in the case of bigger units. Many interesting examples of small-scale downdraft biomass gasifiers are discussed in the literature, for example, Ref. [28].

Fluidized bed gasifiers, as the name goes, are based on the fluidization phenomenon which guarantees a good agent-fuel mixing rate and in consequence a high fuel conversion degree (more than 90%). Thanks to the application of the inert bed material (sand is commonly used) characterized by high heat capacity, it is quite easy to maintain homogenous thermal conditions in the fluidized bed zone. The temperature reaches there 700°C–1000°C [4, 11]. It is crucial to avoid a too high-temperature level due to the risk of melting and agglomeration of the alkaline components of the fuel particles [19], as well as melting of the bed inert particles.

Usually, in this reactor, the biomass is supplied on the side of the device (Figure 15.15) and mixed with the inert material. From the bottom, the gasifying medium is introduced to the reactor under high pressure. This causes a constant movement of the fluidizing bed and improves heat transfer, resulting in rapid pyrolysis. The produced gases and solid particles are directed to the cyclone, where ash and gas are separated.

One important advantage of the discussed gasifier type is that it is possible to apply a fuel mass stream (per installed gasifier body area) much bigger than in fixed or moving bed reactors. Of course, after all, it always depends on the specific subtype of the fluidized bed applied. Two device designs have to be listed here: Bubbling (BFBG) and circulating (CFBG) fluidized bed gasifiers.

Although the BFBGs have a lower fluidization velocity and are characterized by relatively low efficiency, their significant advantage is the possibility of applying a wide range of biomass types [11]. However, nowadays it is still a tall order to conserve acceptably low tar content in the produced gas.

CFBGs, compared to BFBGs, are characterized by a much higher fluidization velocity, which results in the occurrence of a turbulent fluid flow regime. Although it certainly improves the reaction rates, it creates problems with a high amount of solid particles lifted to the freeboard of the gasifier. Due to this fact, their recirculation has to be applied, which complicates the overall process. However, thanks to the recirculation, a higher gasification degree and a significant reduction in the tar content are observed. Furthermore, it can be lowered even more efficiently by using appropriate catalysts [29–31].

Theoretical Background

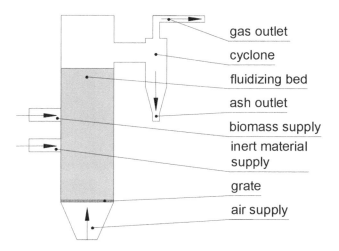

FIGURE 15.15 Construction of a fluidized bed gasifier.

Entrained flow gasifiers are not commonly used for biomass thermochemical treatment, but rather as high-temperature and pressure reactors for coal gasification [11]. Nevertheless, this kind of reactors is included as an additional class in Figure 15.12, due to many studies devoted to biomass gasification in such devices in the last years. Results of interesting analyses can be found, for example, in Refs. [32–34].

15.3.4 Heat Accumulation Systems for RES-Based Technologies

The consistency between energy supply and demand is difficult to achieve, especially in the case of RES-based technologies [35]. To manage this challenge, efficient and reliable storage systems which absorb the excess energy, store it, and deliver upon demand must be implemented. Heat and electricity storage are considered crucial issue in balancing the energy system. Moreover, it allows to integrate numerous energy sources, such as renewables, conventionals, and waste heat from industrial processes [36]. Last but not least, energy storage has a positive impact on the consumption of the primary energy source. Thermal energy storage (TES) methods that may be coupled with RES-based technologies include three groups of technologies: Sensible, latent, and thermochemical.

Sensible TES technologies are the most commonly deployed around the world due to their high development stage, simplicity of operation, and a relatively low price. Their principle of operation is based on heating/cooling of the storage medium, without phase change. The media may be in solid (ceramic, concrete, rocks) or liquid state (water, molten salts) [36]. In each case, the storage-material bed must be insulated to prevent energy losses. Typically, sensible TES has an efficiency between 50% and 98%, capacities between 10 and 50 kWh/t and they work in a wide range of temperatures (from −160°C to 1000°C) [37]. In small-scale energy facilities, water tanks are usually installed [16].

However, there are also other ways to accumulate and use heat, which have recently become more popular in the case of the smallest heating units, like stoves and fireplaces. Figure 15.16 presents a heat accumulation system based on a dedicated ceramic material with high heat capacity ($600 \sim 1000$ [J/(kg K)]), high density ($2500 \sim 3500$ [kg/m^3]), and, as for this type of ceramics, high heat conductivity ($1.5 \sim 3$ [W/(m K)]).

The ceramic heat exchanger presented in Figure 15.16 is constructed from prefabricated modules with an internal channel, which allows to create a complex exhaust gas way from the flue to the

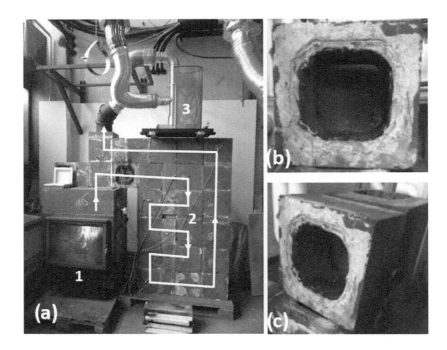

FIGURE 15.16 (a) Didactic experimental station (AGH University) with a heat source (1) and an accumulation heat exchanger (2) a small precipitator, where the line with arrows represents the way of exhaust gas from the combustion chamber to the flue; (b) and (c) ceramic prefabricates.

chimney. Hot gas heat is recovered by the ceramics and slowly released into the heated space. The advantage of the system is that the heat release is slow enough to heat the room for a long time (even 8–12 hours) after the biomass combustion in the fireplace or stove. Of course, the exchanger size has to be adjusted to the nominal power of the heating unit. The second group of TES uses materials that release/absorb significant amounts of energy during their melting and solidification, such as paraffin waxes and inorganic salt hydrates [35]. They usually operate in temperatures between 0°C and 120°C. The amount of latent heat in phase-changing materials is much greater than their sensible heat, so latent TES are characterized by higher mass capacity than sensible TES. The phase-changing process occurs at a constant temperature, so this kind of energy storage is suitable for sensitive applications [36]. They are also used for domestic heating and air-conditioning.

Thermochemical TES exploits sorption and non-sorption processes, which provide the highest energy density among all the mentioned technologies. Sorption accumulators use the energy of sorbent-sorbate binding or hydration–dehydration transformation, whereas the non-sorption storages are based on a reversible exothermic reaction of two chemical substances (chemical looping). The main advantage of thermochemical TES is the fact that they conserve energy without any heat losses as long as desired [36].

15.4 FUNDAMENTALS OF ENERGY CONVERSION AND BALANCE

15.4.1 Fundamentals of Combustion, Pyrolysis, and Gasification Processes

The process of biomass combustion covers a series of subsequent complex stages. Characteristics of the process depend both on the fuel properties and combustion technology. When it comes to a general definition of combustion itself, it is described as a rapid reaction of a fuel and an oxidizer (the most commonly oxygen from the air), resulting in the creation of combustion products, heat release, and the light effect.

Theoretical Background

General process phases that have to be considered in the combustion of both biomass and other solid fuels are

- inert heating of the solid fuel,
- drying,
- thermal decomposition of the fuel components (pyrolysis),
- combustion of the gas product of thermal decomposition,
- combustion of coke in the solid phase,
- further inert heat exchange between the remaining mineral substance and the environment.

Heating and drying aren't associated with any chemical processes. In the case of biomass, the moisture content is a parameter with a very changeable value, depending on the fuel type and conditioning processes of the raw material. Freshly cut wood contains from around 30 to even 60% mass of water, while some unconventional fuel materials, like sewage sludge, can even exceed 90% [38,39]. Evaporation of each kilogram of water from a fuel consumes at least 2260 kJ of energy, which is very difficult to regain in practice [7].

It can, therefore, be stated that high moisture content in biomass constitutes a serious problem. Thus, it is common to apply a pre-drying process before combustion, to remove the maximum amount of water (taking into account the economic issue, as well as the energy consumption) Biomass boilers usually burn fuel with moisture content in the range of 10%–20%. It evaporates when the temperature in the combustion chamber exceeds 100°C. However, the evaporation of other substances is also possible – at a temperature higher than 200°C. Certain low-molecular-weight organic extracts can be released this way as well [7].

Release and combustion of volatiles begin only after moisture evaporation and after reaching the appropriate temperature by fuel particles. The above-mentioned thermal decomposition of fuel, resulting in the release of some gases is generally called devolatilization. However, this term is used rather when there is no accurate information about the process conditions. Thus, it is correct to use it when it comes to the description of biomass combustion in a certain chamber, where conditions are actually mixed (mostly oxidizing, but locally reducing). In case of a reduced or inert atmosphere (no oxidizer access), this process is referred to as pyrolysis.

Biomass pyrolysis is defined as incomplete thermal decomposition resulting in the creation of coke, tar, condensates (liquid substances), and gas. The latter one contains hydrogen, water, carbon monoxide, carbon dioxide, methane, and higher hydrocarbons. When it comes to tar, it has to be said that it contains a large amount of polycyclic aromatic hydrocarbons. Biomass devolatilization starts at a temperature slightly above 200°C (for wood, it is around 220°C). Individual components are characterized by different decomposition temperatures [40,41]. Three main so-called pseudo-components which play a crucial role in biomass thermal decomposition have to be listed

- hemicellulose (decomposition at 220°C–320°C),
- cellulose (320°C–370°C),
- lignin (320°C–500°C).

Hemicellulose and cellulose belong to the group of homocelluloses. They release more volatiles compared with lignin, which gives more solid products (mainly coke). Conventional pyrolysis results in a shift of the process products toward carbon on the C-H-O diagram (Figure 15.17). The consequence is formation of the above-mentioned coke. In the case of so-called fast pyrolysis, products are much closer to the C-H axis, opposite to the oxygen corner. Oxygen concentration is significantly reduced then and the formation of much more liquid hydrocarbons is likely [42,43].

Because the solid residue coming from devolatilization is characterized by certain content of fixed carbon, combustion concerns also this phase. However, it is possible only in the presence of an oxidizer, so it can be said that coke combustion cannot occur from a small biomass particle as long as intensive devolatilization takes place (then combustion goes in the gas phase).

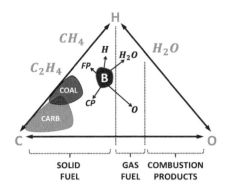

FIGURE 15.17 C-H-O diagram of biomass thermal decomposition (Based on Ref. [7]): "B" – Biomass fuel, "FP" – Fast pyrolysis, and "CF" – Conventional pyrolysis.

When the amount of oxidizer in the process lowers below stoichiometric conditions, characteristics of the chemistry move from direct combustion to the so-called gasification. In contrast to pyrolysis, this process occurs with the contribution of an additional medium-gasifying agent which, among others (according to the previous point), can be oxygen from the air. In the case of gasification, coke and part of the gaseous substrates take part in a set of complex reduction-oxidation reactions which, in contrast to pyrolysis, lead to obtaining only simple gaseous compounds (see Figure 15.4).

Biomass pyrolysis and gasification stages influence the overall combustion process more significantly, compared to coal combustion. It is due to a relatively high volatile content in biomass. Relative significance of gasification varies depending on the combustion technology applied, biomass properties, and process conditions. Intensification of gasification as a stage of biomass combustion in boilers requires strict control of the air mass flow and appropriate localization of its inlets.

A well-known and commonly used solution in modern biomass boilers (like batch boilers or wood-fired gasifying boilers) is partial separation of gasification and post-combustion of its gaseous products through the application of two combustion chambers (sometimes called primary and secondary chamber). Of course, in such devices, it is necessary to implement air staging with at least two steps: The first one for partial combustion of the solid fuel and gasification, and the second one for post-combustion of its products. When it comes to medium and big-scale biomass-fired energy units that work in permanent mode, for example, with a moving grate, different combustion phases can occur simultaneously in different parts of the grate, due to its size and inhomogeneity of the conditions (that is, fields of temperature and oxidizer concentration). It is different in the case of small-scale fixed bed devices, where the division of the combustion phases is ambiguous.

15.4.2 Heat Balance of the Biomass-Fired Heating Units

As in the case of any energy technology, the fundamental criterion for the assessment of biomass-fired energy devices is their total operational efficiency. Depending on the degree of the data detail and the type of technology, direct and indirect methods of efficiency calculation or estimation can be used. Because the first one requires the knowledge of the temporary fuel mass flow rate, it becomes quite problematic when fixed bed devices are considered. In such a situation, determination of the efficiency is possible based on the calculation of the losses in the system and applying the equation as follows:

$$\eta = 100\% - \sum_{i}^{n} L_i, \tag{15.1}$$

where L_i is i-loss of energy considered in the system.

The types and the number of different loss types that have to be taken into account are strictly connected with the type of studied device and assumed thermodynamic boundaries of the system. From the point of view of the biomass-fired units, usually three most important loss sources have to be determined when the thermodynamic system covers only a heating unit (without piping and buffer tank which generate additional transportation and storage losses)

- stack loss L_{SL},
- loss of incomplete combustion in the gas phase L_{IG},
- loss of incomplete combustion in the solid phase that can occur both in slag (L_{IS}) and fly ash (L_{IF}).

Including the above-mentioned elements of the heat balance, the total efficiency of a device can be calculated using the formula

$$\eta = 100\% - (L_{SL} + L_{IG} + L_{IS} + L_{IF})[\%], \qquad (15.2)$$

where subsequent balance elements are expressed in [%].

Stack loss comes directly from the heat lifted with the exhaust gas to the chimney. This heat cannot be used in the system anymore. A popular and useful way to determine this loss in the case of biomass-fired devices is the Siegert method, which is based on semi-empirical equation 15.3 (generalized form).

$$L_{SL} = \sigma \cdot \frac{(T_1 - T_2)}{CO_2}, [\%], \qquad (15.3)$$

where σ – Siegert constant, determined for certain fuel based on its moisture content and average CO_2 emission during combustion [1/K], T_1, T_2 – respectively, the temperature of the exhaust gas (after the last heat exchanger) and temperature of the supplied combustion air [K], CO_2 – molar concentration of the component in the exhaust gas [%].

The loss of incomplete combustion in the gas phase is the consequence of lifting the chemical energy of the combustible gaseous compounds (coming from biomass thermochemical conversion) to the chimney by the exhaust gas. In frames of the methodology of practical determination of this loss, it is often assumed that estimation can be based on the measurement of carbon monoxide emission. This compound prevails in gasification products, and thus, it is their good representative. Loss of incomplete combustion in the gas phase can be calculated based on the following equation:

$$L_{IG} = \frac{V_{EG} \cdot Q_{CO} \cdot [CO]}{Q_B}, [\%], \qquad (15.4)$$

where V_{EG} – volume flow rate of the exhaust gas $\left[\frac{m^3}{s}\right]$, Q_{CO} – carbon monoxide higher heating value $\left[\frac{MJ}{kg}\right]$, $[CO]$ – carbon monoxide molar concentration in the exhaust gas [%], Q_B – biomass calorific value $\left[\frac{MJ}{kg}\right]$.

An unburned organic substance in the solid fraction of the combustion residues also contains unused chemical energy. Depending on whether the unburned solid fuel remains in the combustion chamber or leaves it with the exhaust gas, the effect of the efficiency reduction is called loss of incomplete combustion in the slag or the fly ash, respectively. The loss in slag can be calculated as below:

$$L_{IS} = \frac{\frac{x*A*Q_C}{Q_B}*C_S}{100-C_S}; [\%],\qquad(15.5)$$

where A – mass fraction of the ash in fuel [%], Q_C – elemental carbon higher heating value [kJ/kg], Q_B – biomass calorific value [kJ/kg], C_S – combustible part mass fraction in the slag [%], x – mass fraction of the slag in the total mass of the solid residues from combustion (slag and fly ash) [-].

Analogously, loss of incomplete combustion in fly ash can be calculated as follows:

$$L_{IS} = \frac{\frac{y*A*Q_C}{Q_B}*C_F}{100-C_S}; [\%],\qquad(15.6)$$

where $y = 1-x$, A, Q_C, and Q_B – as in equation (15.5), C_F – combustible part mass fraction in the fly ash [%].

Based on the above-discussed methodology of calculation of the selected losses associated with biomass-fired energy devices, it can be concluded that to find each loss it is required to determine the biomass gross calorific value (GCV).

Of course, it is the best situation when there is a possibility to determine GCV in the course of a proximate analysis (using a calorimeter). Alternatively, when a high accuracy level of the results is not required, it is possible to apply a number of empirical equations that have been proposed in the literature (for example, Refs. [44–46]). Of course, it is still necessary to have data concerning the content of fixed carbon or elemental composition of the fuel (for example, from ultimate analysis). An exemplary approach is presented below [47]:

$$Q_W = 339 \cdot C + 1214.2\left(H_2 - \frac{O}{8}\right) + 104.7 \cdot S - 25.1 \cdot W, \left[\frac{MJ}{kg}\right],\qquad(15.7)$$

where C, H_2, O, S – mass fraction of the certain chemical elements in the fuel [%], W – moisture content in the fuel [%].

Excess air coefficient allows to assess conditions in a reactor during biomass combustion (and gasification). The coefficient informs about the ratio between the real amount of air in the reactor and the amount required to keep the combustion process stoichiometric. In practice, due to the boiler type or technical limitations of the experimental station (range of data that can be obtained with existing equipment), it is common to use indirect methods to calculate temporary excess air coefficient, based on equation 15.8 [47].

$$\lambda = \frac{CO_{2,MAX}}{CO_2},\qquad(15.8a)$$

$$CO_{2,MAX} = \frac{21C}{C+2.37\left(H\frac{O-S}{8}\right)}, [\%],\qquad(15.8b)$$

where $CO_{2,\,MAX}$ – maximum theoretical molar concentration of CO_2 in exhaust gas [%], CO_2 – temporary CO_2 molar concentration in exhaust gas [%],

C, H, O, S – mass fraction of the chemical elements in the fuel [%].

REFERENCES

1. B. Sørensen, *Renewable Energy Its Physics, Engineering, Use, Environmental Impacts, Economy and Planning Aspects*, Academic Press, San Diego, CA, 2000.
2. K. R. Smith, A. Sagar, Making the clean available: escaping India's Chulha trap, *Energy Policy*, 75 (2014), 410–414.
3. M. Khandelwal et al., Why have improved cook-stove initiatives in India failed? *World Development*, 92 (2017), 13–27.
4. *Biomass in Small-Scale Energy Applications. Theory and Practice*, Edited by: M. Szubel, M. Filipowicz, CRC Press Taylor & Francis Group, Boca Raton, FL, 2020.
5. L. Rosendahl, *Biomass Combustion Science, Technology and Engineering*, Woodhead Publishing, Cambridge, UK, 2013.
6. Statistics Data of the International Renewable Energy Agency, www.irena.org (last access: 07.05.2022).
7. P. Basu, *Biomass Gasification and Pyrolysis Practical Design*, Academic Press, Oxford, 2010.
8. S. van Loo, J. Coppejan, *The Handbook of Biomass Combustion and Co-Firing*, EARTHSCAN, London, 2008.
9. T. B. Johansson, H. Kelly, A. K. N. Reddy, R. H. Williams, *Renewable Energy Sources for Fuels and Electricity*, Island Press, Washington, DC, 1993.
10. B. Dubis et al., Biomass yield and energy balance of fodder galega in different production technologies: An 11-year field experiment in a large-area farm in Poland, *Renewable Energy*, 154 (2020), 813–825.
11. Y. A. Situmorang et al., Small-scale biomass gasification systems for power generation (<200 kW class): A review, *Renewable and Sustainable Energy Reviews*, 117 (2020), 1–14 (109486).
12. IRENA, *Rise of Renewables in Cities: Energy Solutions for the Urban Future*, International Renewable Energy Agency, Abu Dhabi (2020), ISBN 978-92-9260-271-0.
13. J. L. Míguez et al., Review of technology in small-scale biomass combustion systems in the European market, *Renewable and Sustainable Energy Reviews*, 16 (2012), 3867–3875.
14. J. P. Wolf, Dong, Biomass Combustion for Power Generation: An Introduction, *Biomass Combustion Science, Technology and Engineering*, Woodhead Publishing Series in Energy, Cambridge, 2013, pp. 3–8.
15. P. Uark, H. Knoef, H. Stassen, Energy from Biomass. A review of Combustion and Gasification Technologies, World Bank Technical Paper, 422 (1999).
16. M. T. Lim et al., Technologies for measurement and mitigation of particulate emissions from domestic combustion of biomass: A review, *Renewable and Sustainable Energy Reviews*, 49 (2015), 574–584.
17. A. Jaworek et al., Particulate matter emission control from small residential boilers after biomass combustion. A review, *Renewable and Sustainable Energy Reviews*, 137 (2021), 1–16 (110446).
18. R. Singh, A. Shukla, A review on methods of flue gas cleaning from combustion of biomass, *Renewable and Sustainable Energy Reviews*, 29 (2014), 854–864.
19. E. Bocci et al., State of art of small scale biomass gasification power systems: a review of the different typologies, *Energy Procedia*, 45 (2014), 247–256.
20. T. Dahou et al., Role of inorganics on the biomass char gasification reactivity: A review involving reaction mechanisms and kinetics models, *Renewable and Sustainable Energy Reviews*, 135 (2021), 1–16 (110136).
21. Y. Xin et al., Kinetic characterizations of biomass char CO_2-gasification reaction within granulated blast furnace slag, *International Journal of Hydrogen energy*, 42 (2017), 20520–20528.
22. H. Qiang et al., Thermal behavior and reaction kinetics analysis of pyrolysis and subsequent in-situ gasification of torrefied biomass pellets, *Energy Conversion and Management*, 161 (2018), 205–214.
23. S. K. Sansaniwala, M. A. Rosenb, S. K. Tyagi, Global challenges in the sustainable development of biomass gasification: An overview, *Renewable and Sustainable Energy Reviews*, 80 (2017), 23–43.
24. G. Scribano, C. Xinwei, T. Manh-Vu, Numerical simulation of the effects of hydrogen and carbon monoxide ratios on the combustion and emissions for syngas fuels in a radiant burner, *Energy*, 214 (2021), 118910.
25. S. M. Alia, S. Varunkumar, Effect of burner diameter and diluents on the extinction strain rate of syngas-air non-premixed Tsuji-type flames, *International Journal of Hydrogen Energy*, 45 (2020), 9113–9127.
26. S. Klayborworn, W. Pakdee, Effects of porous insertion in a round-jet burner on flame characteristics of turbulent non-premixed syngas combustion, *Case Studies in Thermal Engineering*, 14 (2019), 100451.
27. K. K. J. Ranga Dinesh et al., Burning syngas in a high swirl burner: Effects of fuel composition, *International Journal of Hydrogen Energy*, 38 (2013), 9028–9042.

28. A. A. P. Susastriawan, H. Saptoadi, Purnomo, Small-scale downdraft gasifiers for biomass gasification: A review, *Renewable and Sustainable Energy Reviews*, 76 (2017), 989–1003.
29. A. Eschenbacher et al., Catalytic upgrading of tars generated in a 100 kWth low temperature circulating fluidized bed gasifier for production of liquid bio-fuels in a polygeneration scheme, *Energy Conversion and Management*, 207 (2020), 1–14 (112538).
30. M. Junguang et al., Highly abrasion resistant thermally fused olivine as in-situ catalysts for tar reduction in a circulating fluidized bed biomass gasifier, *Bioresource Technology*, 268 (2018), 212–220.
31. M. Xiangmei et al., Biomass gasification in a 100 kWth steam-oxygen blown circulating fluidized bed gasifier: Effects of operational conditions on product gas distribution and tar formation, *Biomass and Bioenergy*, 35 (2011), 2910–2924.
32. K. Xiaoke et al., Effects of operating conditions and reactor structure on biomass entrained-flow gasification, *Renewable Energy*, 139 (2019), 781–795.
33. L. Yan et al. Evolution of PM2.5 from biomass high-temperature pyrolysis in an entrained flow reactor, *Journal of the Energy Institute*, 92 (2019), 1548–1556.
34. Y. Ögren, Development of a vision-based soft sensor for estimating equivalence ratio and major species concentration in entrained flow biomass gasification reactors, *Applied Energy*, 226 (2018), 450–460.
35. R. Rybár, M. Beer, M. Kaľavský, Development of heat accumulation unit based on heterogeneous structure of MF/PCM for cogeneration units, *Journal of Energy Storage*, 21 (2019), 72–77.
36. IRENA (2020), Innovation Outlook: Thermal Energy Storage, International Renewable Energy Agency, available online: https://www.irena.org/publications/2020/Nov/Innovation-outlook-Thermal-energy-storage (last access: 14.04.2022).
37. European Association for Storage of Energy, available online: https://ease-storage.eu/publication/energy-storage-targets-2030-and-2050/ (last access: 11.06.2023).
38. N. Striūgas et al., Estimating the fuel moisture content to control the reciprocating grate furnace firing wet woody biomass, *Energy Conversion and Management*, 149 (2017), 937–949.
39. I. Lopes Motta et al., Biomass gasification in fluidized beds: A review of biomass moisture content and operating pressure effects, *Renewable and Sustainable Energy Reviews*, 94 (2018), 998–1023.
40. Chen et al., Gasification kinetic analysis of the three pseudocomponents of biomass-cellulose, semicellulose and lignin, *Bioresource Technology*, 153 (2014), 223–229.
41. X. Wang et al., High-temperature pyrolysis of biomass pellets: The effect of ash melting on the structure of the char residue, *Fuel*, 285 (2021), 1–11.
42. Y. Wang et al., Characteristics of the catalytic fast pyrolysis of vegetable oil soapstock for hydrocarbon-rich fuel, *Energy Conversion and Management*, 213 (2020), 1–11.
43. Nishu, Catalytic pyrolysis of microcrystalline cellulose extracted from rice straw for high yield of hydrocarbon over alkali modified ZSM-5, *Fuel*, 285 (2021), 1–13.
44. R. Galhano dos Santos, J. C. Bordado, M. M. Mateus, Estimation of HHV of lignocellulosic biomass towards hierarchical cluster analysis by Euclidean's distance method, *Fuel*, 221 (2018), 72–77.
45. A. Dashti et al., Estimation of biomass higher heating value (HHV) based on the proximate analysis: Smart modeling and correlation, *Fuel*, 257 (2019), 1–11 (115931).
46. S. Hosseinpour et al., Estimation of biomass higher heating value (HHV) based on the proximate analysis by using iterative neural network-adapted partial least squares (INNPLS), *Energy*, 138 (2017), 473–479.
47. M. Szubel, W. Adamczyk, G. Basista, M. Filipowicz, Homogenous and heterogeneous combustion in the secondary chamber of a straw-fired batch boiler, *EPJ Web of Conferences*, 143 (2017), 1–11 (02125).

16 Tutorial 7 – Syngas Burner

16.1 EXERCISE SCOPE

This exercise aims to design the model of a simplified syngas burner, as in Figure 16.1. Fuel *(1)* and air *(2)* are delivered to the burner, where combustion occurs. During this process, heat is generated and hot flue gases *(3)* are removed. This model allows us to observe the temperature distribution in the burner domain and the concentration of reagents.

FIGURE 16.1 Cross-section of the burner geometry considered in the exercise with marked locations of (1) fuel inlet, (2) air inlet, and (3) flue gas outlet.

Try to recall from the theoretical part of this chapter (Section 15.3) the details of gasification and pyrolysis technologies and the fundamentals of energy conversion and balance (Section 15.4) for these processes. Pay attention to the possible applications of generated biogas. This exercise allows us to analyze its combustion in a burner under various initial conditions.

This exercise helps to get familiar with

- general rules of geometry preparation for axisymmetric cases in *ANSYS Design Modeler*,
- the *Mesh Metrics* tool in *Ansys Meshing* which allows to locate of mesh elements characterized by a given quality,
- the *Eddy Dissipation* model of combustion and parameters that describe it in *ANSYS Fluent*,
- various tools to supervise the concentration of the reagent in the burner domain in *ANSYS Results*.

Pay special attention to the impact of constants A and B (describing the Eddy Dissipation model) on the temperature changes and oxygen mass fraction in the burner.

All of the images included in this tutorial use courtesy of ANSYS, Inc.

16.2 PREPROCESSING – GEOMETRY

Drag the *Geometry* module *(1)* to the ANSYS Workbench *Project Schematic* as in Figure 16.2. To open the geometry editor, double-click the left mouse button *(LMB)* on A2 cell *(2)*.

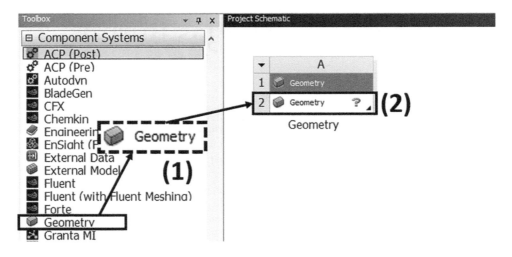

FIGURE 16.2

The *Sketch (1)* tab will be open (Figure 16.3). Firstly, click LMB on the *Select New Sketch Plane (2)* icon. It is essential to note that for axisymmetric geometries, like a burner, the body should be located in the positive quadrant of the *XY* plane. Moreover, the rotation axis must be set on the *X*-axis. To select the *XY* plane, it is necessary to click on the *Z*-axis *(3)* of the global coordinate system. To make the sketching process easier, set the *XY* grid head-on *(4)* and finally choose the *Line (5)* drawing tool.

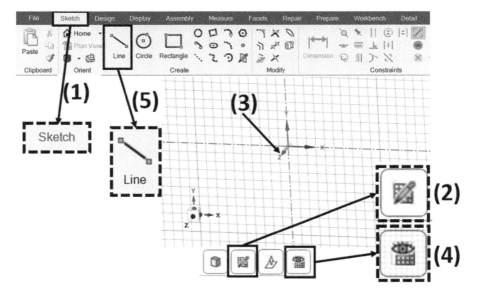

FIGURE 16.3

Using the *Line* tool, draw the burner shape as shown in Figure 16.4 *(1)*. You can set the proper dimensions during the sketching process and check them later, using the *Dimension (2)* tool. Remember that the drawing should be located in the positive quadrant of the *XY* plane and the axis of symmetry must be colinear with the *X*-axis (in this case). Finish your work by choosing the *End*

Tutorial 7 – Syngas Burner

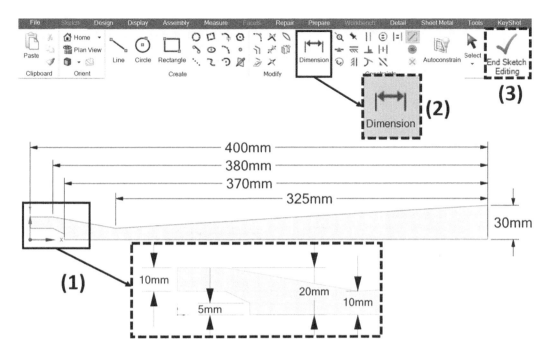

FIGURE 16.4

Sketch Editing icon *(3)*. You will be automatically switched to the *Design* tab and the sketch will be converted to a surface. Close the *SpaceClaim* module.

16.3   PREPROCESSING – MESHING

Select the *Mesh* module *(1)* from the *Component System* ribbon and drag it to the ANSYS Workbench *Project Schematic (2)* as in Figure 16.5. Create an appropriate connection between the geometry and mesh modules *(3)*. Double-click the left mouse button *(LMB)* on *B3* cell *(4)* and open *Mesh*.

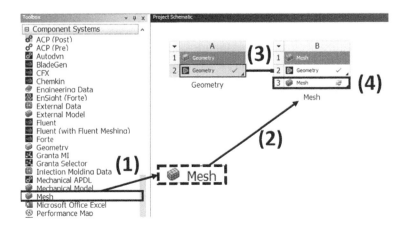

FIGURE 16.5

Find the *Mesh* icon in the project tree *(1)*. In the *Details of Mesh* window, change appropriate basic settings that are required for CFD simulations (Figure 16.6): *Physics Preference – CFD (2)* and *Solver preference – Fluent (3)*. To control the maximum size of mesh elements, change the *Element Size* value from default to *0.5 mm (4)*. Be careful and check your default unit of length. In the *Mesh Defeaturing* options, take into account that the geometric model does not have rounds (*Capture Curvature: No*), but includes constrictions (*Capture proximity: Yes*) that require additional mesh refinement. Generate the mesh by clicking 1× RMB on *Mesh (6)* and selecting the option *Generate Mesh*. It is not required to change the default proximity settings.

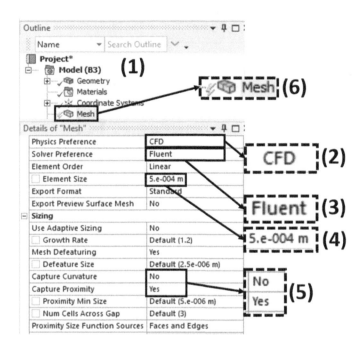

FIGURE 16.6

Find the *Quality* option in *Details of Mesh* (Figure 16.7) and set the *Mesh Metric* parameter as *Skewness (1)*, which describes the difference between the shape of the cell and the shape of an equilateral cell with the same area. The lower skewness, the better, because highly skewed cells decrease accuracy and destabilize the solution. Generally, skewness should not be higher than 0.95. Rate the mesh quality based on the minimum, maximum, and average values of the analyzed parameter *(2)*. In the *Statistics* tab, check the number of mesh *Nodes* and *Elements (3)*. Take a closer look at the generated mesh *(4)*.

Tutorial 7 – Syngas Burner

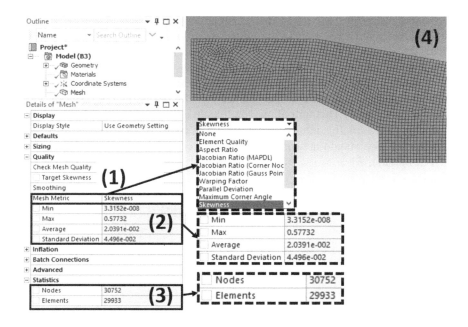

FIGURE 16.7

In the ***Mesh Metrics*** window (Figure 16.8), check the shape of the mesh elements *(1)* and their participation in the total number of mesh elements and their quality. To locate the elements characterized by the worst quality, change the options of the displayed chart. Click ***Controls (2)*** in the ***Mesh Metrics*** window and, in the popup window, change the minimum value of the ***X-Axis*** to 0.4, to display only the worst-quality elements. Also, change the maximum value of the ***Y-Axis*** to 20 (the number of the worst elements is relatively low). There is no ***Apply*** button, just click ***X*** in the corner of the popup window *(3)*. Click on the bars displayed on the chart *(4)* and select them all. The worst-quality elements are now visible in the workspace *(5)*. To check another quality parameter, just select its name in the ***Mesh Metric*** row in the ***Quality*** tab. You can try ***Orthogonal Quality***, which (in simplification) describes the equilaterality of a cell. The value equal to 0 is the worst and means skewed cells (like an obtuse triangle), whereas value 1 is the best and means fully equilateral cells (like a square). It is recommended that the Orthogonal Quality should not be lower than 0.05.

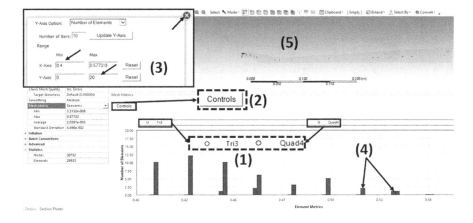

FIGURE 16.8

Use the ***Named Selection*** function to define the names of different objects (like edges or surfaces in the 2D case) as in Figure 16.9. In the ***Edge Selection*** mode *(1)*, click on the proper edge *(2)*. Then, click 1× RMB and, from the dropdown list, choose the ***Create Named Selection (3)*** option (instead, you can just press the *N* button). In a new window, insert the name of the new ***Named Selection (4)***: *inlet_fuel* and confirm by clicking ***OK***. Repeat the same step and name the other input and output: *inlet_air* and *outlet*. The geometry of the burner is axisymmetric, so name the appropriate edge as the ***axis***. This will allow you to define the condition of the axial symmetry in the solver preprocessor. Finally, define all the remaining edges (without assigned names) as ***wall***. All the named selections should appear in the project tree, below the ***Named Selections*** tab *(5)*. Close the ***Mesh*** module.

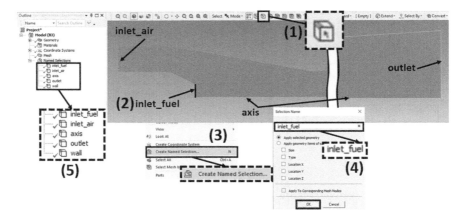

FIGURE 16.9

16.4 PREPROCESSING – SOLVER SETTINGS

Before moving to the next step, save the project in the ANSYS Workbench window. Then, drag the ***Fluent (1)*** module to the ***Project Schematic*** and set the appropriate connection *(2)* between the ***Mesh*** module (B3 cell) and ***Fluent Setup*** (C2) as in Figure 16.10. Click *1× RMB* on *B3* cell *(3)* and choose the ***Update*** option *(4)*, to export the mesh data to the solver. Start the solver by clicking *2× LMB* on the *C2* cell and in a new window, run the solver in the ***parallel*** computing mode according to the available computing power.

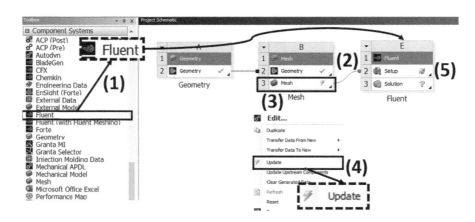

FIGURE 16.10

Tutorial 7 – Syngas Burner

First of all, pay attention to the warning displayed at the bottom of the *console (1)*, as in Figure 16.11, the default task type is not appropriate for an axisymmetric problem. To change it, find the *General* option *(2)* and change the *2D Space* definition from *Planar* to *Axisymmetric (3)*. To display in the *Console (1)* window the most important properties of the imported geometry and mesh, click *1× LMB* the *Check (4)* button. Then, set the *Solver Type* as *Pressure-Based (5)* to analyze fluid flows with a velocity lower than 200 m/s. To provide steady-state simulation, choose the *Time* option as *Steady (6)*. As the burner works in a vertical configuration and the geometry edge defined as the axis lies on the *X*-axis, set the *Gravity (7)* vector along the *X*-axis to *−9.81 (8)*.

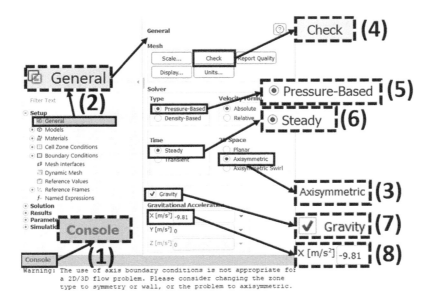

FIGURE 16.11

Move to the *Models* option *(1)* to select appropriate mathematical models of the physical phenomenon occurring in the analyzed case (Figure 16.12). Firstly, activate the *Energy Equation (2)* for heat transfer. Do this by ticking the available option in the popup window *(3)* and confirm with *OK*. Secondly, choose the *Viscous (4)* option to choose the turbulence model for fluid flow. In the case of a mesh without an inflation layer, the *k-epsilon (5)* turbulence model is the most appropriate. Activate the *Realizable* mode *(6)* and *Scalable Wall Functions (7)* to properly describe the phenomenon occurring in the boundary layer. Confirm all changes by clicking *OK*.

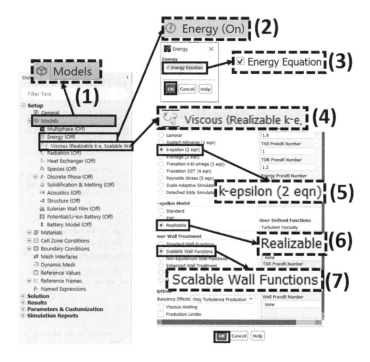

FIGURE 16.12

Being still in the *Models* tab *(1)*, find the *Species (2)* option and set the appropriate combustion model settings (Figure 16.13): Model *Species Transport (3)* with *Volumetric* reactions *(4)*. Choose the type of turbulence and chemical interactions described by the *Eddy Dissipation (5)* model. Make sure that the *Chemistry Solver* is set to *None-Direct Source (6)*. Change the default reaction mix from the *mixture template* to the *wood-volatiles-air (7)*, to describe the composition of the fuel from biomass gasification. If necessary, the user may define the new mixture composition.

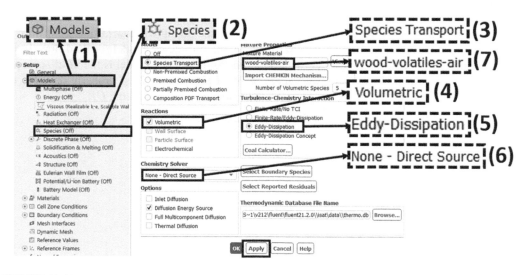

FIGURE 16.13

Tutorial 7 – Syngas Burner

Note that the ***radiation*** model should be activated to properly describe the temperature distribution inside the burner. In this case, the ***Discrete Ordinate*** model would be the most suitable one, because it includes heat absorption in gas particles. Nevertheless, to make this exercise easier and quicker to compute, the radiation model is not included.

Now let's focus on the ***Materials (1)***. The displayed list should include ***Mixture: wood-volatiles-air*** defined in the previous step. The ingredients of this mixture can be previewed under its name, as in Figure 16.14. To display more information, click 2× LMB on ***Mixture (2)***. Edit the details of the ***Eddy-Dissipation (3)*** chemical ***Reaction***.

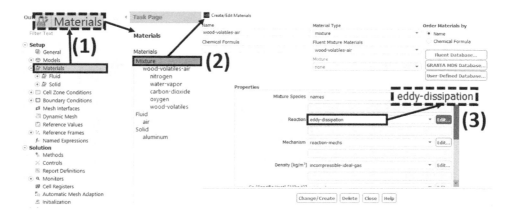

FIGURE 16.14

In the newly opened window (Figure 16.15), it is possible to decrease/increase the ***Total Number of Reactions (1)*** and specify their ***reactants (2)*** and ***products (3)***. For each substance, ***Stoichiometric Coefficient*** and ***Rate Exponent*** are calculated. In the case of the ***Eddy-Dissipation*** model, ***Mixing Rate*** is controlled by the constants A and B ***(4)***. Most often, the default values should not be changed, although there are models that require "calibration" with the use of A and B. In the

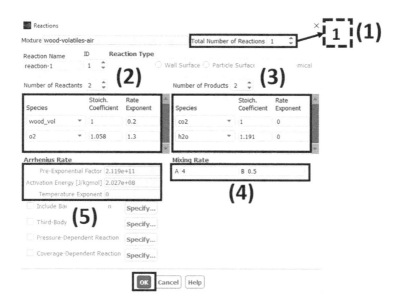

FIGURE 16.15

Reactions window, there is also an ***Arrhenius rate*** section *(5)*, which defines kinetic parameters of the reaction – note that in the ***EDM*** model, this section is not active. Generally, the kinetic parameters of decomposition are determined experimentally. To prepare a detailed combustion model, it is necessary to find their exact values in the literature (instead of using ***Fluent*** defaults).

Find the ***Cell Zone Conditions*** option *(1)* and check if the geometry surface *(2)* type is defined as ***fluid*** *(3)* as in Figure 16.16. Then, click ***Edit*** *(4)* and in the popup window, set the fluid material as ***wood-volatiles-air*** *(5)*. Confirm all changes by clicking ***Apply***.

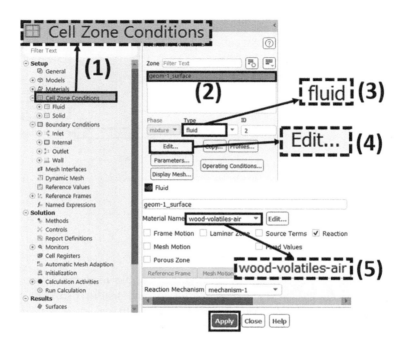

FIGURE 16.16

It's time to define the boundary condition (Figure 16.17). Click ***2× LMB*** on the ***Boundary Conditions*** *(1)* option and select the ***axis*** *(2)*. Its ***Type*** should be set as ***axis*** *(3)* by default. You can open the details window for condition by clicking ***Edit*** *(4)*. Nevertheless, in the case of an ***axis***, there are no parameters to modify *(5)*.

Tutorial 7 – Syngas Burner

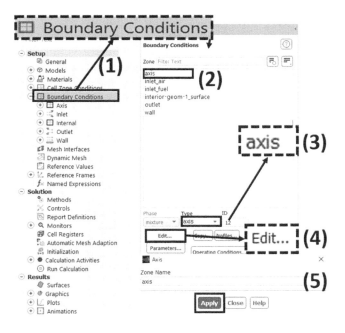

FIGURE 16.17

Move to the next boundary condition: *inlet_air* **(1)**. Check if its **Type** is set as **velocity-inlet (2)** as in Figure 16.18. Open the details window by clicking **2× LMB** on the zone name or by selecting the Edit option. In the **Momentum** tab, insert the inlet **Velocity Magnitude** value as 3 m/s **(3)**. **Turbulence** should be described by **Intensity and Viscosity Ratio (4)** (default values are OK). In the **Thermal** tab, set the **Temperature** as **300 K (8)**. In the last active tab, **Species** define the mass fractions of each substance at the inlet. In the case of *inlet_air*, there is only air with **23.15%** of oxygen and the remaining part of nitrogen.

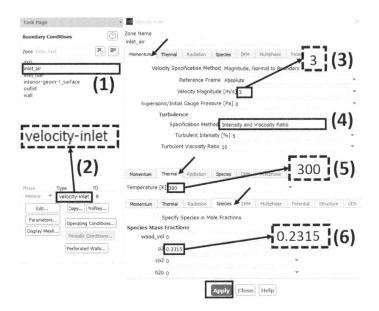

FIGURE 16.18

The second inlet boundary condition, *inlet_fuel (1)*, should also be defined as ***velocity-inlet (2)*** as in Figure 16.19. Set the value of ***Velocity Magnitude*** as *3 m/s (3)* and describe the ***Turbulence (4)*** by ***Intensity*** (default value of *5%* is OK) and ***Hydraulic Diameter***. In this project, the inlet channel is circular and has a *5 mm* radius, so the hydraulic diameter should be *10 mm*. The fuel ***Temperature*** should be equal to the air temperature *(300 K) (5)*. Specify the fuel composition as *72%* of ***wood-volatiles (6)***, 15% of ***carbon dioxide (7)***, and the remaining part as nitrogen.

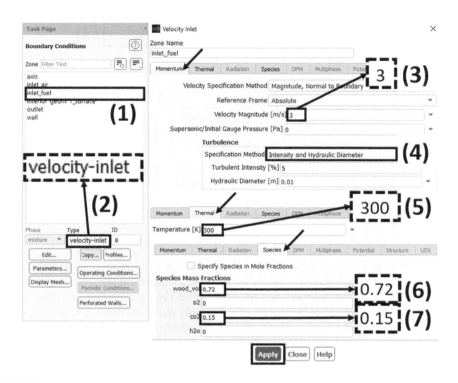

FIGURE 16.19

Move to the ***Outlet (1)*** (Figure 16.20). Its default type is ***pressure-outlet (2)***, which describes gauge pressure instead of outlet velocity. In the ***Momentum*** tab, set the ***Gauge Pressure*** as *0 Pa (3)* and define ***Turbulence (4)*** by ***Intensity*** (5%) ***and Hydraulic Diameter*** (60 mm – based on the geometry dimensions). In the ***Thermal*** tab, leave the default value of backflow temperature because the reversed flow is not expected and ***Apply*** all changes.

Tutorial 7 – Syngas Burner

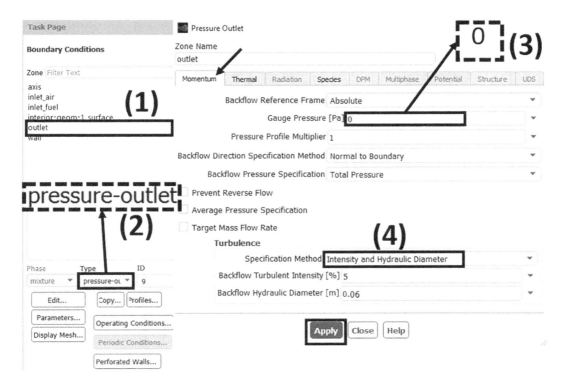

FIGURE 16.20

The last boundary condition is *wall (1)* (Figure 16.21). Its type is set by default as *wall (2)* with *Heat Flux (3)* equal to 0 *(4)*. This is fine in the case of thermally insulated surfaces. Confirm these selections by clicking *Apply*.

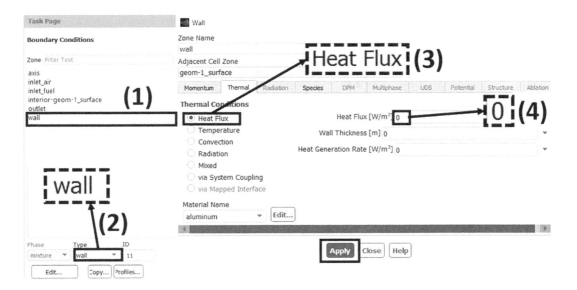

FIGURE 16.21

When the physics for the model has already been determined, the next step is to define numeric **Methods (1)** of the solution process. Set the algorithm of pressure-velocity coupling as **Coupled (2)**, which ensures a stable computation process. Check if the **Spatial Discretization** for each turbulence parameter is based on the **First Order Upwind** scheme, whereas all the other parameters are discretized with higher-order schemes, as in Figure 16.22. Use the **Pseudo Transient (3)** mode for calculations and activate the combined discretization of all the components of the mixture **(4)**.

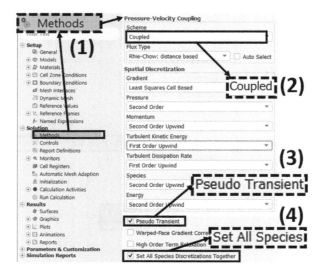

FIGURE 16.22

To supervise the computation process, create at least one **Report definition (1)**. In this case, it is reasonable to supervise the CO_2 concentration at the outlet, which gives information about the combustion process (Figure 16.23). Instead of setting a concentration monitor, the flue gas temperature or the balance of heat fluxes may be observed. To create a **New** monitor **(2)** click **2× LMB** on the **Report Definition** option and choose from the dropdown list **Surface Report** with the **Mass-Weighted Average** type **(3)**.

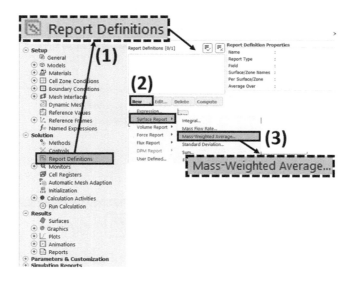

FIGURE 16.23

Tutorial 7 – Syngas Burner

Change the Report Definition *Name* to CO$_2$-out *(1)* and select *Species (2)* and *Mass fraction of CO$_2$ (3)* as *Field Variable*. As the name implies, this parameter should be calculated at the *outlet (4)*. Moreover, check if the *report Plot* option *(5)* is active to monitor the changes in CO$_2$ concentration in real time, during calculations (Figure 16.24). Confirm settings with the *OK* button.

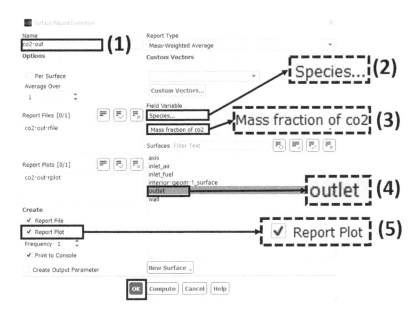

FIGURE 16.24

In each simulation, it is necessary to initialize the solution process (Figure 16.25). Find the *Initialization* option *(1)* and select the *Hybrid Initialization (2)* option. Perform a hybrid solution initialization by clicking the *Initialize* button *(3)*. Pay attention to the information displayed in the console at the end of the initialization process. The solver has automatically set non-zero concentrations of reactants and combustion products. This operation is required to initialize whenever the combustion model is set as *Eddy-Dissipation*.

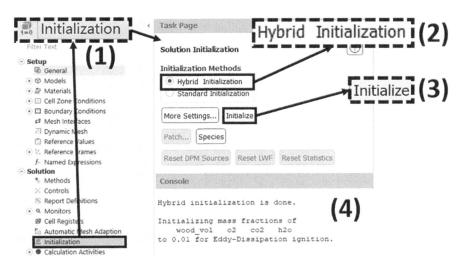

FIGURE 16.25

It is time to **Run Calculation (1)**. Set the **Number of Iterations** to 200 *(2)* as in Figure 16.26 – it should be enough to reach the convergence. Finally, click the **Calculate** button *(3)*.

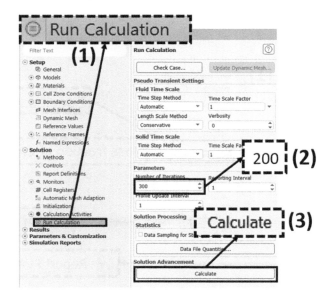

FIGURE 16.26

The level of calculation accuracy may be observed on the **Scaled Residuals** chart *(1)* as in Figure 16.27. The curves which describe the error in an individual solution for each parameter, should be heading down until the convergence is reached. The information that the calculation is complete will appear in the solver console and in the popup window. Then, change the residual graph window *(1)* to the monitor graph window *(2)*. Check if the CO_2 concentration at the **outlet** has reached a stable value after the calculations.

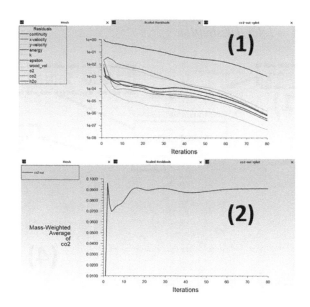

FIGURE 16.27

Tutorial 7 – Syngas Burner

Save the *Case* and *Data* files (Figure 16.28) under the name: *velo_3* in the newly created folder. NOTE: The *Case* files and the corresponding Data files must be in the same location! Then, repeat the calculations for the inlet speeds of air and fuel equal to *5 m/s* (do not perform initialization). Re-export the *Case* and *Data* files to the same folder as previously, with the name *velo_5*. Perform further calculations sequentially for both inlet velocities *3* and *5 m/s*, but with modified constants *A* and *B* of the combustion model *(A = 3, B = 1)* to consider their impact on simulation results. Re-export the relevant files as *velo_3_modified* and *velo_5_modified*. Close the solver and save the project from the *ANSYS Workbench* module.

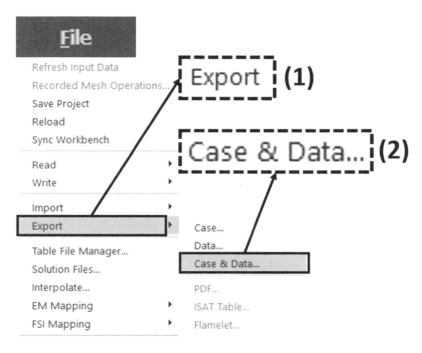

FIGURE 16.28

TABLE 16.1
Solver Settings for Four Variants of Biogas Burner Simulation

Characteristic	Variant			
	I	II	III	IV
Air and fuel inlet velocities [m/s]	3	5	3	5
Constants A and B		default		A = 3 B = 1
Exported files names	velo_3	velo_5	velo_3_modified	velo_5_modified

16.5 POSTPROCESSING

Find the *Results* module *(1)* in *Component Systems* and drag it to the *Project Schematic*, as in Figure 16.29. DO NOT export data from the *Solution* cell to the *Results* (you should run the "empty" postprocessor module). Click *2× LMB* on the *D2* cell to launch the module *(2)*.

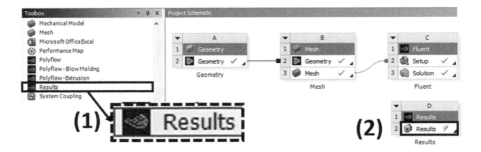

FIGURE 16.29

Now it is time to import the data on the four analyzed variants (Figure 16.30). To do this, open the *File* dropdown list and choose the *Load Results (1)* option. Then, find the location of your folder which contains the exported case and data files. Select all the files in *.dat* format *(2)* and *Open (3)*. After a successful import, the four wireframes of the geometry should be visible – one for each case *(4)*.

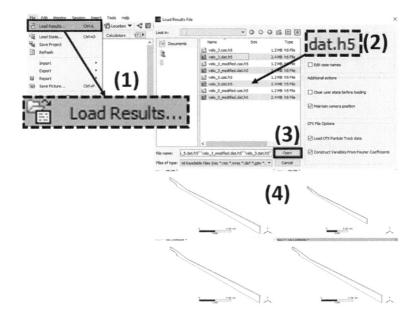

FIGURE 16.30

To display the CO_2 concentration inside the burner, select the *Contour (1)* tool and insert *Name (2)*. In *Details of Contour* specify *Locations* as *periodic 1(3)*. Then, choose the *Variable* which will be displayed: CO_2 *Mass fraction (4)* and its *Range: Global (5)*. In the *# of Contours* cell increase the number of isolines to *50 (6)*. Confirm these settings by clicking *Apply* and enjoy the first contours (Figure 16.31). As you can see, the differences in the results presented by contours are barely visible.

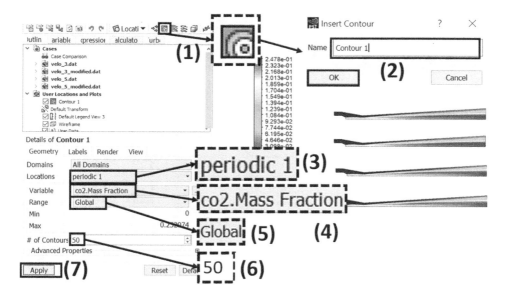

FIGURE 16.31

To present the results more legibly, use the charts showing the linear distribution of the selected results on the axis of symmetry. Therefore, an appropriate line should be generated, as in Figure 16.32. Open ***Location (1)*** and select ***Line (2)*** from the dropdown list. ***Name*** your location ***(3)*** and define its details: Line defined by ***Two Points (4)***. Specify the first point as ***(0.0, 0.0, 0.0)*** and the second one as ***(0.4, 0.0, 0.0)***. These coordinates represent the beginning and the end of the symmetry line. Check if the ***Line Type*** is set as ***Cut (5)*** and ***Apply*** your selections.

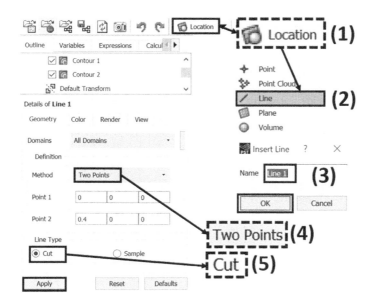

FIGURE 16.32

Find the **Chart** icon *(1)* and add a new chart (Figure 16.33). You can change its **Name** if necessary *(2)*. In the **Details of Chart** window, find the **Data Series** tab *(3)* and insert the **Location** of a data series at the newly created **Line1** *(4)*. In the **X-Axis** tab, define **Variable** as the **X** *(5)* coordinate and set **Y-Axis Variable** as CO_2 **Mass fraction** *(6)*. Confirm your settings with **Apply** and the chart will be displayed.

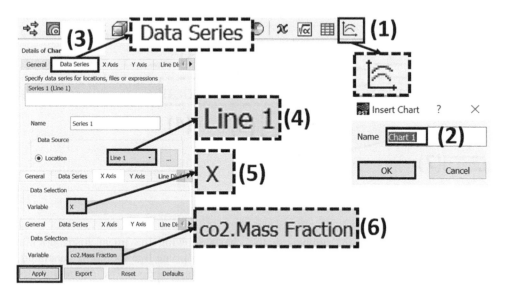

FIGURE 16.33

Using the procedure described above, prepare similar charts presenting the temperature changes, changes in the kinetic energy of turbulence, and oxygen mass fraction alongside the symmetry *(Line1)*. Then, answer the following questions:

What is the relationship between the change in the velocity and the change in the kinetic energy of turbulence and what does it result from?
What is the result of the modifications of the constants A and B?

To calculate the temperature of the flue gases at the outlet, use **Function Calculator** *(1)*. Define the parameter as **massFlowAve** *(2)* at the **outlet** *(3)* as in Figure 16.34. Select **All cases** *(4)* to display the results for all the variants simultaneously. Finally, set the **Temperature** *(5)* as the analyzed parameter. Click the **Calculate** button *(6)* and the results will be displayed *(7)*. Using this function, calculate the CO_2 concentration at the outlet for each case.

Tutorial 7 – Syngas Burner

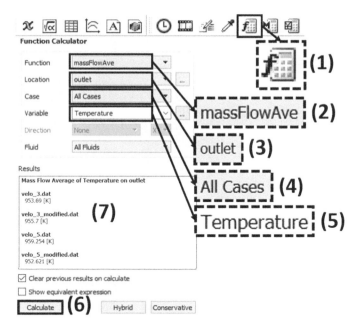

FIGURE 16.34

17 Tutorial 8 – Particulate Matter Separation in Cyclone

17.1 EXERCISE SCOPE

This tutorial presents the procedure of modeling the discrete phase transport in the domain representing particulate matter (PM) separation system – a small cyclone, as the one in Figure 17.1. The device has one inlet *(1)* and two outlets. The bottom one *(2)* allows to extract PM to separate from the system. The air cleaning of dust leaves the cyclone through the top outlet *(3)*.

Try to recall from the theoretical part (see Chapter 15) the principle of PM separation in cyclones. It is based on the centrifugal force and the gravity influence on the particles. The model considered in this exercise allows us to observe trajectories of different particles (different diameters) inside the cyclone and to assess the influence of their size on the separation efficiency.

Among others, this exercise will teach you how to

- create selection names directly in **SpaceClaim Direct Modeler** instead of using **Ansys Meshing**,
- define global mesh settings in the **Fluent Mesh** module,
- apply PM injectors in **Discrete Phase Model** in **Ansys Fluent**.

Pay attention to the settings of the renormalization group (RNG) turbulence model mode for the swirl flow, the settings of the interactions between the continuous and discrete phase, as well as the way of exporting particle history data from the solver.

All of the images included in this tutorial use courtesy of Ansys, Inc.

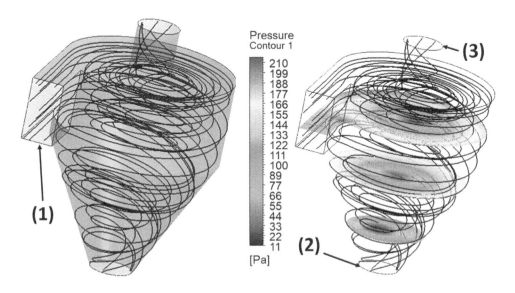

FIGURE 17.1 Geometry of the cyclone considered in the exercise with the visualization of exemplary results.

17.2 PREPROCESSING – GEOMETRY

Fluent Meshing will be used to generate a mesh in frames of the exercise. It is part of the *Fluid Flow (Fluent with Fluent Meshing) (1)* module. Find this module on the *Analysis Systems* list, then drag and drop it in the *Project Schematic* window *(2)*, as in Figure 17.2. Open *Ansys SpaceClaim Direct Modeler* (*ASCDM*) – click *2× RMB* on *B2* (*Geometry*) cell *(3)*.

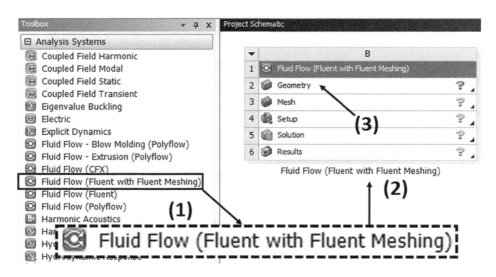

FIGURE 17.2

In *ASCDM*, go to the *Sketch (1)* tab and activate the *Line (2)* tool to create a 2D sketch. First, select the appropriate sketching plane – click *New Sketching Plane (3)*. Now the sketching plane selection can be done by moving the mouse cursor on the triad axis that is normal to the required plane – in this case the *Z (4)* axis. Then, click the axis *1× LMB*. Press the *V* key to set the normal view. Prepare the sketch according to *(5)* in Figure 17.3. Then, click *End Sketch Editing (6)*.

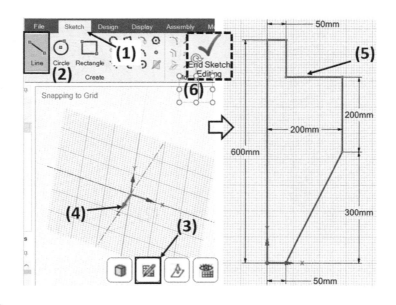

FIGURE 17.3

Tutorial 8 – Particulate Matter Separation in Cyclone

A new surface should appear in the *Structure* window (*LHS* of the *GUI*). It can be revolved now to create the cyclone body. Select the *Pull (1)* tool from the top toolbar (Figure 17.4). Click *Select (2)* and select the surface that is going to be revolved *(3)*. Click the *Revolve (4)* option and select the rotation axis – the longest surface edge. Click and hold *LMB* on the small yellow arrow, and turn it to any angle. Then, insert **360°** and press *ENTER* to get a full rotation.

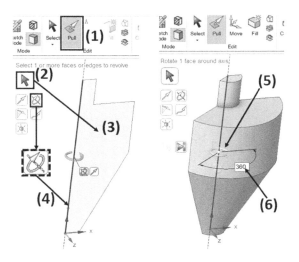

FIGURE 17.4

Go back to the *Sketch (1)* tab. Click the *Circle (2)* tool. The sketching plane selection mode should activate automatically. Select the top face *(3)* of the cyclone body. If encounter a problem with the selection of the sketching plane, use the *Select New Sketching Plane (4)* function. Set the normal view using the *V* key or the *Plane View (5)* function. Create two circles: One *(6)* with a **100 mm** diameter (equal to the top outlet diameter) and the other *(7)* with a **106 mm** diameter, as in Figure 17.5. Then, click *End Sketch Editing* as previously.

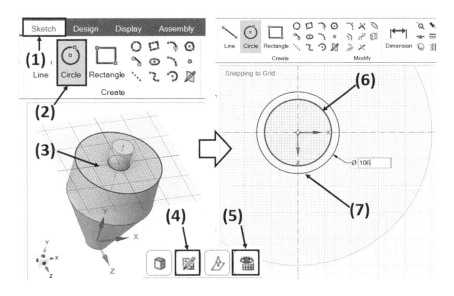

FIGURE 17.5

Apply the ***Pull (1)*** tool to pull down the wall of the outlet channel inside the cyclone body. Click ***Select (2)***, then click **1× *LMB*** on the cylinder created by the two sketched circles. Click and hold ***LMB*** on the small yellow arrow ***(3)*** and pull it down a bit. Then, insert the length of ***150 mm*** and press ***Enter***. A narrow slot, as ***(5)*** in Figure 17.6, should appear. Check whether the ***Structure*** window contains any surfaces. If so, they have to be removed (***1× RMB -> Delete***). Only one solid body representing the cyclone should be visible there.

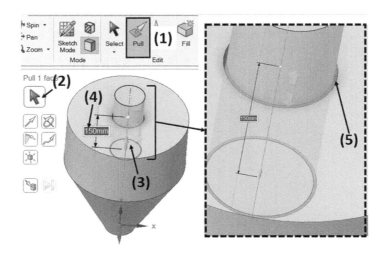

FIGURE 17.6

The third sketch of the inlet channel has to be created now. In the ***Design*** tab, (next to the ***Sketch*** tab) find and click the ***Plane (1)*** tool, as in Figure 17.7. Move the mouse cursor on the ***Z (2)*** axis of the triad and click ***1 × LMB***. A new plane ***(3)*** appears. Now, it has to be moved away appropriately. Click ***Move (4)*** (next to ***Pull***) and select any edge of the new plane ***(5)***. A new triad of the ***Move*** tool appears. Click and hold ***LMB*** on its Z axis, and drag it a bit. Insert the dimension of ***300 mm (7)*** and press ***Enter***. You can deactivate or completely remove the plane from the ***Structure*** window.

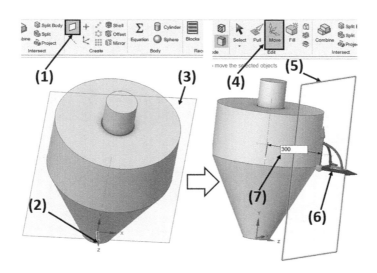

FIGURE 17.7

Tutorial 8 – Particulate Matter Separation in Cyclone

Again, go to the **Sketch** *(1)* tab and activate the **Rectangle** *(2)* tool. Select the created plane as the sketching plane *(3)*. Set the view *(4)*. Create a rectangle with dimensions as in *(5)*, Figure 17.8. This rectangle will be used to pull the inlet channel. You can also use the **Move** function to place the sketch. Remember to click **End Sketch Editing** at the end.

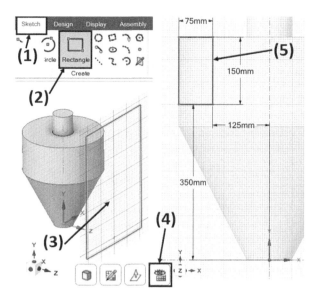

FIGURE 17.8

Use the **Pull** *(1)* tool again. Click **Select** *(2)* and mark the surface *(3)* created automatically from the rectangle sketch, according to Figure 17.9. Click the **Up To** *(4)* function and mark the upper part of the cyclone's external surface *(5)*. Complete cyclone geometry should look like *(6)*. You can remove the plane (**Delete**) *(7)* from the **Structure** window.

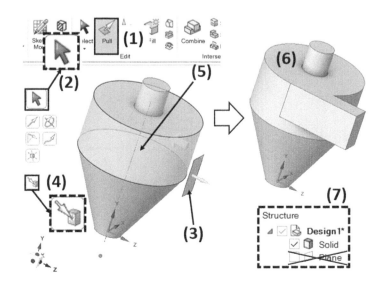

FIGURE 17.9

Face names (*Named Selections*) that will be used during the meshing process have to be defined now. Start with the *inlet*. Go to the *Groups (1)* tab below the *Structure* window. Select (*1× LMB*) the inlet surface *(2)*, as in Figure 17.10. Then, click *Create NS (3)*. A new item with a default name appears below the *Named Selection* tab *(4)*. Change the name to *inlet (5)*.

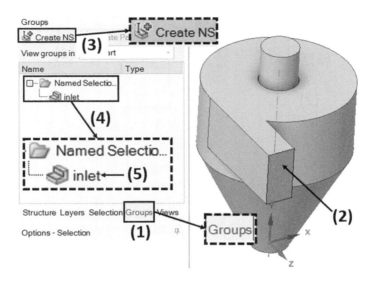

FIGURE 17.10

In the same way as the *inlet* (Figure 17.10) add all the other required *Named Selections* (Figure 17.11): *Outlet-top (1)*, *outlet-bottom (2)*, and *wall (3)*. In the latter case, it is important to select all of the external surfaces that are neither outlets nor an inlet, including three surfaces in the narrow slot region *(4)* that are a bit more difficult to mark. To hide a face, click *RMB* on it and select the *Hide* option. Another possibility to look inside the geometry is to use the *Ctrl + scroll* button combination. Close *ASCDM*. Then save the whole project in the *Ansys Workbench* (*AWB*) main window (*File -> Save*).

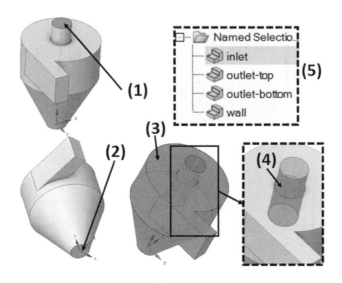

FIGURE 17.11

Tutorial 8 – Particulate Matter Separation in Cyclone

17.3 PREPROCESSING – MESHING

The check mark has appeared in the **Geometry B2** cell (Figure 17.12). It means that the cyclone geometry file is already defined. Now, it is time to create a polyhedral mesh using the **Fluent Meshing** approach. Click **2× LMB** in the **Mesh C3 (1)** cell to launch this module.

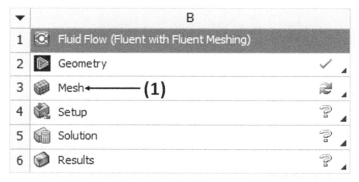

Fluid Flow (Fluent with Fluent Meshing)

FIGURE 17.12

In the **Fluent Launcher** window (as in Figure 17.13), it is possible to predefine settings, both for **Fluent Meshing** and the **Fluent** Solver. In the **Parallel (Local Machine) (1)** section, the number of parallel partitions for meshing and for the solution process can be determined individually. The **AWB** student license allows running the solution with four parallel processes. Thus **4** should be set both for **Meshing Processes and Solver Processes (2)**. Click **Start** to launch **Fluent Meshing**.

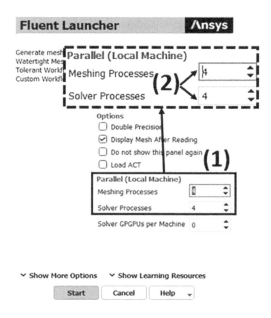

FIGURE 17.13

The first step in **Fluent Meshing** is importing the geometry *(1)*. Pay attention to the subsequent tabs in the workflow – all of them are incomplete tasks, thus, lightning thumbnails appear there (as in Figure 17.14). The **File Name** cell includes a geometry file path. In the case of creating a project in the **Analysis Systems** mode, as in this case, the existing path should be correct. Click **Import Geometry** *(2)* and wait until the process is over to continue.

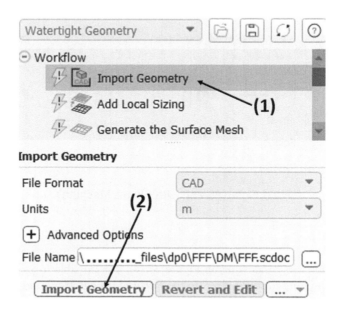

FIGURE 17.14

The **Add Local Sizing** *(1)* tab allows to define different cell sizes individually for selected regions (bodies, faces, edges) of the geometry. The selection of the appropriate region takes place based on **Named Selections**. In this simple case, there is no need to use this tool. Thus, set **No** in cell *(2)* (according to Figure 17.15) and **Update** *(3)* the project to continue.

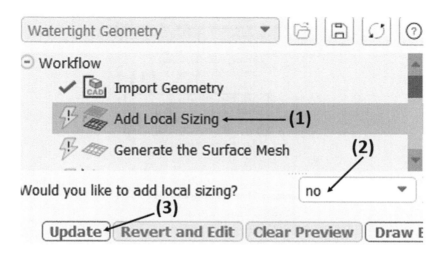

FIGURE 17.15

Tutorial 8 – Particulate Matter Separation in Cyclone

The surface mesh is generated in the same way as the first one. Then, this mesh is the basis for creating a volume mesh. The surface mesh settings are defined in the **Generate the Surface Mesh (1)** tab. These settings are the global ones (for the surface). Change **Maximum Size (2)** to **0.005**. Pay attention to the cell size visualization at the model surface *(3)*, as in Figure 17.16. **Size Functions** should be set to **Curvature & Proximity (4)**. **Curvature Normal Angle** should be set to **16 (5)**. Click **Generate Surface Mesh (6)** to continue.

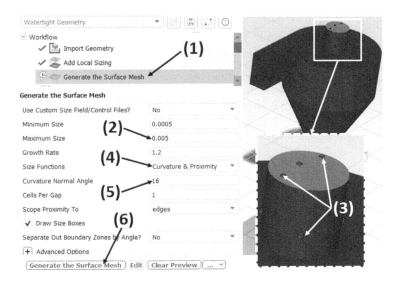

FIGURE 17.16

The created mesh is for now triangular *(1)*, as in Figure 17.17. Later, the mesh will be turned into a polyhedral one. The surface mesh summary in the **Console**, below the project workspace window, includes information about the quality of the created mesh. To improve the mesh quality, click **1 × RMB** on the **Generate Surface Mesh (2)** tab (**Workflow** list) -> **Insert Next Task (3)** -> **Improve Surface Mesh (4)**. Set **Face Quality Limit (5)** to **0.7** (for skewness correction) and click **Improve Surface Mesh (6)** to continue.

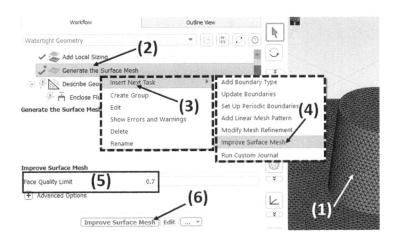

FIGURE 17.17

The Describe Geometry (1) tab allows to predefine certain domain features, according to Figure 17.18. In the *Geometry Type* section select, *the geometry consists of only fluid regions with no voids (2)*. For *Change all fluid-fluid boundary types...* select *No (3)*. Share topology application is not required *(4)* here. Click *Describe Geometry* to continue *(5)*.

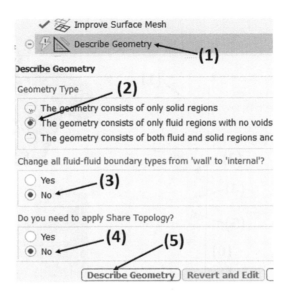

FIGURE 17.18

The *Update Boundaries (1)* tab includes a list of the model boundaries (Figure 17.19). Based on *Selection Names* defined in *Ansys SpaceClaim Direct Modeler*, the boundary types in the table *(2)* have been set automatically. Check whether all the boundaries are assigned correctly and click *Update Boundaries (3)* to continue.

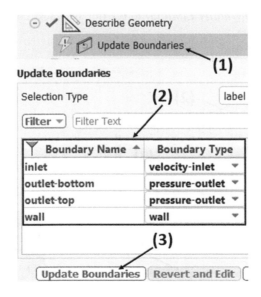

FIGURE 17.19

The Update Regions (1) tab allows to control and confirm the region types. Although the default name of the created geometry according to Figure 17.20 is *solid*, its type (*Region Type* cell) is *fluid*, so it is correct. Click *Update Region (2)* to confirm and go to the *Add Boundary Layers (3)* tab. The inflation layer in the near-wall-regions can be defined here. In this exercise it is not required, so set *no (4)* next to the *Add Boundary Layer?* inscription and click *Update (5)* to go ahead.

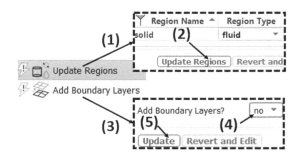

FIGURE 17.20

The last step in the *Fluent Meshing* module is creating a volume mesh based on the surface mesh that has been already created. In the *Generate the Volume Mesh (1)* tab (Figure 17.21) set *Fill With → polyhedra (2)* – this is the most appropriate mesh type for this case. Set the *Growth Rate* to *1.2 (3)* and *Max Cell Length* to *0.008 m (4)*. Now click *Generate the Volume Mesh (5)* to create it.

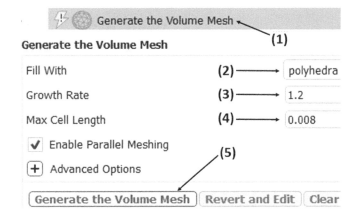

FIGURE 17.21

When meshing is completed, its summary is displayed in the *Console (1)* – as previously for the surface mesh. Table *(2)* (Figure 17.22) collates the data concerning the worst cells. The number of the created cells *(3)*, processing time *(4)*, and the worst quality indicator value *(5)* are available in the table.

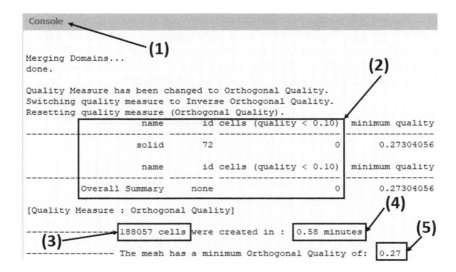

FIGURE 17.22

The Clipping Planes section of the top toolbar allows us to analyze the mesh cells inside the object. Activate (if not active) the *Insert Clipping Plane (1)* option. The currently active cross-section is visible in the project workspace *(2)*. Note (Figure 17.23) that the mesh cells are polyhedral. The *Limit in… (3)* function allows to switch the plane (*X, Y, Z*). Slider *(4)* moves the plane along the axis.

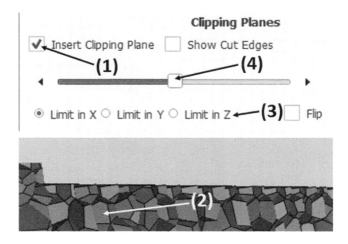

FIGURE 17.23

Grid cell quality distribution can be controlled using the *Display (1)* drop-down list (Figure 17.24) → *Plot (2)* → *Cell Distribution… (3)*. *Options (4)* allows to analyze the cell quality or size. The quality indicator can be set using *Quality Measure… (5)* options. If an assembly model is considered, the appropriate geometry part can be selected in the *Cell Zones* window. Click *1× LMB* on *Solid (6)* and *Plot (7)* to display results. You can test different options in the *Cell Distribution…* window and then go to the solver using the *Switch to Solution (8)* button in the top left corner of the *Fluent Meshing* GUI.

Tutorial 8 – Particulate Matter Separation in Cyclone

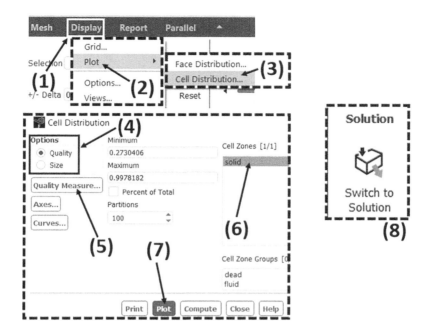

FIGURE 17.24

17.4 PREPROCESSOR – SOLVER SETTINGS

In the *Ansys Fluent* solver, go to the *General (1)* tab (Figure 17.25). Have a look at the available tools. *Display (2)* allows to show and hide model surfaces and interior. *Check (3)* enables the display of the model geometry parameters in the *Console*. *Report Quality (4)* allows controlling the mesh quality indicators. Now activate *Gravity (5)* and based on the model orientation *(6)* set the earth acceleration vector $Y = -9.81$ $[m/s^2]$ *(7)*.

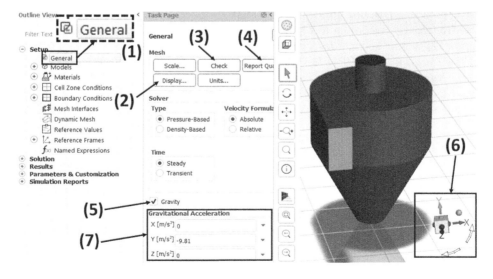

FIGURE 17.25

Click *2× LMB* on the *Models (1)* tab to activate appropriate models/transport equations. Modify the turbulence modeling settings – click *2× LMB* on the *Viscous… (2)* tab. In the *Viscous Model* window, change the settings of the *Model (3)* list to the *k-epsilon (4)* model which is appropriate for the cases with coarse mesh on the model walls (Figure 17.26). In the *k-epsilon Model* section *(5)*. Select the *RNG (6)* mode, which is often recommended for highly swirled flows. In the *RNG Options* section *(7)* below, activate the *Swirl Domain Flow (8)* option. Then click *OK (9)* to continue.

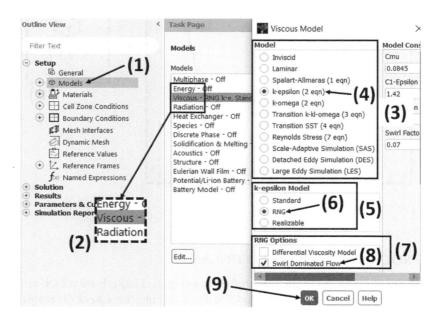

FIGURE 17.26

Particulate matter (PM) carried by air flowing into the cyclone will be modeled using the simplest approach – **Discrete Phase Model (DPM)**. Select (*2× LMB*) **Discrete Phase-Off (1)** from the *Models* list (according to Figure 17.27). In the *Discrete Phase Model* window, in the *Interaction (2)* section activate *Interaction with Continuous Phase* and *Update DPM Sources Every Flow Iteration*, with *DPM Iteration Interval* of *10*. *Tracking Parameters (3)* control the *Maximum Number of Steps* – the maximum length of the particle path in the domain before it vanishes and the *Step Length Factor*, which indirectly determines the number of time steps required for a particle to traverse a computational cell. Default settings are fine here. Go to *Injections… (4)* to continue.

Tutorial 8 – Particulate Matter Separation in Cyclone

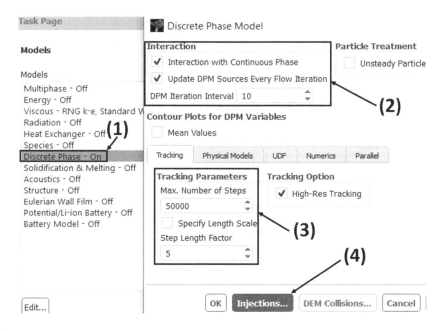

FIGURE 17.27

The *Injections* list is empty now, as in Figure 17.28. Click the *Create (1)* button to add a new injector. Insert the *pm1* name in the *Injection Name (2)* field. Set *Injection Type* to *Surface (3)*. Click *1× LMB* on the inlet in the *Injection Surfaces (4)* window. *Particle Type* should be set to *Inert (5)* (there is no chemical reaction or even heat transfer in the model). Select *ash-solid* from the *Material (7)* list. Now define the particle properties below *(8)*. Set the *Diameter* to *0.000001 [m]*. If you don't want to set velocity vector components individually, activate the *Inject Using Face Normal Direction (9)* option. Then, go back to the properties and set *Velocity Magnitude* to *10 [m/s]*. The *Total Flow Rate* should be *0.00001 [kg/s]*. Now, click *OK* to continue.

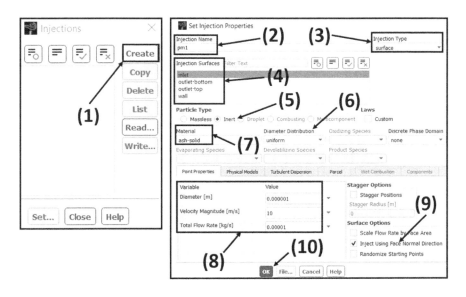

FIGURE 17.28

The Defined *pm1 (1)* injector has appeared on the *Injections* list, as in Figure 17.29. There is no need to create further injectors step by step if they are similar to the already existing ones, as in this model. Click *1× LMB* on the *pm1* injector on the list and *Copy (2)*. Change the injector name to *pm2.5 (3)*. Change the particle *Diameter* to *0.0000025 [m] (4)*. Click *OK*. The second injector should appear on the *Injections* list. Analogously, add the third injector – *pm10* with the particle *Diameter* of *0.00001 [m]*. Finally, three injectors should be visible on the *Injections (5)* list. Then, *Close (6)* the *Injections* window.

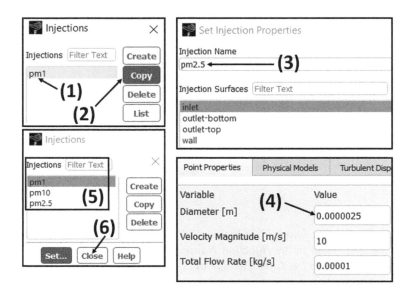

FIGURE 17.29

Open the *Materials (1)* tab in the project tree (Figure 17.30). Open the *Inert Particle* tab below and click *2× LMB* on *ash-solid (2)*. In the displayed window, change *Density* to *2000 [kg/m³]* and click *Change/Create (4)*. *Close (5)* the window.

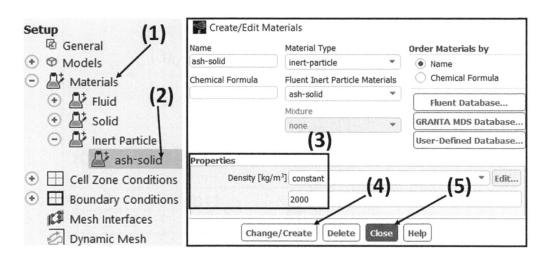

FIGURE 17.30

Tutorial 8 – Particulate Matter Separation in Cyclone

Open the **Cell Zone Conditions (1)** tab in the project tree to make sure that the domain type is **Fluid (2)**, according to Figure 17.31. Click **2× LMB** on **Cell Zone Conditions (1)** and select *fluid* from the **Type** drop-down list. Click **2× LMB** on **Boundary Conditions (3)** (BCs) in the project tree. Click **1× LMB** on **inlet (4)**. Note that BC **Type (5)** is *velocity inlet*. Click **Edit (6)** to open the **Velocity Inlet** settings. Change **Velocity Magnitude (7)** to **10 [m/s]**. In the **Turbulence** section below, change the **Specification Method** to **Intensity and Hydraulic Diameter (8)**. Set **Hydraulic Diameter** to **0.109 [m] (9)**. Then, go to the **DPM (10)** tab to continue.

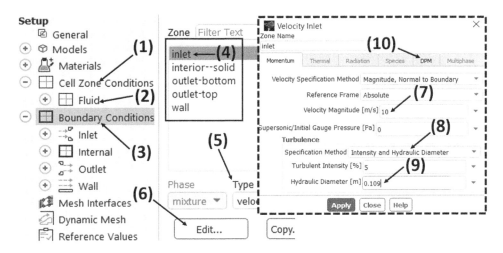

FIGURE 17.31

The **DPM (1)** tab includes a description of the particle behavior when reaching a certain BC. The Default setting of **Discrete Phase BC Type** (Figure 17.32) is *escaped*. Change it to *reflect* **(2)** – it prevents PM reversed flow through the injection surface. Click **Apply (3)** and **Close (4)**. Now click **2× LMB** on the **outlet-bottom** BC. In the **Pressure Outlet** window, change the **Turbulence Specification Method** (see Figure 17.31) to **Intensity and Hydraulic Diameter**, and set the **Diameter to 0.1 [m] (5)**. **Apply** and **Close** the window. Then define analogously the **outlet-top** BC.

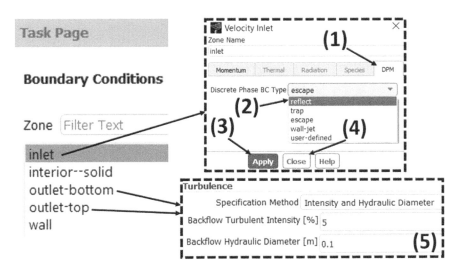

FIGURE 17.32

Click **2× LMB** on the **Methods (1)** tab. In the **Pressure-Velocity Coupling** (Figure 17.33) section, set the **SIMPLE (2) Scheme**. Change **Spatial Discretization** of the **Momentum** equations to **First Order Upwind (3)**. Click **2× LMB** on the **Initialization (4)** tab. Switch **Initialization Methods** to **Standard Initialization (5)**. Select **inlet (6)** from the **Compute** the list. It means that initialization parameters will be read from this BC. Click **Initialize** to continue.

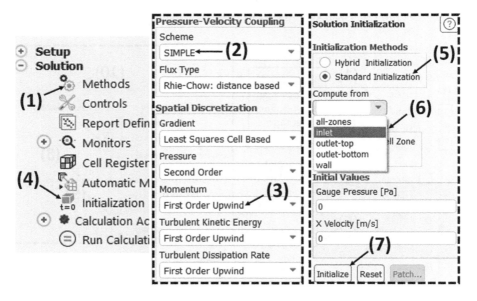

FIGURE 17.33

Click **2× LMB** on the **Run Calculation (1)** tab in the project tree. Set the **Number of Iterations** to **250 (3)** and click **Calculate (3)** to start the solution process (Figure 17.34). Solution information will be displayed in the **Console**. The change in the residual level of the solved transport equations can be tracked using the chart displayed in the workspace window.

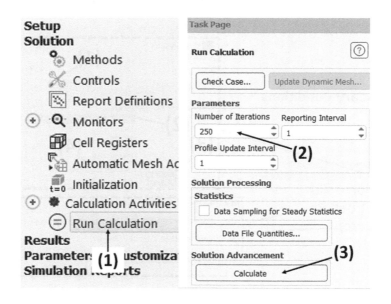

FIGURE 17.34

17.5 POSTPROCESSING

Have a look at the **Console** now (Figure 17.35). The *iter (1)* column refers to the current iteration number. After several iterations, the title of the subsequent equation columns *(2)* appears. One discrete phase iteration per each ten continuous phase iterations has been set, thus after each tenth iteration the particle tracking report is displayed *(3)*, where ***tracked*** = refers to the total number of injected particles and ***escaped*** = refers to the particles leaving the domain through any outlet. Residual values can be read in columns *(4)*. When the residual for each equation is equal to or lower than the convergence criteria, the solver displays information that the ***solution is converged (5)*** and ends the simulation.

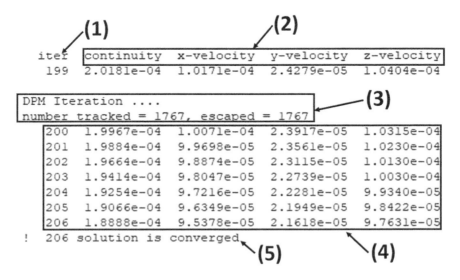

FIGURE 17.35

To analyze the computed particle trajectories, in the ***Results (1)*** tab of the project tree (Figure 17.36) open the ***Graphics (2)*** tab and find the ***Particle Tracks (3)*** tool. To add ***particle-tracks-1 (4)*** as in Figure 17.36, click ***2× LMB*** on the ***Particle Tracks*** tab (project tree). In the ***Particle Tracks*** window, find the ***Color by (5)*** section. Select ***Particle Variables*** from the upper list and ***Particle Residence Time*** from the lower list. Now set ***Skip*** to ***50 (6)*** to reduce the number of the displayed trajectories. Now, select ***pm1*** (the smallest particles) from the ***Release from Injections*** list *(7)*. Click ***Save/Display (8)*** to generate the visualization in the workspace window.

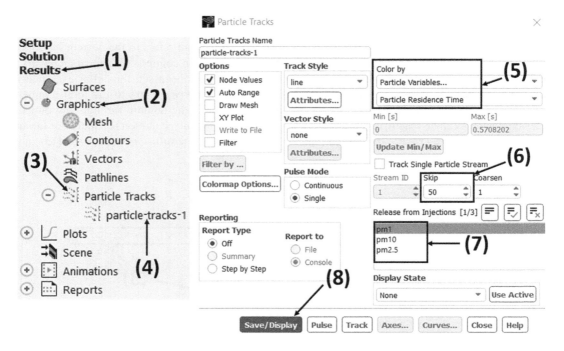

FIGURE 17.36

Based on the above-presented procedure (see Figure 17.36), you can display the particle trajectories for two remaining particle sizes (***pm2.5, pm10***). Pictures *(a)–(c)* in Figure 17.37 present particle flow paths through the cyclone for three particle dimensions considered in the model: *1 μm, 2.5 μm*, and *10 μm*. The fewer particles that escape through the top outlet *(1)*, the more efficient is the operation of the cyclone. Unfortunately, in the case of the smallest particles, a large percentage of them is not captured and they flow out through the top outlet. Efficiency is much better in the case of bigger particles. All the biggest (*10 μm*) particles escape through the bottom outlet *(2)*. Furthermore, there is a clear relation between the particle diameter and the particle residence time in the cyclone. Can you describe it using the legend *(3)*? You can also show all three injections in one picture.

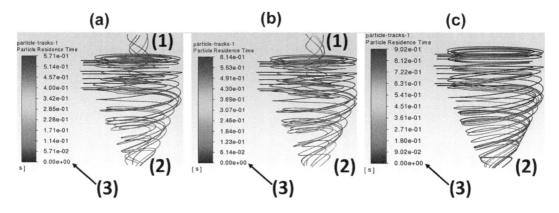

FIGURE 17.37

Tutorial 8 – Particulate Matter Separation in Cyclone 367

If you want to check the mass balance of the continuous phase in the domain, open the *Reports (1)* tab (Figure 17.38) and click *2× LMB* on the *Surface Integrals* tool. In the *Surface Integrals* window, select *Mass Flow Rate (2)* from the *Report Type* drop-down list. Then, mark the *inlet*, *outlet-bottom*, and *outlet-top* surfaces in the *Surfaces (3)* window. Click compute *(4)* to display the report in *Console (5)*.

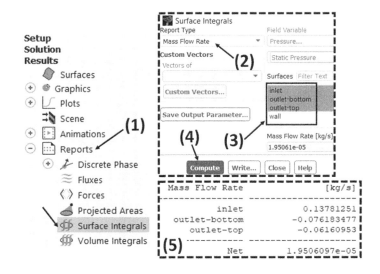

FIGURE 17.38

If the DPM trajectories are also to be displayed in the *Ansys Results* module, appropriate data files have to be exported first. Click *File (1)* (top bar) → *Export (2)* -> *Particle History Data… (3)*, according to Figure 17.39. In the displayed *Export Particle History Data* window, select *pm1 (4)* from the *Injections* list. Set *Skip* to *50 (5)* – every 50th particle will be tracked to reduce the amount of the saved data. Insert the file name, for example, *pm1 (6)*, and set its directory *(7)*. Then, click *Write (8)* to export the data. Of course, if you want, you can save *Particle History Data* for any other particles (diameters) or even for all the injections together.

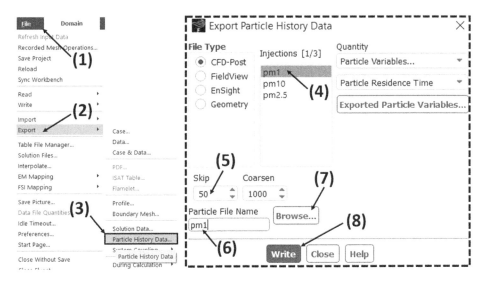

FIGURE 17.39

Close **Ansys Fluent** and save the project in the **Ansys Workbench** main window (**File → Save**). Then, click **2× LMB** on **A6 (1)** (**Results**) cell to launch the postprocessor module. Any geometry face can be displayed by marking it below the solid tab in the **Outline** window (project tree), as in Figure 17.40. Mark **wall (2)** and click it **2× LMB**. Go to the **Details of wall** window below. You can make the wall semitransparent in the **Render (3)** tab. To do it, set **Transparency (4)** to the **0.5–0.7** range. Confirm by **Apply (5)**. The edited face **(6)** is transparent now. In the **Color** tab, it is possible to change the wall color or even fill it with a specific **Variable**.

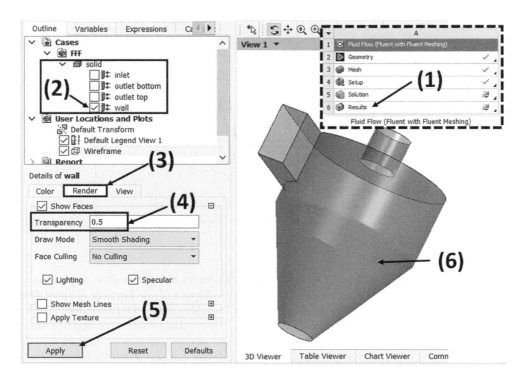

FIGURE 17.40

Click **File (1) → Import (2) → Import FLUENT Particle Track File (3)** and import the previously saved file (**pm1.xlm**). The DPM trajectories should appear automatically **(4)**, as in Figure 17.41. If not, just mark the data **(5)** in the **Outline** window. Click **2× LMB** on **FLUENT PT for Ash Solid (6)**. Set the **Color (6)** option to black (**1× LMB**). Confirm by **Apply (7)**.

Section planes allow to display field variables in determined locations. Deactivate wall visibility and DPM trajectories **(1)**. Click **Location (2)** drop-down list → **Plane (3)**. In the **Definition (4)** section, select the appropriate plane (**ZX**) from the **Method** list and set the distance **Y = 0.1 [m]** just below. **Apply (5)** settings. A new plane should appear in the project workspace **(6)**. When required, the plane visibility can be deactivated, in the same way as in the case of the other objects in the **Outline** window.

Tutorial 8 – Particulate Matter Separation in Cyclone 369

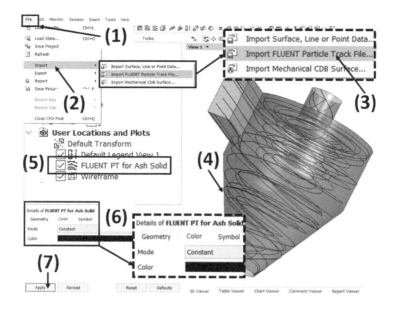

FIGURE 17.41

Analogously to the first section plane (see Figure 17.42), add two other planes (Figure 17.43), on heights *Y=0.25 [m]* and *Y=0.4 [m], respectively.* Of course, any other planes can be created too. They can be easily activated and deactivated and used to generate contour or vector variable fields.

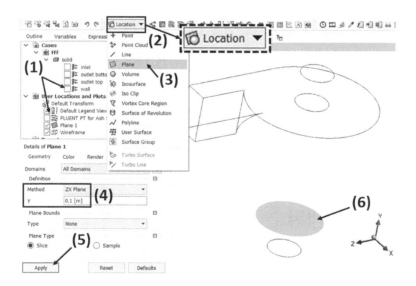

FIGURE 17.42

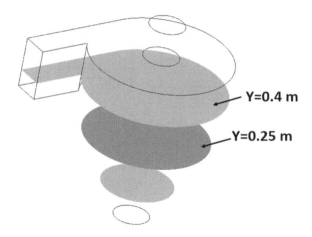

FIGURE 17.43

Among the visualization tools *(1)* available in the top toolbar (Figure 17.44), find and click **Contour**. You don't have to change the default name. Click **OK** *(2)*. Go to the **Details of Contour 1** window and find the **Location** list. Click three dots *(3)* next to it. Hold the **Ctrl** key and select all the three created planes *(4)*: **Plane 1**, **Plane 2**, and **Plane 3**. Click **OK**. Go back to the **Details of Contour 1** window, find the **Variable** list, and set **Pressure** with **Local Range** (just below) *(5)*. Increase **# of Contours to 75** *(6)* and **Apply** *(7)* settings.

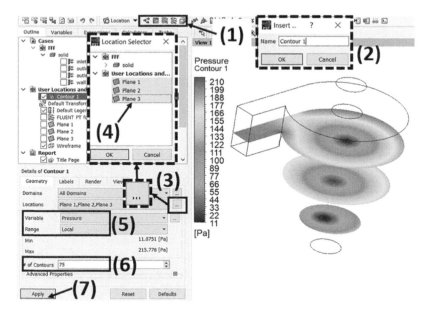

FIGURE 17.44

The option that is very useful in the case of discrete phase modeling with the **DPM** approach is the particle track animation. If at the end of a simulation, there are any incomplete particles and the continuity (or any other) equation is unstable (no convergence), an animation allows us to find very quickly the area in which the particles 'stop' for some reason. The problem comes quite frequently

Tutorial 8 – Particulate Matter Separation in Cyclone

from the poor mesh quality in this region, so it can be easily fixed. Let's practice running animations now (Figure 17.45). Activate displaying DPM trajectory *(1)* again and click *2× LMB* on *FLUENT PT for Ash Solid* to open the *Details of FLUENT PT for Ash Solid* window. Go to the *Symbols (2)* tab, mark the *Show Symbols (3)* option, and *Apply (4)*. Find and click the clock thumbnail *(6)* on the top toolbar. In the displayed *Animations* window, mark the DPM trajectory (*FLUENT PT for...*) *(7)* on the *Select one or more...* list. Before running the animation, it is recommended to reduce its speed using the slider *(8)*. Then run the animation *(9)*. If you want to keep the animation, you can mark the *Save Movie (10)* option. The *Infinity (11)* mode cannot be active then. In *Options*, you can also adjust quality and resolution.

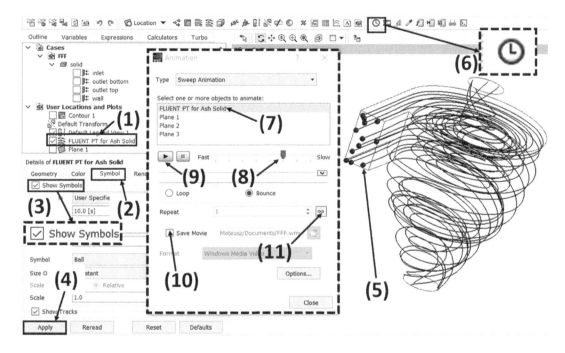

FIGURE 17.45

18 Tutorial 9 – Accumulation Heat Exchanger

18.1 EXERCISE SCOPE

This tutorial is dedicated to the simulations of the two accumulation heat exchangers presented in Figure 18.1. Both of them consist of the same number of prefabricated elements but are arranged differently. In the course of the simulation, it is possible to assess the influence of the internal channel shape on the heat accumulation process.

Try to recall from the theoretical background (Section 15.3.4) the methods of thermal energy storage which may be coupled with RES-based technologies. This exercise allows us to examine one of them, based on heat accumulation in ceramic elements.

Among others, this exercise can teach you how to

- split geometry into smaller bodies in *SpaceClaim Direct Modeler* to generate an excellent quality mesh in *ANSYS Meshing*,
- define properties of a given material and create interfaces to run a simulation of the model with an inconsistent mesh in *Ansys CFX*,
- calculate the thermal power of a heat exchanger and visualize the flow field of flue gas in *Ansys Results*.

Pay special attention to the way of geometry division which makes it possible to generate a fully structured mesh.

All of the images included in this tutorial use courtesy of ANSYS, Inc.

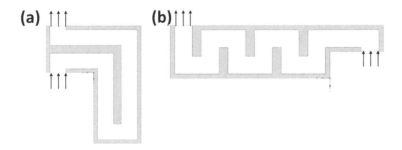

FIGURE 18.1 Geometries of the analyzed accumulation heat exchangers with marked inlets and outlets: (a) Case 1, (b) case 2.

18.2 PREPROCESSING – GEOMETRY

Drag the ***Geometry*** module *(1)* to the ANSYS Workbench ***Project Schematic***, as in Figure 18.2. To open the geometry editor, double-click the left mouse button *(LMB)* on the ***A2*** cell *(2)*.

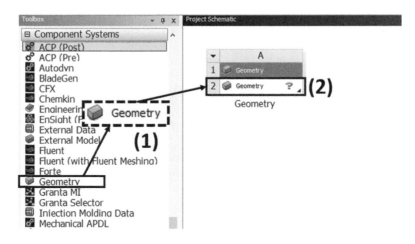

FIGURE 18.2

The ***Sketch*** tab will open (Figure 18.3). Firstly, click LMB on the ***Plane View*** *(1)* to view the ***XZ*** grid head-on. Then, draw the external shape of the accumulation heat exchanger using the ***Line*** tool from the ***Create*** ribbon *(2)*. Start drawing from the coordinate system origin *(3)* and set the dimensions as shown in the picture. You can check the sketch dimensions with the ***Dimension*** tool *(4)*. Finish your work by switching to the ***Design*** tab *(5)*.

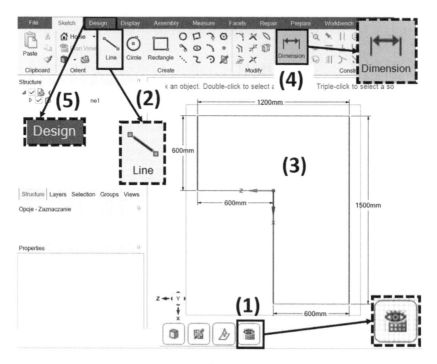

FIGURE 18.3

Tutorial 9 – Accumulation Heat Exchanger

In the ***Design*** tab click on the ***Pull*** tool *(1)* from the ***Edit*** ribbon, which will allow you to create a 3D object from the previously prepared sketch (Figure 18.4). Firstly, select the surface created from the sketch *(2)*. Then, pull the sketch downwards (along the *Y*-axis) and set the thickness to ***150 mm***. After that, return to ***Sketch*** mode *(3)*.

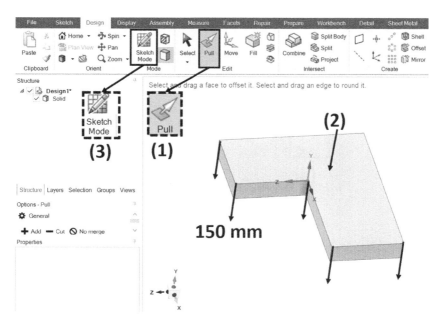

FIGURE 18.4

The main body is ready, so it is time to prepare the internal channels. Start by choosing ***Select New Sketching Plane** (1)*. It should be placed in the ***XZ*** plane, so click on the upper surface of the created geometry. Draw a shape of the channel using the ***Line*** tool *(2)*. Set the dimensions as shown in Figure 18.5 *(3)*.

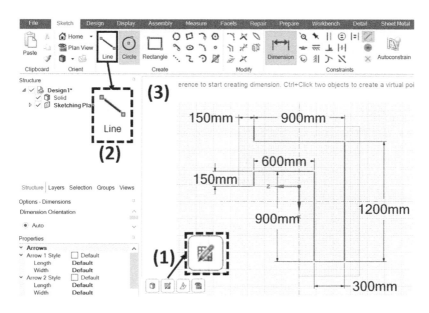

FIGURE 18.5

Now it is time to design the cross-section of the internal channel, which will be swept by the prepared trajectory. *Select New Sketching Plane (1)* in the *YZ* and set your view perpendicularly to it *(2)*. Using the *Line* or *Rectangle (3)* tool from the *Create* ribbon, draw a rectangle of dimensions *100 mm × 200 mm*. As shown in Figure 18.6, one of its sides should be located on the *Z* axis, whereas the other side *50 mm* from the geometry edge. Finish your sketch by clicking on the *End Sketch Editing icon (4)* to be switched automatically into the *Design* mode.

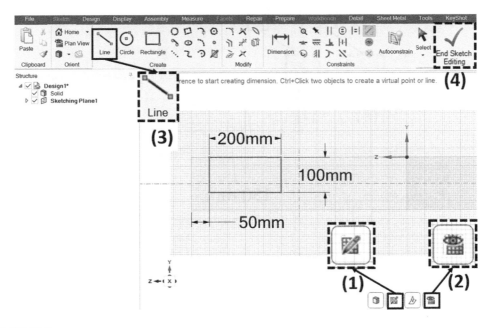

FIGURE 18.6

In this step, the geometry of the internal channel will be created. Click on the *Pull* tool *(1)* and choose the *Sweep* option *(2)*. Firstly, select a trajectory to sweep along *(3)* as in Figure 18.7, and confirm by pressing *Enter*. Then, pick the rectangular surface *(4)* created in the last step. Ensure that the

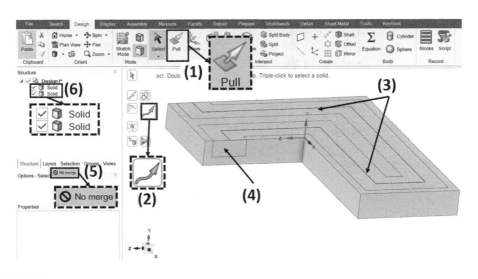

FIGURE 18.7

Tutorial 9 – Accumulation Heat Exchanger   377

No merge mode *(5)* is active to prevent unification of the newly created body and the existing one. Click *Enter* to generate the sweep operation. As a result, two bodies should be visible in the project *Structure (6)*. By clicking on them, you could see that they are overlapping in the space where fluid in the internal channel should be located.

To remove the solid from the internal channel, use the *Combine* tool *(1)* from the *Intersect* ribbon. Ensure that the *Cutter (2)* option is active and choose the receiver body *(3)* as the target object. Then, select the internal channel body as the cutter object *(4)* and again click on the fluid, so that the overlapping geometry can be deleted. This operation will result in two bodies that are completely separate *(5)* as shown in Figure 18.8. Check which body represents the solid and which represents the fluid in the internal channel. You can do this by clicking on the *V* sign in the square located to the left of the body name. Then, change the names of these bodies to solid and fluid, respectively *(5)*.

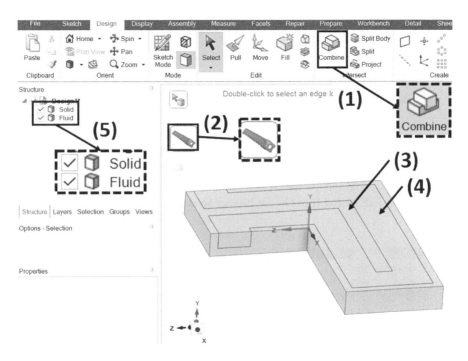

FIGURE 18.8

To provide smooth meshing in the next step, the geometry should be sliced into smaller cuboids. To do so, select edges that outline the fluid and copy them *(1)*. Then, hide the solid and fluid body *(2)* for better visibility. Select the *Curves* from the outline and *Pull* them *(3)* downwards for any dimension *(4)*. A new *Surface* object will appear in the Project *Structure (5)* as in Figure 18.9.

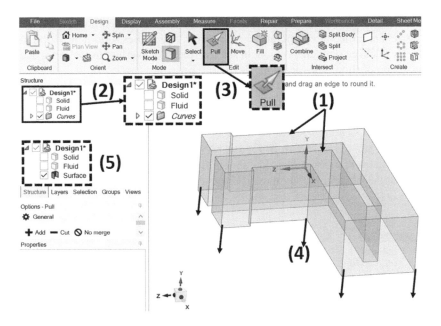

FIGURE 18.9

Use the generated surfaces to split the geometry into smaller bodies. Choose the option ***Split Body*** *(1)*, and select the fluid and the solid domain as the target bodies *(2)*, which will be sliced. Ensure that the ***Cutter*** *(3)* option is active and use the created surfaces as cutting planes *(4)*. Numerous bodies will be created as in Figure 18.10 and listed in the Project ***Structure***. Find the ***Surfaces*** object there and delete it *(6)*.

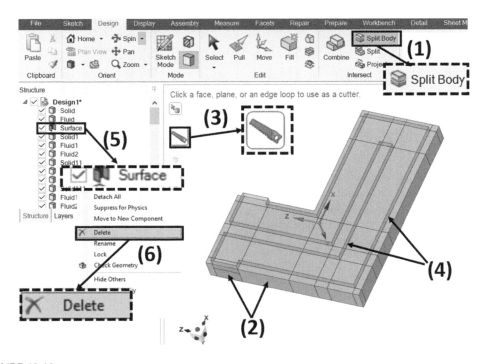

FIGURE 18.10

Tutorial 9 – Accumulation Heat Exchanger

In the previous step, numerous cuboids were created. Nevertheless, not all of them are necessary. Use the **Combine** tool *(1)* to merge smaller elements from the sliced geometry *(2)* into bigger ones *(3)* as shown in Figure 18.11.

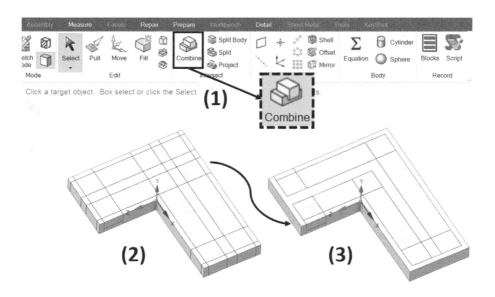

FIGURE 18.11

The final part of the geometry preparation is to create two components from the bodies visible in the Project **Structure**. Select all the bodies representing fluid *(1)* and choose the **Move to New Component** option *(2)* from the popup window. The Properties **Share** the topology *(3)* of the component (Figure 18.12). Repeat this step with solid bodies to obtain only two components in the final version of the geometry.

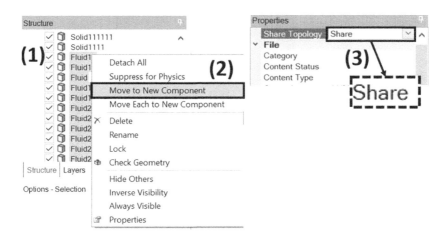

FIGURE 18.12

18.3 PREPROCESSING – MESHING

Select the *Mesh* module *(1)* from the *Component System* ribbon and drag it to the ANSYS Workbench *Project Schematic (2)* as in Figure 18.13. Create an appropriate connection between the geometry and mesh modules *(3)*. Double-click the left mouse button *(LMB)* on the *B3* cell *(4)* and open *Mesh*.

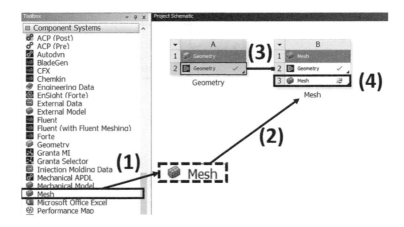

FIGURE 18.13

Find the *Geometry (1)* icon in the project tree and check if there are two separate bodies: solid and fluid (Figure 18.14). Then, move to the *Mesh (2)* option. In the *Details of Mesh* window, change the appropriate basic settings that are required for CFD simulations: *Physics Preference – CFD (3)* and *Solver preference – CFX (4)*. To limit the maximum size of the mesh elements, change the *Element Size* value from default to *1 mm (5)*. Be careful and pay attention to the general length unit!

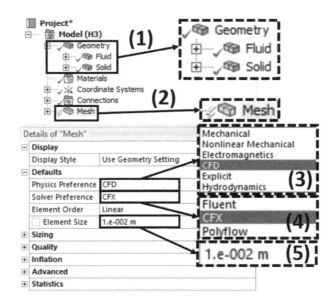

FIGURE 18.14

Tutorial 9 – Accumulation Heat Exchanger

Use the ***Volume Selection*** mode *(1)* and select both: Fluid and solid bodies *(2)* as in Figure 18.15. Click 1× (right mouse button) RMB and choose the option ***Insert (3)*** → ***Method (4)***. In the ***Method*** row, select the appropriate option – ***MultiZone (5)*** in this case. Generate the mesh by clicking *1×* ***RMB*** on the ***Mesh (6)*** and selecting the option ***Generate Mesh (7)*** from the popup window.

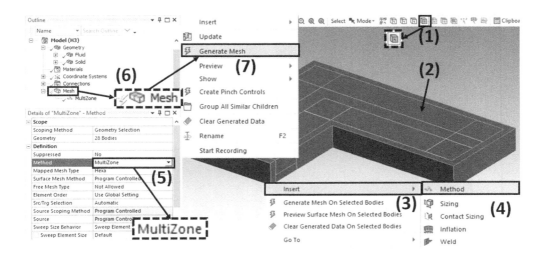

FIGURE 18.15

Find the ***Quality*** option in ***Details of Mesh*** and set the ***Mesh Metric*** parameter as ***Skewness (1)***. It describes the difference between the shape of the cell and the shape of an equilateral cell with the same volume. For most applications, skewness should not exceed 0.95. Rate the mesh quality based on the minimum, maximum, and average values of the analyzed parameter *(2)*. In the ***Statistics*** tab, check the number of mesh ***Nodes*** and ***Elements (3)***. Look at the mesh appearance (Figure 18.16) – it consists of cubic elements with an edge length of *1 mm (4)*.

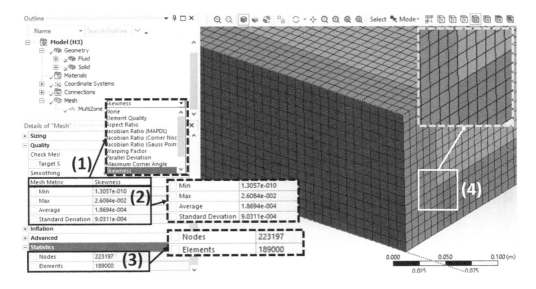

FIGURE 18.16

This regular shape of mesh elements results in an excellent mesh quality (Figure 18.17). To achieve this, the geometry should be divided into smaller parts *(1)*, as described in the previous part of this exercise. Without this initial division, the mesh will contain some elements with an irregular shape *(2)*. They will negatively influence the overall mesh quality if there are more of them.

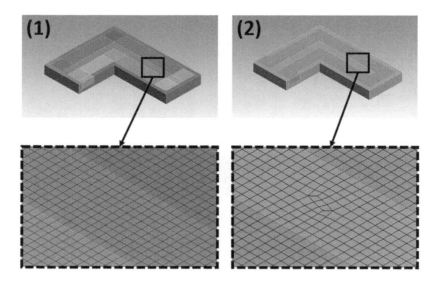

FIGURE 18.17

Evaluate mesh *Quality* in a more detailed way. In the *Mesh Metrics* window, check the shape of the mesh elements *(1)* and their participation in the total number of mesh elements and quality (Figure 18.18). To locate the elements characterized by the worst quality, change the options of the displayed chart by clicking on the *Controls (2)* button in the *Mesh Metrics* window. In the popup window, change the minimum value of the *X-Axis* to 0.02 and the maximum value of the *Y-Axis* to 60 (the number of the worst elements is low). There is no *Apply* button, just click *X* in the corner of the popup window *(3)*. Click on the last bar displayed on the chart *(4)*. The worst-quality elements are now visible in the workspace *(5)*. Then, you can check another quality parameter.

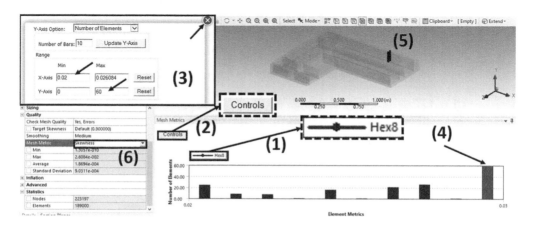

FIGURE 18.18

Tutorial 9 – Accumulation Heat Exchanger

Named Selection is the function that allows to assign names to different objects to make them easier to find in the solver. Remember: Use the *Face Selection* mode to name a surface and use the *Volume Selection* mode to name a body *(1)*. Select all fluid elements as in Figure 18.19, click *LMB* and then choose the *Create Named Selection (3)* option from the dropdown list. You can also just press the *N button* from the keyboard. Then, insert the name for the new *Named Selection (4)*: *fluid* and confirm by clicking *OK*.

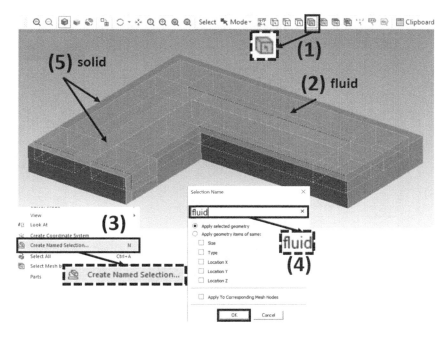

FIGURE 18.19

Repeat the same step and name other surfaces *(1)*: *Inlet*, *outlet*, *wall* (side and bottom surfaces of geometry), and *symmetry* (surfaces located on the upper symmetry plane) – see Figure 18.20. Their names should appear in the project tree, below the *Named Selections* tab *(2)*. When you click on it, you can see the related object highlighted in red in the workspace.

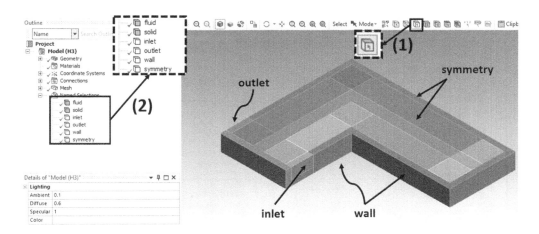

FIGURE 18.20

Because the geometry consists of two separate parts, an interface should be created between them (Figure 18.21). It allows to transfer information from one domain to another during calculations. To make the settings in the solver easier, it is important to name appropriate surfaces properly, using *Named Selection*. Firstly, find the *Fluid* domain in the project tree *(1)* and click RMB on it. From the popup window, choose the *Suppress Body (2)* option to hide the fluid for physics. Using *Surface Selection (3)*, choose all the surfaces where solid and fluid bodies are in direct contact and then name them as *interface_s-f (4)*, which means that the data is transferred from the solid body to the fluid domain. Then, unsuppress the fluid body and suppress the solid. Name all the surfaces where the fluid is in direct contact with the solid as *interface_f-s (5)*. Restore the solid body by clicking *1× RMB* on its name in the project tree *(1)*, then select *Unsuppress* from the popup window list.

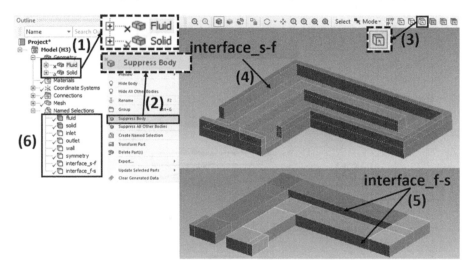

FIGURE 18.21

18.4 REPROCESSING – SOLVER SETTINGS

Drag the *CFX (1)* module to the *Project Schematic* in Workbench and set an appropriate connection *(2)* between the *Mesh* module (B3 cell) and *Fluent Setup* (C2) as in Figure 18.22. Click *1× RMB* on *B3* cell *(3)* and choose the *Update* option *(4)* to create the input file with the generated mesh data and export it to the solver. Start *CFX* by clicking *2× LMB* on the *C2* cell.

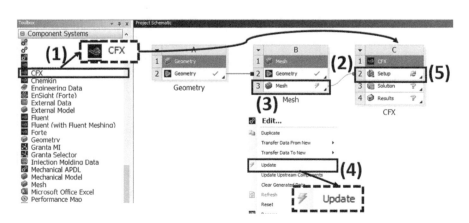

FIGURE 18.22

Tutorial 9 – Accumulation Heat Exchanger

Before you start the model setup, make sure that the ***Automatic Default Interfaces*** option is activated. Click ***2xLMB*** on ***General*** (in the ***Case Option*** tab) at the top of the project tree. And mark all checkboxes below the ***Auto Generation*** inscription. Then ***Apply*** the settings and close the window.

Now assign solid and fluid domains to the appropriate geometry elements (Figure 18.23). To do so, click the ***Domain*** icon *(1)* and insert ***Name*** as *Fluid* and confirm ***OK (2)***. In the project window, a new tab will appear. In ***Basic Settings (3)***, set the ***Location*** of the new domain as *fluid (4)* – this is a named selection assigned in the ***Mesh*** module. The filling of the internal channel should be highlighted in green. Choose the ***Domain Type*** as ***Fluid (5)*** and set the ***Material*** as ***Air Ideal Gas (6)***. This is a simplification that gives satisfying results during an analysis of the relation between the shape of a heat accumulator and the effectiveness of heat transfer. Nevertheless, in a more detailed case, the exhaust gases should be defined as a mixture. Activate ***Buoyant Option (7)*** to investigate the impact of gravity. Set the appropriate values of gravitational acceleration in each direction – in this case, *−9.81 m/s²* in the *X* direction and *0* in *Y* and *Z* directions. Define also reference density as *1.26 kg/m³ (7)*.

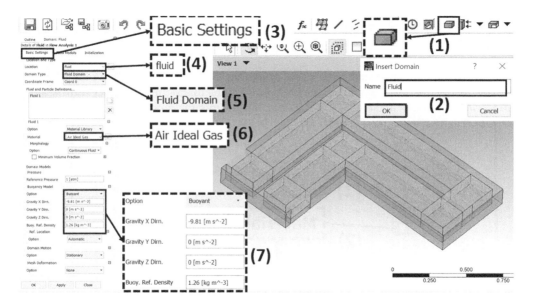

FIGURE 18.23

Now it is time to select appropriate mathematical models of physical phenomena occurring in the analyzed case (Figure 18.24). Switch to the ***Fluid Models (1)*** tab and activate the ***Thermal Energy (2)*** for heat transfer. ***Thermal Energy*** is used for incompressible low-speed flows. Then, choose the turbulence model for fluid flow as ***k-Epsilon (3)*** with ***Scalable Wall Functions (4)***, which is suitable for meshes without an inflation layer. ***Apply*** all the changes *(5)*. Remember that each operation in ***CFX-Pre*** should be confirmed by ***Apply***, not ***OK***. Generally, in the case of exhaust gases of high-temperature, the ***thermal radiation*** model should be included to obtain more reliable results. The analyzed case is a simplification made to compare two different geometries of an accumulation heat exchanger, so the application of radiation is not necessary.

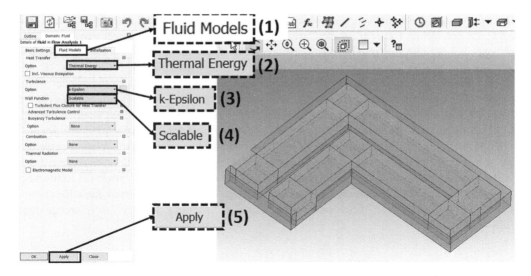

FIGURE 18.24

The accumulation heat exchanger is made of *akubet*, which is not specified in the *CFX* database. New material should be defined manually (see Figure 18.25). In the project *Outline (1)*, find the *Materials (2)* option and click *1× RMB* on it. Then, from the dropdown list, choose *Insert → Material (3)* and set its name as *akubet (4)*.

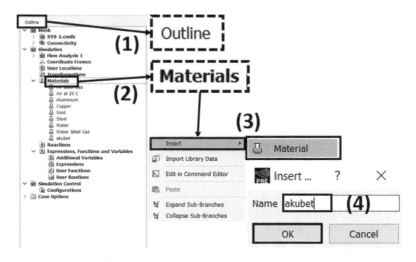

FIGURE 18.25

In *Basic Settings (1)*, set the material type as *Pure Substance (2)* and its *Thermodynamic State* as *Solid (3)*. Switch to the *Material Properties (4)* tab to describe *molar mass (5)*, *density (6)*, and *specific heat capacity (7)* as in Figure 18.26. For these properties, define constant values as

$$M = 22.3 \left[\frac{kg}{k\,mol}\right], d = 2556 \left[\frac{kg}{m^3}\right], c_p = 670 \left[\frac{J}{kg\,K}\right].$$ Confirm all changes by *Apply (8)*.

Tutorial 9 – Accumulation Heat Exchanger

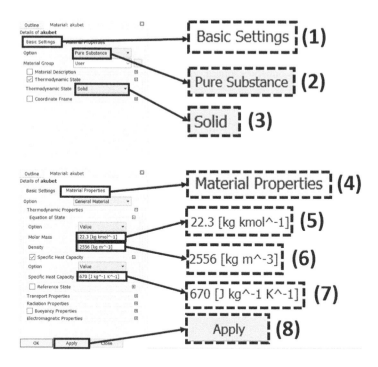

FIGURE 18.26

When the new material is created, insert a new *Domain (1)* and name it *Solid (2)* as in Figure 18.27. In *Basic Settings (3)*, set the *Location* of the new domain as *solid (4)* – it should be highlighted in green. Choose *Domain Type* as *Solid (5)* and set the *Material* as *akubet (6)*. In the *Solid Models* tab *(7)* activate the *Thermal Energy* equation *(8)* and confirm all changes by clicking *Apply (9)*.

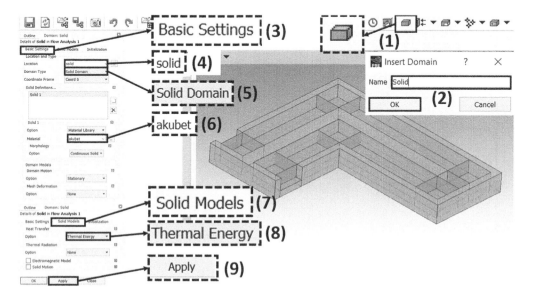

FIGURE 18.27

To provide data transfer between solid and fluid parts of the geometry, the interface should be defined (Figure 18.28). In the project *Outline (1)*, find the *Interfaces (2)* option. Click *1× RMB* on the default interface name listed below and from the dropdown list choose *Edit (3)*. In case when there is no default interface, it can be created by clicking on the *Domain Interface* icon *(4)* and inserting a new *Name (5)*.

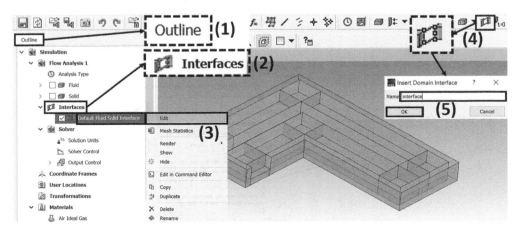

FIGURE 18.28

Basic Settings (1) allows to set the interface type as *Fluid Solid (2)*. In the next step, define *Interface Side 1* as *interface_f-s* located in the *Fluid* domain *(3)*. The second side of this interface is located in the *Solid* domain on the surfaces named *interface_s-f (4)*. *2x LMB* on this option and in the popup window create a new interface (Figure 18.29). Finally, *Apply (5)* inserted changes.

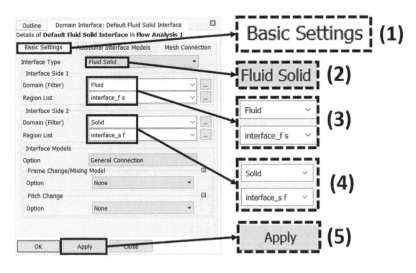

FIGURE 18.29

Tutorial 9 – Accumulation Heat Exchanger

It's time to define *Inlet*, the first boundary condition (Figure 18.30). Firstly, click on the *Boundary* icon *(1)* and select its general location *in Fluid (2)*. Insert the *Name* as *Inlet (3)* and then move to the *Basic Settings* tab *(4)*. Set *Boundary Type* as *Inlet (5)* and specify its *Location* as *inlet* named selection *(6)*. Move to the *Boundary Details* tab *(7)* and choose the option of the inlet definition as *Static Pressure* with *0 Pa* of *Relative Pressure (8)*. Turbulence should be described by *Intensity* (a default value of *5%* is OK). Set the *Static Temperature* to *600 K (10)* and *Apply* all changes *(11)*.

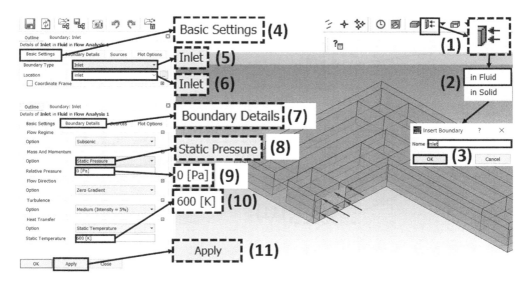

FIGURE 18.30

Repeat the previous procedure and add the outlet boundary condition as in Figure 18.31 *(1–3)*. In *Basic Settings (4)*, select the *Boundary Type* as *Opening (5)* and locate it on the *outlet* surface *(6)*. Move to the *Boundary Details (7)* and define the outlet by *Opening Pressure (8)* equal to *−10 Pa*. The reversed flow of exhaust gases is expected, so in the *Heat Transfer* option describe the *Opening Temperature* as an average temperature at the outlet. To do so, insert the expression: *areaAve(T)@outlet* and *Apply* all changes.

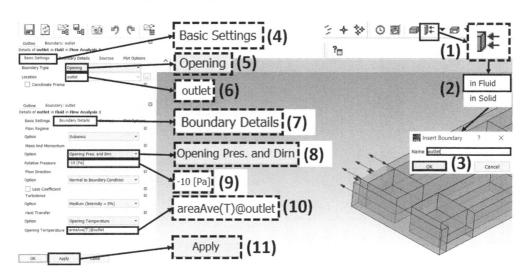

FIGURE 18.31

Add another Boundary condition, this time *in* the *Solid* domain (Figure 18.32): the wall *(1–5)* located on surfaces named as *wall (6)*. Define the *Boundary Details (7)* as *Heat Transfer Coefficient (8)* equal *12 W/(m² K) (9)* and set the *Outside Temperature* to *293 K (10)* – the temperature of thermal comfort in residential buildings. *Apply* changes *(11)*.

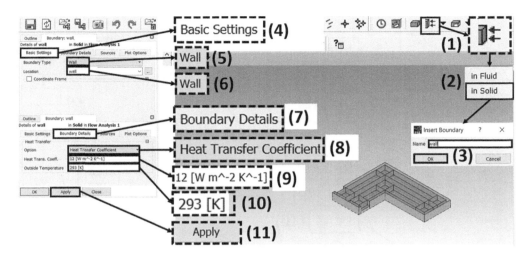

FIGURE 18.32

The last two boundary conditions are located on the symmetry plane. Assign the appropriate surfaces to them as shown in the picture. *Symmetry_s (1)* is defined in the solid domain, whereas *symmetry (2)* is located in the fluid (compare with Figure 18.33).

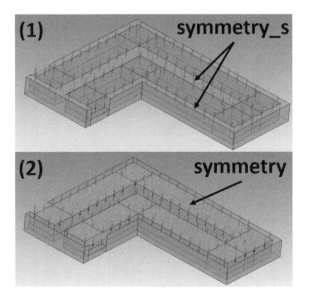

FIGURE 18.33

Tutorial 9 – Accumulation Heat Exchanger

All the inserted boundary conditions should be visible in project *Outline (1)* below the *Fluid* and *Solid* domains *(2)* as in Figure 18.34. When the physics for the model physics is determined, the next step is to define some details of the solution process in *Solver (3)*. Find the *Solver Control (4)* and the *Output Control (5)* options. Choose the first one *(4)* and move to the next step.

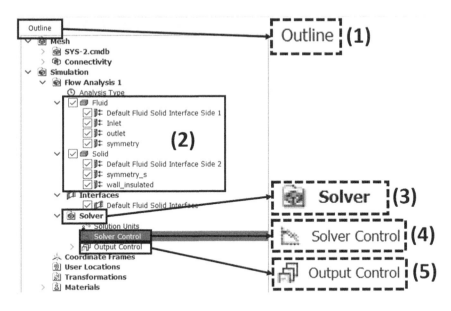

FIGURE 18.34

In *Solver Control (1)*, change some *Basic Settings (2)* like: *Advection Scheme* to *Upwind (3)* and *Turbulence Numerics* to *First Order (4)*. Set the *Max. Iterations* value to *500 (5)* – it should be enough to reach the convergence with *Residual Target RMS* (Root Mean Square) equal 10^{-4} *(6)*. Confirm all changes *(7)* (Figure 18.35).

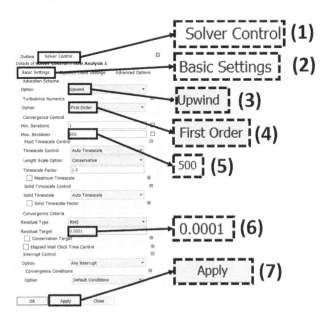

FIGURE 18.35

To supervise the computation process, create at least one *Output Control (1)* as in Figure 18.36. Switch to the *Monitor (2)* tab and add a *New* monitor point *(3)*. Define it by *Expression (4)* which calculates the average temperature of exhaust gases at the outlet from the heat exchanger. Insert formula: *areaAve(T)@outlet (5)* and *Apply (6)*. The plot of the outlet temperature will be printed into the console during calculations and the shape of the curve will indicate whether the final temperature has been stabilized. After finishing this step, exit the *CFX-Pre* module.

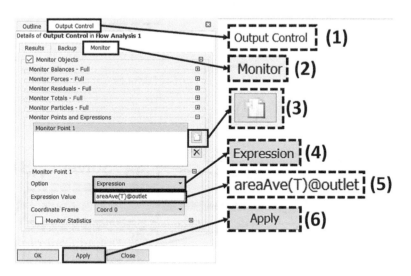

FIGURE 18.36

It is time to start the calculations (Figure 18.37). *Save* your project from Workbench *(1)* and then click *2× LMB* on the *Solution* cell *(2)*. The popup window will appear, where *Run Mode* should be set as *Local Parallel* to reduce the calculation time. Set the number of *Partitions* to *4 (4)* and finally *Start Run (5)*.

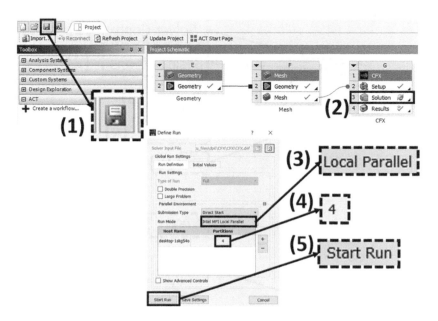

FIGURE 18.37

Tutorial 9 – Accumulation Heat Exchanger   393

The level of the accuracy of the computation may be observed on the chart, as in Figure 18.38 *(1)*. It is possible to change the displayed data by clicking on appropriate tabs *(2)*. In the last one, the monitor points created by the user may be found. The information that the calculation is complete will appear in the solver console *(3)* and the popup window.

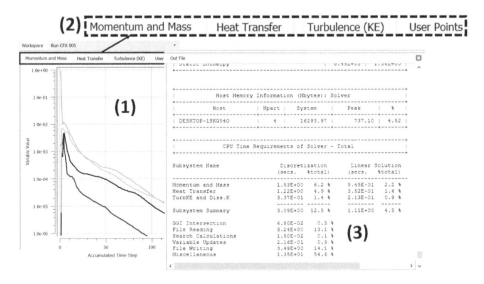

FIGURE 18.38

18.5 POST-PROCESSING

There is no need to add the *Results* module from *Component Systems* because it is included in the *CFX* module. Click *2× LMB* on the *Results* cell *(1)* to launch the post-processing mode (Figure 18.39).

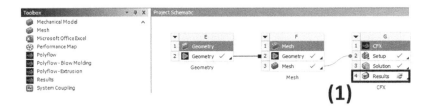

FIGURE 18.39

Select the *Contour (1)* tool and insert a *Name (2)* as in Figure 18.40. In the *Details of Contour* window, select *All Domains (3)* and specify *Locations* as the all symmetry plane. To do this, click on the three dots icon *(4)* and select the necessary surfaces. Confirm your choice by clicking *OK* or Enter. Then, choose the *Variable* which will be displayed: *Temperature (5)* and its *Range: Global (6)*. The *# of Contours* cell increase the number of temperature isolines to *25 (7)*. Confirm the settings by *Apply (8)* and enjoy the first contour. Deactivate the generated temperature distribution by clicking the tick in the project *tree window (9)*. Create the second contour – visualize the local temperature changes of density in the fluid domain.

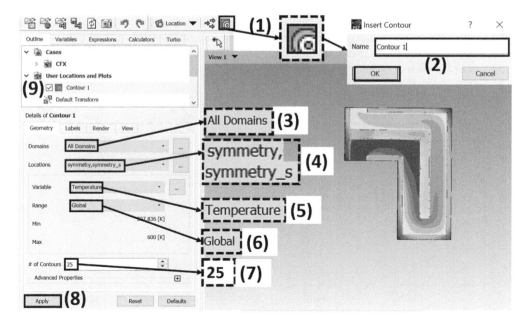

FIGURE 18.40

Now generate **Vectors** to visualize the velocity distribution *(1–2)* as in Figure 18.41. Specify the **Domain** as *fluid (3)* and display vectors on the symmetry plane *(4)*. Choose the **Sampling** type as **Equally Spaced** *(5)* to obtain a more legible visualization. Set the **# of Points** as *2000 (6)* and confirm by clicking **Apply** *(7)*. Like in the case of contours, it is possible to hide the generated vectors *(8)*.

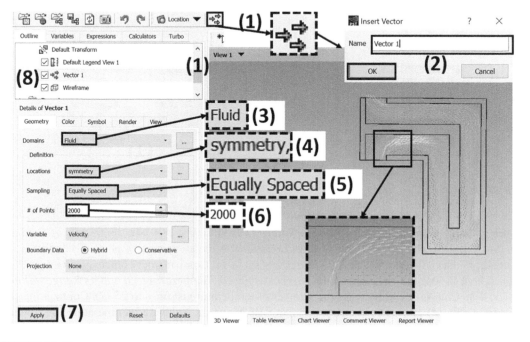

FIGURE 18.41

Tutorial 9 – Accumulation Heat Exchanger

Another way to visualize the fluid movement is to use streamlines (Figure 18.42). Find the *Streamline* icon *(1)* and confirm the default *Name (2)*. Define the *Domain* as fluid *(3)* and set the *inlet (4)* as the face from which the streamlines start. Insert *# of Points* as 25 *(6)* and click *Apply (7)* to create a new visualization. Hide the created streamlines.

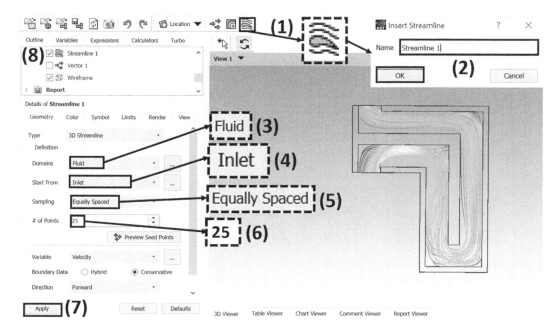

FIGURE 18.42

To calculate the average temperature of a wall with an accumulation heat exchanger (Figure 18.43) insert a new *Expression (1)* and *Name* it *Twall (2)*. Define the formula as: *areaAve(T)@wall (3)* and *Apply (4)*. The result will be displayed in the *Value* row *(5)*. Similarly, try to calculate the thermal power of a heat exchanger. Firstly, define the necessary variables, that is, mass flow *(m)* as *massFlow()@inlet*, and temperature difference *(dT)* between the *inlet* and the *outlet* as *areaAve(T)@outlet-areaAve(T)@inlet*. Then, create the final expression for thermal power *(P)* using the previously defined parameters: *m * 1004.4 [J*kg^–1*K^–1] * dT*.

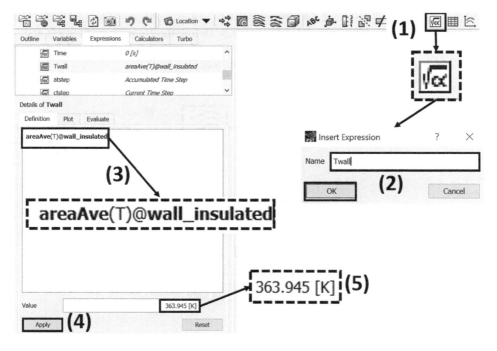

FIGURE 18.43

18.6 ADDITIONAL EXERCISE

Create the second geometry of an accumulation heat exchanger as shown in Figure 18.44. It consists of the same number of prefabricated elements, but the shape of the internal channel is different. Analyze the new geometry as described in this manual and compare the results. Which geometry provides better heat accumulation?

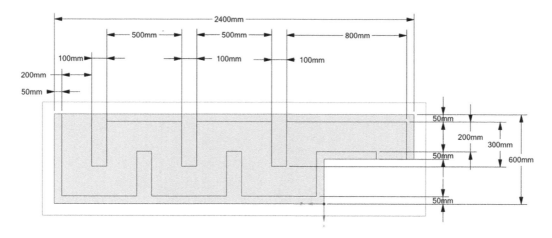

FIGURE 18.44

Part VI

Geothermal-Energy-Based Systems

19 Theoretical Background

19.1 DEVELOPMENT OF TECHNOLOGY

Man has been harnessing geothermal energy for centuries. Wherever hot water gushes out of the ground, people have been using it for bathing, washing, or cooking. What is more, some sources confirm that geothermal springs have been even used for medical purposes [1], like in New Zealand or North America. The latter application of geothermal energy was also well-known by ancient civilizations of Greece, Rome, and also Japan [1]. Furthermore, thanks to the neighborhood of Vesuvius and its influence on the underground water, the inhabitants of today's Naples area have been able to introduce geothermal heating to their houses [2]. As reported in Ref. [1], the first modern application of commercial district heating took place in the USA (Boise, Idaho) and it happened as early as the end of the 19th century. Also at that time arose an idea of applying geothermal energy for big-scale electricity generation.

19.2 STATISTICAL DATA

Figure 19.1 shows the installed capacity of geothermal-energy-based systems in the top ten countries around the world in 2021. It is worth mentioning that the same ten countries have dominated this ranking since 2014, as the possibilities of gathering geothermal energy are the highest close to the areas of volcanic activity [4]. The first three places belonged to the USA, Indonesia, and the Philippines, with 3889 MW, 2277 MW, and 1928 MW installed capacity, respectively. Until 2017,

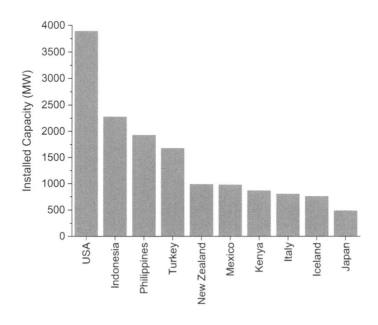

FIGURE 19.1 Top ten countries with the highest installed capacity of geothermal-energy-based systems in 2021. (Based on Ref. [3].)

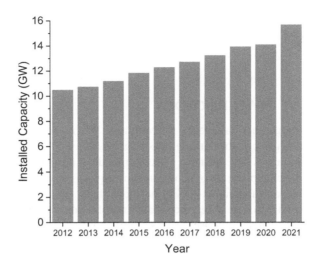

FIGURE 19.2 Cumulative installed capacity of geothermal-energy-based systems in the world. (Based on Ref. [3].)

the third place had been occupied by Indonesia, but in 2018 this country overtook the Philippines. Moreover, it is predicted that in the next decade, Indonesia will become a world leader because it has got the largest geothermal reserves in the world [5]. The fourth place belongs to Turkey, with 1677 MW of geothermal installations. All the remaining countries: New Zealand, Mexico, Kenya, Italy, Iceland, and Japan have less than 1 GW of installed capacity. Nevertheless, in Iceland, New Zealand, Kenya, and the Philippines, geothermal energy is the key renewable source, which covers a significant share of electricity and heat demand [3].

Figure 19.2 presents the installed capacity of geothermal-energy-based systems in the world during the 10 years from 2012 to 2021. Generally, in the analyzed period, the cumulative installed capacity constantly increased from 10 GW in 2010 to nearly 16 GW in 2021. The most significant net addition was observed in 2021 with 1571 MW. The global average cost of geothermal-energy-based systems constantly fluctuated over the 10 years. In 2012, the total installed cost of 1 kW was 5317 USD and increased to 4468 USD in 2021. The year-to-year variations are caused by the relatively small market for geothermal power and specified on-site conditions [4].

Figure 19.3 presents the electricity generation from geothermal sources in 2019 for ten top countries. This year, the global electricity production from geothermal technologies was about 92 TWh. The 20% share belongs to the USA, which takes the first place in the rank. They are followed by Indonesia (14 TWh) and the Philippines (11 TWh). All the other countries are characterized by energy generation lower than 10 TWh. The last place is occupied by Japan, where energy production is 2800 GWh. All the countries presented in this rank are located in tectonically active regions, where the occurrence of high-/medium-temperature geothermal resources is abundant.

19.3 CLASSIFICATIONS AND CHARACTERISTICS

Most of the heat of the Earth's core comes from nuclear reactions taking place in the core at the Sun's surface temperature – around 5500°C. Of course, it changes with depth, and the temperature of the ground layers reachable for exploitation is much lower. However, thermal energy occurring in the upper 50 km of the crust is still about 5.4×10^9 EJ [6].

Nowadays, the most common application of the discussed energy is for geothermal heat pump power supply. Apart from space heating, geothermal is used in a wide range of different areas of life, starting with greenhouse and aquaculture pond heating or agricultural drying, through many

Theoretical Background

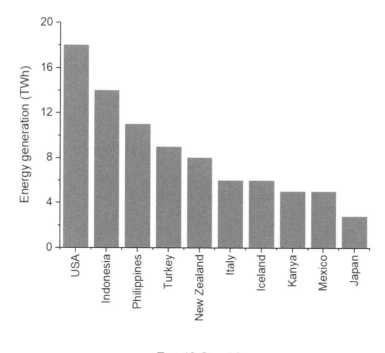

FIGURE 19.3 Top ten countries with the highest electricity generation from geothermal-energy-based systems in 2019. (Based on Ref. [3].)

industrial applications and power generation, finishing with bathing and swimming or even snow melting [7].

Depending on the purpose for which it is to be used, different temperature sources of geothermal energy are required. This constitutes the base for the general classification of geothermal system types, which can be divided into three groups, characterized by different potentials.

Low-temperature geothermal systems use the ground to a depth of about 100m and usually work at temperatures up to 10°C. Such systems can be used to power heat-pump-based space heating or to prepare domestic hot water.

Medium-temperature geothermal systems use geothermal water available even at a depth of 3km and work at temperatures around 100°C. They can be applied for recreational and medical purposes, as well as a power source for geothermal heating plants. In this case, the probe operates in sedimentary rocks, like sandstones, dolomites, or limestones.

The third group covers high-temperature systems, including so-called enhanced geothermal systems (EGSs) that get heat from hot dry rocks (HDR), like granites and basalts (up to 6km into the ground). EGSs work at temperatures close to 200 and higher, which allows them to produce steam used for electricity generation.

When it comes to small-scale energy applications of low-temperature geothermal systems, like in the case of households, two kinds of heat exchangers for drawing heat from the ground are in use. The horizontal ground collector uses energy accumulated in the top ground layers. It is installed at 120–150cm depth, so there is no need for an expensive well. However, the whole installation requires a lot of space to provide enough power to supply the heat pump. In contrast, vertical probes use heat from deeper ground layers by U-shape, meander, or spiral plastic probes, which can reach even 200m down the ground.

In the case of the solutions based on groundwater, different types of well systems can be applied. Based on Ref. [6], four general designs can be distinguished:

- single-well extraction system (SWES), where only production well extracting hot water is applied to supply the heat exchange station (then water is discharged directly to the surface reservoir),
- two-well circulation system (TWCS), adding to SWES an individual injection well (injecting cold water to the ground water reservoir),
- standing column well system (SCWS) – using the same well for the extraction of hot groundwater and the injection of cold water,
- single-well circulation system – combining SCWS and CWTS by applying only one well to extract and inject water, but the inlet and outlet streams are separated in the groundwater reservoir by a sealing system.

Injecting cold water into the reservoir serves two purposes – it helps to avoid gradual exhaustion of groundwater (what can potentially happen in the case of the SWES system) and environmental changes and, at the same time supplies the medium which can store heat in the reservoir. Of course, solid material in an appropriately designed heat accumulator can perform the same function as water. Nowadays, this solution is considered as one allowing to increase the efficiency of renewable (especially solar) energy use. Four types of thermal energy storage systems can be listed here

- hot-water thermal energy store (HWTES),
- borehole thermal energy store (BTES),
- aquifer thermal energy store (ATES),
- gravel-water thermal energy store (GWTS).

According to Ref. [8], BTES (and ATES) are currently among the most commonly used systems. Borehole thermal energy storage (BTES), as presented in Figure 19.4, is a closed-loop system that is capable of accumulating large amounts of heat for a long time [9]. BTES uses the ground as a

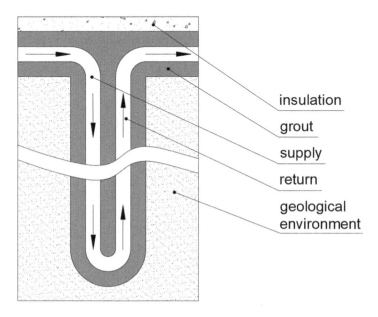

FIGURE 19.4 Construction of a single borehole.

Theoretical Background

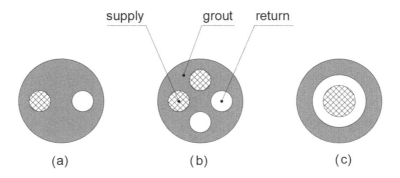

FIGURE 19.5 Three types of tube arrangement in BTES: (a) Single U-tube, (b) double U-tube, and (c) coaxial tubes.

thermal battery that seasonally stores or abstracts the heat. The main element of this facility is an array of ground heat exchangers installed vertically or horizontally. Application of numerous boreholes instead of a single one provides more efficient heat storage [10]. The whole array is usually shaped as a cylinder with similar dimensions of a borehole diameter and depth [11], but there are also known efficient hexagonal or quadratic arrangements [12].

Each borehole has a 100–150 mm diameter and length of up to 350 m. The inner tube is usually made of polyethylene material [12] due to its beneficial properties, such as low density, corrosion resistance, flexibility, and low price [12]. However, for extreme operating conditions, steel pipes are preferred [13]. The inner tube is surrounded by ground. Additionally, a small upper part of the pipe is reinforced and insulated to protect the construction from unfavorable weather conditions, mechanical damage, and thermal losses [12]. The tubes of the ground heat exchanger may be shaped in many ways. The most popular solutions are single U, double U, or coaxial (see Figure 19.5). The first one is characterized by the lowest investment cost, whereas the second one provides the highest reliability [12].

Heat exchanger tubes are surrounded by rock, sediment mass, or groundwater. These materials are characterized by high volumetric heat capacities, so they are suitable for thermal storage applications [11]. The presence of water in porous rock increases the heat transfer efficiency between the heat exchanger and the solid ground [12]. The working fluid is pumped through pumps. The most popular fluids are water (in hot climates) and water-glycol solutions (in climates where freezing is possible). Additional additives are required to inhibit corrosion or eliminate biocides [12].

The direction of working fluid flow is reversed seasonally to store or abstract the heat from the ground. When the working fluid has a higher temperature than the surrounding rocks, then the heat is transferred to the geological environment. In the opposite situation, when the working fluid is cooler than the surroundings, the heat is extracted to the surface [12]. Finally, the working fluid is transferred to another heat exchanger or heat pump located on the surface [12].

The BTES are often combined with conventional and hybrid solar-thermal installations [14], air conditioning systems, or industrial facilities which produce excess heat. The stored energy may be used for space heating/cooling or district heating applications [15]. This kind of system provides long-term heat storage and is especially attractive in strongly seasonal climates. Despite the high investment costs [16], BTES reaches high values of coefficient of performance COP=4÷8 [12].

19.4 FUNDAMENTALS OF ENERGY CONVERSION AND BALANCE

The fact related to geothermal energy is that 99% of the Earth's mass has a temperature above 1000°C. The heat is generated in the internal part of the Earth and transferred to its surface.

As it has already been mentioned, it is assumed that as a result of radioactive decay, heat is generated throughout the interior of the Earth. The most important radioactive elements are ^{235}U (decay rate $9.7 \cdot 10^{-10}$ y^{-1}), ^{238}U (decay rate $1.5 \cdot 10^{-10}$ y^{-1}), ^{232}Th (decay rate $5.0 \cdot 10^{-11}$ y^{-1}), and ^{40}K (decay rate $5.3 \cdot 10^{-10}$ y^{-1}). These isotopes are present in different concentrations in different geological formations. They are more abundant in the granite-containing continental shields than in the ocean floors. Most of the material containing radioactive elements is concentrated in the upper part of the Earth's crust. In the lower half of the crust (total depth around 40 km), the radiogenic heat production is believed to be fairly constant at a value of about 2×10^{-7} W/m^3.

The heat generation value inside the upper crustal rocks, typically obtained from three radioactive elements, Uranium, Thorium, and Potassium, is of special interest [17]

$$Q = 10^{-3} \rho (9.52 C_U + 2.56 C_{Th} + 3.48 C_K), \tag{19.1}$$

where Q is heat production (W/km^3), ρ is the rock density (kg/m^3), and C_U, C_{Th}, and C_K are Uranium, Thorium, and Potassium contents (ppm), respectively. The units for all numeric heat production constants are in W/kg.

Geothermal heat flow can be treated as a "truly" renewable energy source because only its part, equal to 2.4×10^{-10} per year, is irreversibly lost due to geothermal heat dissipation into space [18].

According to Figure 19.6, the following mechanisms are the most important for the heat transfer to the Earth's surface, depending on the layer [19].

Major compositional and rheological divisions as well as the relative proportion and type of heat flow from each division are illustrated. Although conduction is the major form of heat flow in the crust, some heat flow occurs by convection (in the areas of circulating crustal fluids) and by advection (with the rise of magma below active volcanoes).

The temperature profile can be illustrated in Figure 19.7 (Ref. [19])

The figure illustrates the Earth's geothermal gradient. Also shown is the solidus or the temperature at which rocks begin to melt. The geothermal gradient is the highest near the surface, indicative of conductive heat flow, and changes gradually with depth, indicating a combination of convective

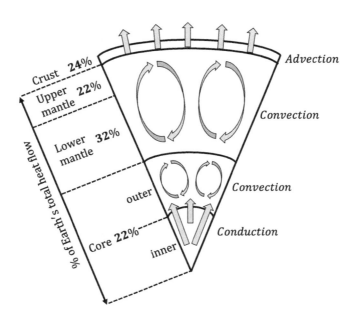

FIGURE 19.6 Heat transfer mechanisms within the Earth, along with the amount of heat flow in each layer.

Theoretical Background

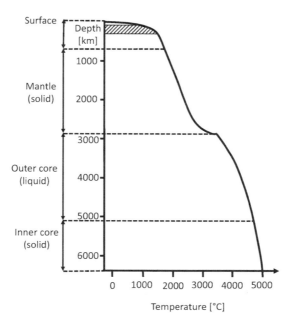

FIGURE 19.7 Temperature profile of the Earth's layers. Temperature increases with depth.

and conductive heat flow. The hatched layer marks the asthenosphere, where the temperature of the solidus and that of the Earth are close, resulting in rheologically weak rock. The layer above the asthenosphere is the lithosphere, where the temperatures of the Earth and solidus are further apart, making for rheologically strong rock: To derive temperature as a function of depth for the range of 0–10 km, a steady state is assumed without the occurrence of heat advection processes, such as magmatism, intense erosion, or hydrothermal convection. All the above-mentioned phenomena can occur only for a short time when an equilibrium thermal regime of the continental crust is considered.

The temperature profile $T(z)$ up to the depth z ca. Therefore, 10 km can be calculated with a simplified equation (where constant thermal conductivity, a depth-dependent temperature field, and suitable boundary conditions for the continental crust are expected as input).

$$T(z) = T_0 + q_0 \cdot \frac{z}{k} - Q \cdot \frac{z^2}{2k}, \tag{19.2}$$

where T_0 is the surface temperature, q_0 is the surface heat flux, z is the depth, k is the thermal conductivity, and Q is the heat production [17]. The surface temperature T_0 is on average a few degrees above zero, in the range of −25°C to +30°C, the heat flux q_0 is in the range of 0–200 mW/m² (see Figure 19.8), thermal conductivity k is on average 3 W/(m K) in the range of 0 to above 6 W/(m K), and Q heat generation is approximately 200 W/km³ and varies in the range of 0–300 W/km³. More detailed information can be found, for example, in Ref. [17].

The available heat calculated at the depth of 1000 m can be obtained from the following expression:

$$Q = \rho c_p V_c \cdot (T_z - T_r) \cdot 10^{-18}, \tag{19.3}$$

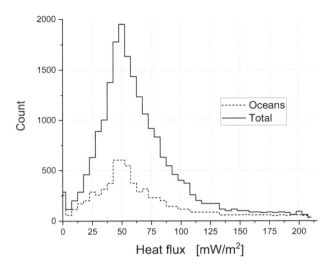

FIGURE 19.8 Heat flux distribution based on measurements. The continental (light gray) and oceanic (medium gray) data are presented. The mean values for continents and oceans are indicated. Fluxes above 210 mW/m2 are included in the highest bin [20].

where density ρ is equal to 2550 kg/m³, and specific heat capacity C_p of the rock is constant and equal to 1000 J/(kg K), V_c is the volume of rock, T_z is the temperature at the depth interval, and T_r is the base or reference temperature:

$$T_r = T_0 + 80°C. \tag{19.4}$$

Of course, the above-mentioned parameters depend on specific geological conditions and are different for different regions. More information can be found in Ref. [17].

Various parameters of the rock heat conduction, permeability, etc., cause variation in the heat flux reaching the Earth's surface. The distribution is presented in Figure 19.8 illustrates this heat flux.

It is very important for sustainable use of geothermal energy to develop accurate models, able to predict the interaction between the reservoir and the heat exchange system used for transferring energy from and to the ground. Otherwise, geothermal energy is considered to be highly risky. Moreover, the concomitance of high water content and high temperatures strongly influences the energy balance in the soil and may result in a phase change. Figure 19.9 presents a classification of geothermal resources as a function of their temperature level and water mass flow rates available. Correct identification of a geothermal system is imposed by a large variability range of parameters characterizing various reservoirs [21].

The complexity of the geothermal system is evident, especially when building detailed models describing its behavior. For example, low permeability shallow formations present a low level of model complexity, supercritical systems are characterized by a very complex thermofluid behavior.

For some energy systems, for example, heat pumps and ground heat storage, a very shallow layer of the ground is utilized. Thus, the knowledge of the ground temperature profile concerning time and depth is very important. Examples of these applications are buried ground heat exchangers connected to ground source heat pump systems or energy storage systems, Earth-to-air heat exchangers for heating and cooling of buildings and agricultural greenhouses, etc. Many factors such as slope orientation, terrain, solar radiation, wind, rain, etc., can affect the ground thermal behavior to a more or less important extent.

Theoretical Background

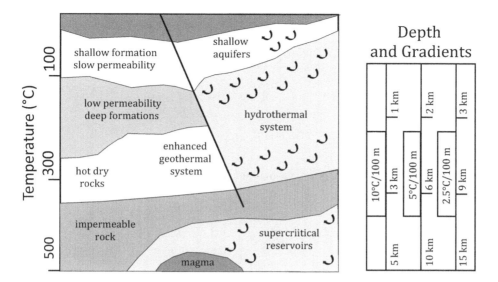

FIGURE 19.9 Simplified classification of geothermal resources [21].

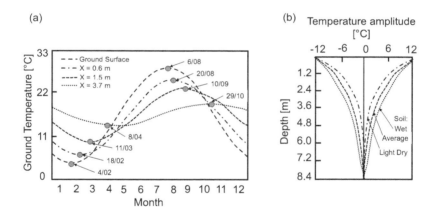

FIGURE 19.10 (a and b) General view of the shallow ground temperature profiles illustrating dependence on the time and depth.

We can assume that the temperature profiles at the shallow ground depth are as presented in Figure 19.10.

Since little measured data on ground temperature are available, several models have been developed for its estimation as a function of time and depth. Determination of this temperature is generally based on the solution of the transient one-dimensional heat conduction problem in the ground and depends on the type of boundary condition at the ground surface. According to Ref. [19], the following calculation scheme can be applied.

The equation describing the transient, one-dimensional heat conduction in the ground is

$$\frac{dT}{dt} = \alpha \frac{dT^2}{dz^2}. \tag{19.5}$$

Here, the ground is assumed to be a semi-infinite medium with constant physical properties.

The following initial and boundary conditions can be applied:

$$T(z,0) = \overline{T}_s \left(at\ t = 0 \right), \tag{19.6}$$

$$T_s = T(0,t) = \overline{T}_s - A_s \cos(\omega t - \phi_s) at\ z = 0, \tag{19.7}$$

$$T(\infty,t) = \overline{T}_s\ at\ z = \infty. \tag{19.8}$$

The solution of function $T(z, t)$ expressing ground temperature at time t (day or hour) and depth z is the following:

$$T(z,t) = \overline{T}_s - A_s \exp\left(\frac{-z}{\delta}\right) \cdot \cos\left(\omega t - \phi_s - \frac{z}{\delta}\right), \tag{19.9}$$

where $\overline{T}_s$ is the annual average ground surface temperature, equivalent to the undisturbed ground temperature, A_s is the annual amplitude of the ground surface temperature, δ is the damping depth of annual fluctuation of the ground temperature which can be calculated from

$$\delta = \sqrt{\frac{2\alpha}{\omega}}, \tag{19.10}$$

where α is the thermal diffusivity, $\omega = \dfrac{2\pi}{365}$ (radian/days) is the angular frequency, and ϕ_s is the phase angle.

To calculate parameters $\overline{T}_s$, A_s, and ϕ_s at the ground surface ($z=0$), the energy balance is applied as a boundary condition. A detailed description of the boundary conditions is presented in Figure 19.11.

Figure 19.11 represents the main heat flux contributions at the surface of the ground, including the conduction heat flux into the ground, the convective heat flux transferred between the surface and the ambient air, the short-wave global solar radiation (radiant flux) absorbed by the ground surface, the long-wave radiant flux exchanged with the surroundings (sky), and the latent heat flux of evaporation.

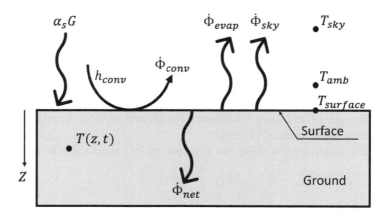

FIGURE 19.11 Boundary condition applied in the calculations.

Net heat flux transferred by conduction across the ground surface $\dot{\Phi}_{net}$ is given by the equation

$$\dot{\Phi}_{net} = -k \frac{\partial T}{\partial z}\bigg|_{z=0} = \dot{\Phi}_{conv} + \alpha_s S - \dot{\Phi}_{sky} - \dot{\Phi}_{evap}, \qquad (19.11)$$

where $\dot{\Phi}_{conv}$ is the sensible heat flux exchanged between the air and the ground surface, $\alpha_s S$ is the short-wave global solar radiation (radiant flux) absorbed by the ground surface, $\dot{\Phi}_{sky}$ is the long-wave radiation (radiant flux) emitted by the ground surface to the sky, and $\dot{\Phi}_{evap}$ is the evaporation heat exchange flux that depends on several factors, including ground surface and ambient air temperatures, wind speed, surface cover, soil moisture content, and air humidity.

A description of the above-mentioned terms and a detailed equation can be found in Ref. [19].

The presented equation is widely used, for example, in dynamic simulation software like TRNSYS, where valid ground temperature profiles are used among others for heat pump calculations, ground heat exchangers, and heat losses in buildings.

REFERENCES

1. P. Allahvirdizadeh, A review on geothermal wells: Well integrity issues, *Journal of Cleaner Production*, 275 (2020), 1–21 (124009).
2. S. Carlino et al., The geothermal exploration of Campanian volcanoes: Historical review and future development, *Renewable and Sustainable Energy Reviews*, 16 (2012), 1004–1030.
3. Statistics Data of the International Renewable Energy Agency, www.irena.org (last access: 07.05.2022).
4. IRENA (2017), *Geothermal Power: Technology Brief*, International Renewable Energy Agency, Abu Dhabi.
5. Indonesia Set to Become World's 2nd Largest Geothermal Power Producer, available from: https://www.indonesia-investments.com/news/todays-headlines/indonesia-has-become-world-s-2nd-largest-geothermal-energy-producer/item8775 (last access: 11.06.2023).
6. C. Siyuan, Quantitative assessment of the environmental risks of geothermal energy: A review, *Journal of Environmental Management*, 276 (2020), 1–17 (111287).
7. J. W. Lund, T. L. Boyd, Direct utilization of geothermal energy 2015 worldwide review, *Geothermics*, 60 (2016), 66–93.
8. G. Liuhua, Z. Jun, T. Zipeng, A review on borehole seasonal solar thermal energy storage, *Energy Procedia*, 70 (2015), 209–218.
9. J. Wołoszyn, Global sensitivity analysis of borehole thermal energy storage efficiency on the heat exchanger arrangement, *Energy Conversion and Management*, 166 (2018), 106v119.
10. S. Lanini et al., Improvement of borehole thermal energy storage design based on experimental and modelling results, *Energy and Buildings*, 77 (2014), 393–400.
11. H. Skarphagen et al., Design considerations for borehole thermal energy storage (BTES): A review with emphasis on convective heat transfer, *Geofluids*, 26 (2019), 393–400.
12. M. Reuss, The use of borehole thermal energy storage (BTES) systems, *Advances in Thermal Energy Storage Systems - Methods and Applications*, edited by L uisa F. Cabeza, Woodhead Publishing, Cambridge, UK, 2015, 117–147
13. D. Banks, *An Introduction to Thermogeology: Ground Source Heating and Cooling*, 2nd edition, John Wiley & Sons, Chichester, 2012.
14. M. Aldubyan, A. Chiasson, Thermal study of hybrid photovoltaic-thermal solar collectors combined with borehole thermal energy storage systems, *Energy Procedia*, 141 (2017), 102–108.
15. B. Welscha et al., Environmental and economic assessment of borehole thermal energy storage in district heating systems, *Applied Energy*, 216 (2018), 73–90.
16. T. Schmidt, D. Mangold, H. Müller-Steinhagen, Central solar heating plants with seasonal storage in Germany, *Solar Energy*, 76 (2004), 165–174.
17. A. Aghahosseini, C. Breyer, From hot rock to useful energy: A global estimate of enhanced geothermal systems potential, *Applied Energy*, 279 (2020), 115769.
18. B. Sørensen, *Renewable Energy. Its Physics, Engineering, Use, Environmental Impacts, Economy and Planning Aspects*, Elsevier Academic Press, Cambridge, 2004.

19. D. R. Boden, *Geologic Fundamentals of Geothermal Energy*, CRC Press, Boca Raton, FL, Routledge Handbooks Online: https://www.routledgehandbooks.com/pdf/doi/10.1201/9781315371436-4/1649967686734 (last access: 14.04.2022).
20. A. M. Hofmeister, R. E. Criss, Earth's heat flux revised and linked to chemistry, *Tectonophysics*, 395 (2005), 159–177.
21. A. Carotenuto et al., Models for thermo-fluid dynamic phenomena in low enthalpy geothermal energy systems: A review, *Renewable and Sustainable Energy Reviews*, 60 (2016), 330–355.

20 Tutorial 10 – Borehole Heat Exchanger

20.1 EXERCISE SCOPE

The goal of this exercise is to design the model of a simplified borehole heat exchanger, as in Figure 20.1. It consists of a single U-tube *(1)* filled with a working fluid *(2)*. The tube is placed in a grout casing *(3)*, which separates the well from the ground *(4)*. This model allows us to observe the changes in the ground and fluid temperature profile with the increasing borehole depth.

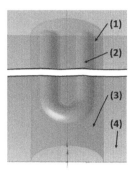

FIGURE 20.1

Try to recall from the theoretical part of this chapter: The construction details (Section 19.3) and fundamentals of energy conversion and balance (Section 19.4) in the case of borehole heat exchangers. Temperature distribution strongly depends on the well depth and initial fluid parameters. The considered model allows us to study this relation.

Among others, this exercise can teach you how to

- create multipart geometry in ***Ansys Design Modeler*** with ***Extrude*** and ***Sweep*** tools,
- generate an inconsistent mesh for geometry without shared topology in ***Ansys Meshing***,
- create a new material with defined properties and, optionally, describe the increase of the ground temperature with the borehole depth with the ***Expression*** option in the ***Ansys Fluent*** solver,
- conduct a simple variant analysis,
- display a chart of temperature distribution depending on borehole depth, and other types of temperature and velocity visualizations in the ***Ansys Results*** module.

Pay special attention to the impact of the non-shared topology in the geometry on the computational grid characteristics and specification of the interfaces in the ***Ansys Fluent*** solver.
All of the images included in this tutorial use courtesy of ANSYS, Inc.

20.2 PREPROCESSING – GEOMETRY

Drag the ***Fluid Flow (Fluenet)*** module *(1)* from the ***Analysis Systems*** tab to the ***ANSYS Workbench Project Schematic*** as in Figure 20.2. The default geometry editor is **SpaceClaim**, but in this case, we will use *Design Modeler*. To open this geometry editor, click the right mouse button *(RMB)* on the *Geometry* A2 cell *(2)*, and from the dropdown list choose the *New Design Modeler Geometry (3)* option.

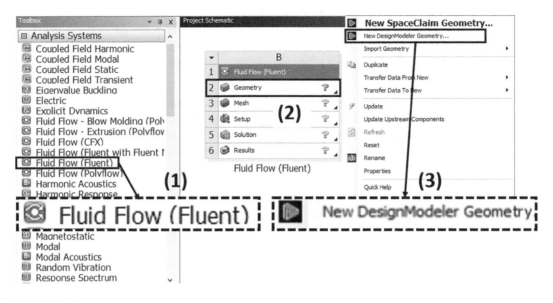

FIGURE 20.2

The *Design Modeler* in the *Modeling* mode will be opened (Figure 20.3). Before starting sketching, it is necessary to choose a proper plane for a new sketch. Find the *Tree Outline* window, select *ZX Plane (1)*, and then create a new sketch *(2)*. Click *LMB* on *Sketch1 (3)* from the *Tree Outline* and switch the mode from Modeling to *Sketching (4)*. To view the sketching plane head-on, click on the *Look at Face/Plane/Sketch* icon *(5)*. You will see the *Z* and *X* axes *(6)* and also a linear scale *(7)* with a default length unit. To zoom in, use a scroll. It would be convenient to take *1 m* as the maximum value on the scale.

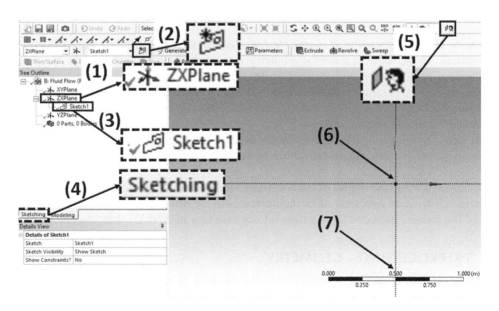

FIGURE 20.3

Tutorial 10 – Borehole Heat Exchanger

In the *Sketching* mode, click on *Draw (1)* to see the drawing tools. Firstly, choose the *Rectangle* option *(2)* and start your drawing from the horizontal axis. Then, select the *Dimensions* tab *(3)* and the *General* option *(4)*. It will allow you to choose the proper dimensions for the previously prepared sketch *(5)*: The rectangle should be *3.2 m* long and *1.6 m* wide, located symmetrically to the vertical axis, as in Figure 20.4.

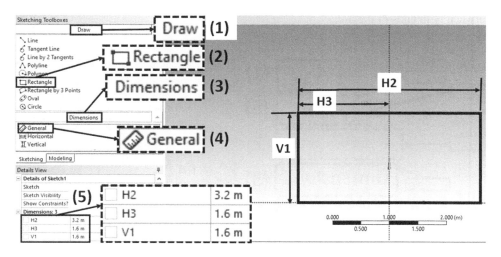

FIGURE 20.4

To create the main body, switch the mode to *Modeling (1)* and select the *Extrude* tool *(2)*. In *Details View (3)*, choose *Sketch1* as the base *Geometry* for extrusion. Check if the *Operation* is *Add Material* and *Direction* is *Normal*. Set the *Extent Type* as *Fixed* with a *12.64 m* value. After that, click *RMB* on the *Extrude1 (4)* in *Tree Outline* and choose the *Generate* option from the dropdown list. As a result, the 3D geometry should be visible *(5)* as in Figure 20.5. Moreover, the lightning icon near the operation name should change to a *green* tick.

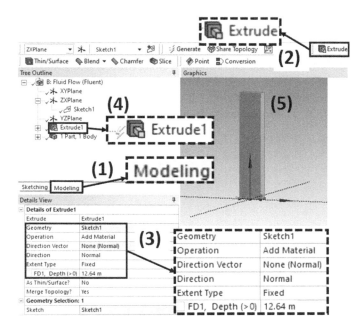

FIGURE 20.5

The next step in the geometry preparation is to draw a semi-circular shape on the same plane (*ZXPlane*) as previously but in another sketch. Add a new sketch, as shown before, and switch the mode to *Sketching*). From the *Draw* tab, *(1)* select the *Arc by Center* tool *(2)* and start drawing from the center of the coordination system as in Figure 20.6. To set the proper *Dimensions (3)*, use the *Radius* tool *(4)* and set the value to 0.32 m *(5)*. Remember that in *Design Modeler*, each sketch for extruding, cutting, or slicing should be closed. To do so, use the *Line* tool *(6)* and draw a horizontal line between both ends of the arc.

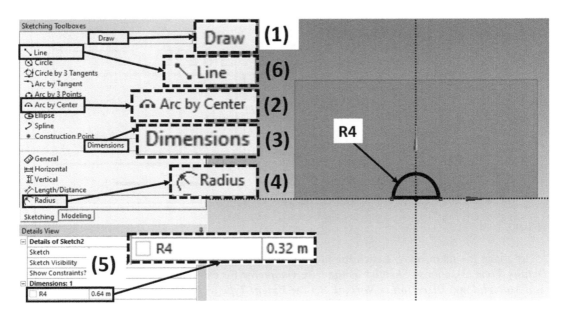

FIGURE 20.6

Now, it is time to extract the external pipe casing from the main body (Figure 20.7). To do this, switch the mode to *Modeling (1)* and again use the *Extrude* tool *(2)*. In the *Details View* window *(3)*, select *Geometry* as the newly prepared *Sketch2*. This time, use *Slice Material Operation* so that the main body will be divided into two parts along the selected geometry. The *Direction* of this operation is *Normal* to the sketch and the *Extent Type* should be *Through All* bodies. Finalize the operation by clicking *RMB* on *Extrude2 (4)* in *Tree Outline* and choose the *Generate* option from the dropdown list. Now, the geometry should consist of *2 Parts* and *2 bodies (5)* which are totally separate.

Tutorial 10 – Borehole Heat Exchanger

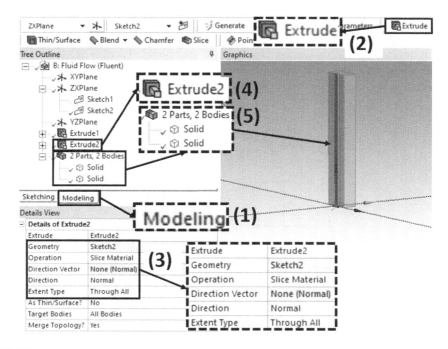

FIGURE 20.7

Now, prepare the shape which will be later used as a trajectory for the pipe. Firstly, you have to prepare a new sketch in *YZ Plane* and switch the mode to *Sketching*. After that, find the *Draw* tab *(1)* and select the *Line* tool *(2)*. Draw two lines, parallel to the longer axis. Using the *Dimensions* tab *(3)* and the *General* tool *(4)*, set the length of the line to 12 m and locate them 0.16 m away from the axis (*0.32 m* from each other) as in Figure 20.8. The distance between the bottom of the geometry and the lines should be set to *0.64 m* *(5)*. The last sketching step is to connect the ends of the lines with the *Arc by Tangent* tool *(6)*.

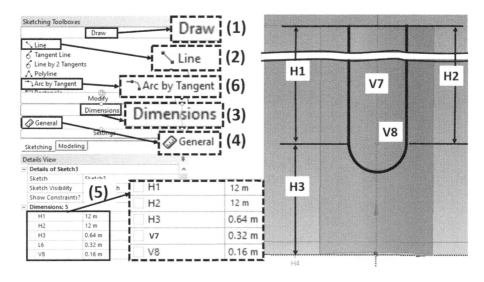

FIGURE 20.8

To create the geometry of pipes, it is necessary to sketch their cross-section shape at one end of the trajectory (Figure 20.9). In this step, create a new plane for sketching. Choose the **Selection Filter** as **Faces (1)** and click on the upper surface of the cuboid. Then, select the new plane icon **(2)**, and in **Details View (3)**, set the **Type** as **From Face**. Make sure that the **Base Face** is **Selected**. Finish the operation by clicking **RMB** on the operation name **(4)** and choosing the **Generate** option from the dropdown list (the same procedure as in **Extrude** operation).

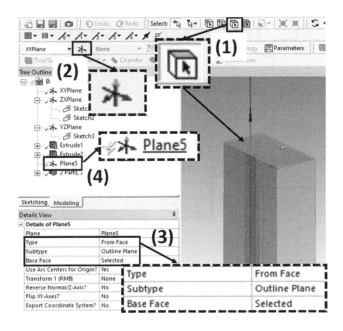

FIGURE 20.9

Firstly, prepare the sketch with the internal shape of the pipe, to sweep it by the previously prepared trajectory. **Draw (1)** an **Arc by Center (2)** and set its **Dimension (3)** with **Radius (4)** tool to **0.08 m (5)**. Remember to close the sketch with **Line (6)**. Locate the center of the arc **0.16 m** away from the edge of the internal body (as in Figure 20.10), using the **General (7)** dimension tool.

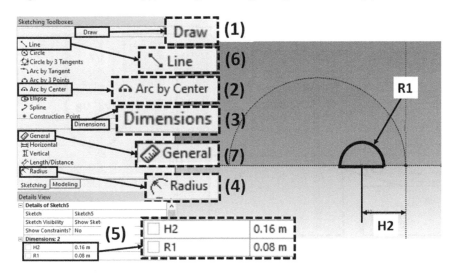

FIGURE 20.10

Tutorial 10 – Borehole Heat Exchanger

The pipe filling will be generated in the **Modeling** mode *(1)* using the **Sweep** tool *(2)*. In **Details View** *(3)*, set the previously prepared sketch as **Profile** and the U-shaped drawing as **Path**. Select the Operation type as **Slice Material** to extract the pipe from the casing body. **Generate** the operation as in previous steps *(4)*. As a result, you should receive *3 Parts* and *3 Bodies (5)* as in Figure 20.11.

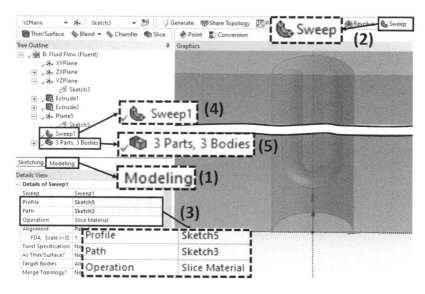

FIGURE 20.11

To create the external shape of the pipe, make a new sketch and locate it on the user-created plane (the same where the inner shape was drawn). Sketch two arcs with a common center and two lines to close the sketch *(1)*. Set the radius of the inner arc to *0.08 m* and of the outer one to *0.1 m*, so that the pipe will be *2 cm* thick. Locate the center of the arcs *0.16 m* away from the inner body edge *(2)* as in Figure 20.12. Finally, use the **Sweep** tool again. The last sketch will be set as **Profile** and the U-shaped sketch as **Path**. Set the **Operation** type as **Slice Material** and **Generate** the operation. The *4 Parts* and *4 Bodies* should appear *(3)*. Close the **Design Modeler** module.

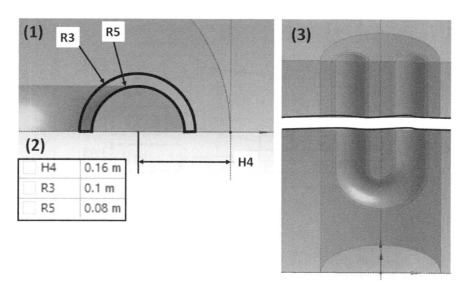

FIGURE 20.12

20.3 PREPROCESSING – MESHING

Move to the *Mesh* module by double-clicking on the *A3* cell *(1)*, as in Figure 20.13.

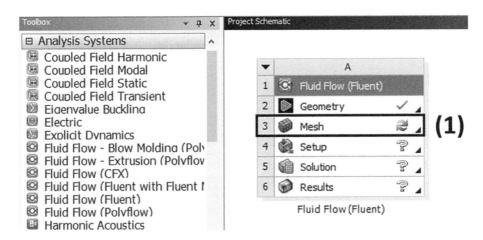

FIGURE 20.13

Find the *Geometry (1)* icon in the project tree and check if there are four separate bodies as in Figure 20.14. Then, click on the *Mesh (2)* option. In the *Details of Mesh* window, some basic settings have been automatically made: *Physics Preference – CFD (3)* and *Solver preference – Fluent (4)*. It is due to the usage of the whole *Analysis System* dedicated to the *Fluent* solver in Workbench, instead of separate modules. Change the global *Element Size (5)* to 0.16 m, to limit the maximum mesh element size.

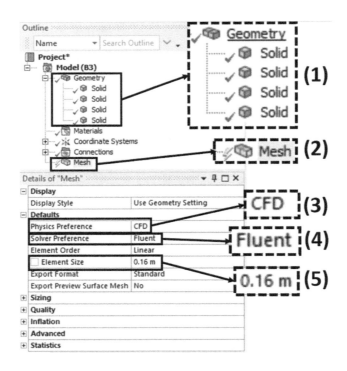

FIGURE 20.14

Tutorial 10 – Borehole Heat Exchanger

Use the ***Volume Selection*** mode *(1)* and click with Ctrl click on the bodies representing fluid filling, pipe walls, and internal casing *(2)* as in Figure 20.15. Next, click *1× RMB* and choose the option ***Suppress Body*** *(3)* from the dropdown list to hide them. As a result, in the ***Project*** tree the green tick should be visible near the element *(4)* which is visible in the main window *(5)*.

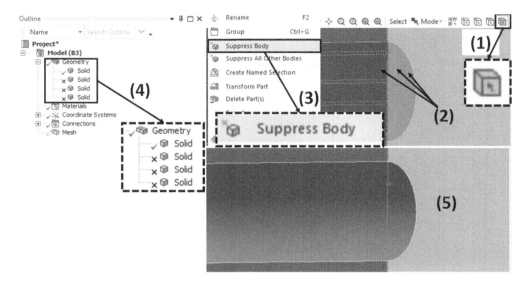

FIGURE 20.15

Still in the ***Volume Selection*** mode, *(1)* select the remaining body *(2)* and click *1× RMB*. From the dropdown list, select ***Insert*** *(3)* → ***Method*** *(4)*. In ***Details of Method***, change the ***Method*** to the ***Sweep*** option *(5)*, which allows to generate hexahedral mesh through the body with a topologically constant cross-section (Figure 20.16). In this method, the ***Source*** surface (user-defined or automatically applied) is first meshed. Then, this mesh pattern is swept through the whole body to the ***Target*** surface. Usually, the ***Sweep*** method allows to maintain a high quality of mesh and reduces the number of mesh elements at the same time.

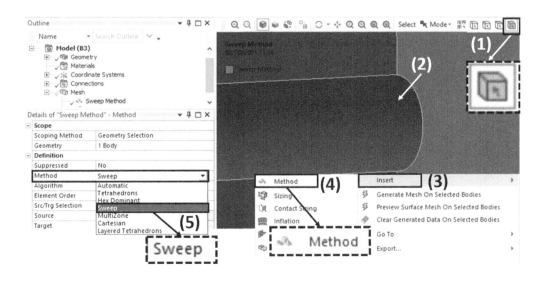

FIGURE 20.16

To increase the number of mesh elements along the curved edge, click on the *Edge Selection* mode *(1)*. Then, select the proper geometry edge *(2)* and click 1× *RMB* on it. From the dropdown list, choose *Insert (3)* → *Sizing (4)*. In *Details of Sizing*, change the *Type* to *Number of Divisions (5)* and set this value to 18 *(6)*. This will result in 18 cells distributed equally along the selected edge (Figure 20.17).

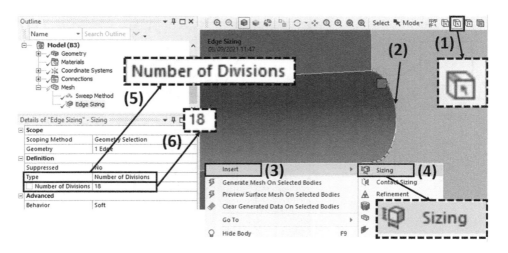

FIGURE 20.17

Again, click on the *Volume Selection* mode *(1)* and select the visible body *(2)*. Click *1× RMB* and choose the *Unsuppress All Bodies* option *(3)* to see the whole geometry (Figure 20.18). Now let's focus on the mesh dedicated to the pipe. The thickness of the pipe wall is *2 cm* and the mesh in each body should be a minimum of 2 elements thick, so the elements should not be bigger than 1 cm. To insert this condition, local sizing will be used. Do it like in the previous step – select the part which represents pipes *(4)* and then click *1× RMB*. From the dropdown list, choose *Insert → Sizing*. In *Details of Body Sizing*, set the *Type* as *Element Size (5)* with a maximum dimension of *1 cm* (pay attention to the units!) *(6)*. Following this procedure, insert local sizing to the fluid body *(7)* with *Element Size* equal to *2 cm*.

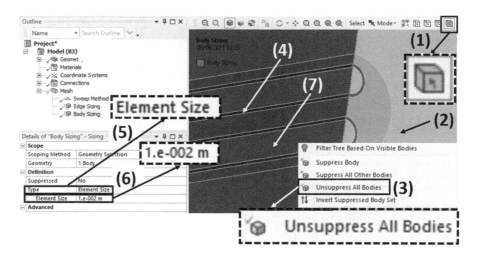

FIGURE 20.18

Tutorial 10 – Borehole Heat Exchanger

Before mesh generation check if all the necessary settings are inserted: ***Sweep Method, Edge Sizing***, and twice ***Body Sizing (1)*** as in Figure 20.19. Finding ***Mesh (2)*** in the ***Project*** tree and click 1× ***RMB*** on it. Select the ***Generate Mesh (3)*** option and wait until the procedure is finished (the lightning icon near the ***Mesh*** option will change into a green tick). Then, find the ***Statistic*** tab ***(4)*** and check the total number of ***Nodes*** and ***Elements*** in your mesh (it should be lower than ***500 000***). Take a look at the mesh appearance ***(5)*** – the mesh nodes generated for each part do not meet the boundaries – the mesh topology is inconsistent. This type of mesh is often used when there are significant differences in size between solid and fluid geometry elements.

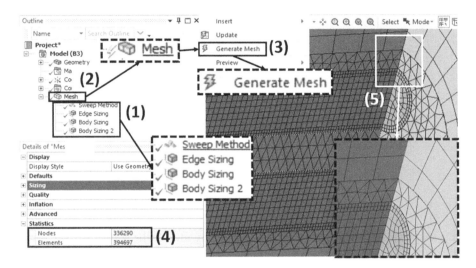

FIGURE 20.19

To analyze the mesh better, create a ***Section Plane (1)*** by clicking on its icon in the ***Insert*** tab. Then, draw a line that will slice the geometry into two parts. It may be activated or deactivated just by clicking the tick near the section plane name ***(2)***. To make the view clearer, choose the ***Show Whole Elements*** option ***(3)*** which displays partially sliced elements in their entirety as in Figure 20.20. Compare the size and shape of elements in each body. In the end, deactivate the created cross-section.

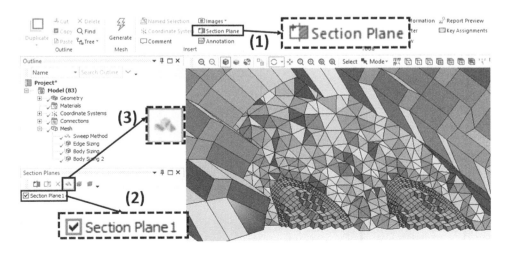

FIGURE 20.20

Click on the **Mesh (1)** option and, in **Details of Mesh**, find the **Quality** option. Set the **Mesh Metric** parameter as **Skewness (2)**. It describes the difference between the shape of the cell and the shape of an equilateral cell with the same volume. The lower skewness, the better. The upper limit recommended for skewness is 0.95 – higher skewed cells may decrease accuracy and destabilize the solution. Rate the mesh quality based on the minimum, maximum, and average values of the analyzed parameter *(3)*. In the **Mesh Metrics** window (Figure 20.21), check the shape of the mesh elements *(4)* and the participation of cells with particular quality in the total number of mesh elements *(5)*.

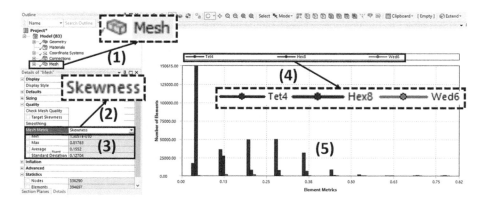

FIGURE 20.21

To locate the elements characterized by the worst quality, change the options of the displayed chart. Click **Controls (1)** in the **Mesh Metrics** window and the popup window, change the minimum value of **X-Axis** to display only the worst-quality elements. Change this parameter to value **0.75**. Also, change the maximum value of **Y-Axis** to **100** (the number of the worst elements is relatively low). There is no **Apply** button, just click **X** in the corner of the popup window *(2)*. By clicking on the bars displayed on the chart *(3)* it is possible to display the worst-quality elements in the workspace *(4)* as in Figure 20.22. Now check another quality parameter, **Orthogonal Quality**, which (in simplification) describes the equilaterality of the cell. The value equal to 0 is the worst and means skewed cells, whereas value 1 is the best, and means fully equilateral cells (like a cube). Generally, **Orthogonal Quality** should not be lower than 0.05. To check the minimum, maximum, and average value of this parameter, just select the **Orthogonal Quality** parameter in the **Mesh Metric** row in the **Quality** tab *(5)*.

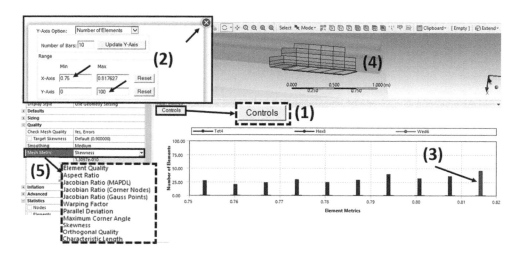

FIGURE 20.22

Tutorial 10 – Borehole Heat Exchanger

Named Selection is the function that allows to assign names to different objects to make them easier to find in the solver. Remember: To name a surface use the *Face Selection* mode and to name a body use the *Volume Selection* mode *(1)*. Choose the external body *(2)* and click *1× RMB* on it. From the dropdown list choose the *Create Named Selection (3)* option or just press the N button from the keyboard. Then, insert the name for the new *Named Selection* in the popup window *(4)*: ground and confirm by clicking *OK*. Repeat this procedure to name fluid, pipe, and grout bodies (Figure 20.23). They should be visible in the *Named Selections (5)* option in the *Project* tree.

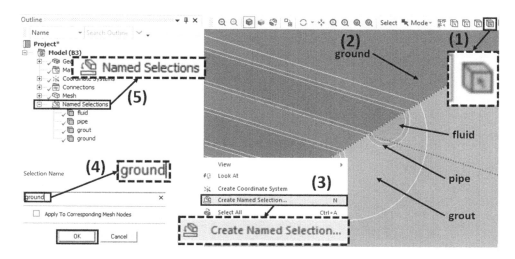

FIGURE 20.23

Now it is time to name surfaces (Figure 20.24), so switch to the *Face Selection* mode *(1)*. The procedure is the same as in the case of naming bodies. Firstly, name the *inlet* (1 surface) and *outlet* (also 1 surface) *(2)*. Then using *Ctrl*, select all the faces on the front side of the geometry (7 surfaces). They create a symmetry plane of the analyzed case, so name them *symmetry (3)*. Now select the remaining side surfaces and the ones located at the top of the geometry (7 of them), and set their name as *surrounding (4)*. This named selection is connected with two assumptions: First – the ground and the ambient air are in thermal balance, second – the length of the borehole is low so the increase in the ground temperature due to the depth may be neglected. Finally, select 2 surfaces located at the bottom of the geometry and name them *bottom (5)*.

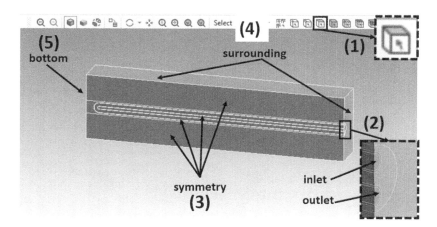

FIGURE 20.24

Because the geometry consists of four separate parts with four separate meshes, the interfaces must be created in the solver. They are surfaces where the information from one domain is transferred to the next one. To make this easier, it is important to properly name these surfaces using *Named Selection* (Figure 20.25). Let's start from the middle of the geometry: Suppress all the bodies except the fluid. Name the convex surfaces (3 of them) as *interface_1-2 (1)*, which means that the data are transferred from body 1 (*fluid*) to body 2 (*pipe*). Then, unsuppress the pipe body and suppress the fluid. Firstly, create the *interface_2-1 (2)* including three concave surfaces which are in direct contact with the fluid domain. Next, move to the *interface_2-3 (3)* located on the three convex surfaces. When these interfaces are named, make only the grout casing visible.

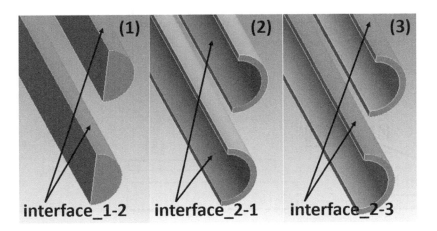

FIGURE 20.25

Continue naming interfaces (Figure 20.26): Three concave surfaces create an *interface_3-2 (1)*, whereas one convex surface should be named *interface_3-4 (2)*. The last step is to name the concave surface located in the ground domain as *interface_4-3 (3)*. The given names should appear in the project tree, below the *Named Selections* tab. After clicking on one of the names, all the related objects should be highlighted in red in the workspace. Double-check the correctness of all the named selections – it is essential for further operations in the solver. Finally, unsuppress all the bodies.

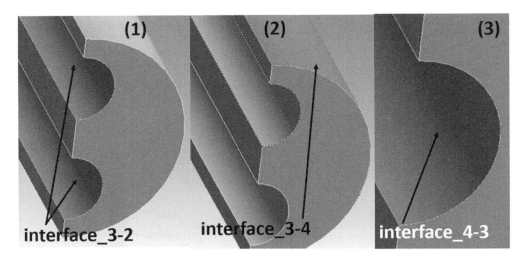

FIGURE 20.26

Tutorial 10 – Borehole Heat Exchanger

Before moving to the next step of the simulation, find the **Connections (1)** tab in the **Project** tree, click *1× RMB* on **Contacts**, and select the **Delete** option *(2)* from the dropdown list (Figure 20.27). **Contacts** are an automatic alternative to tackle interfaces between volumes. In this case, we named interfaces manually so this tool should be disabled. Close the **Mesh** module.

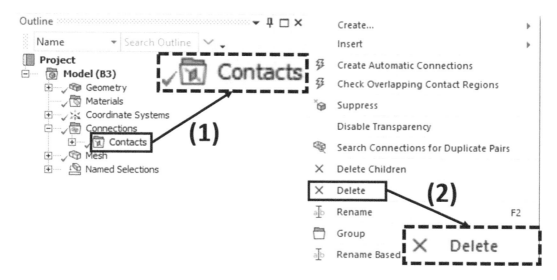

FIGURE 20.27

20.4 PREPROCESSING – SOLVER SETTINGS

Click *1× RMB* on cell *A3 (1)* with **Mesh** and choose the **Update** option *(2)* to export the generated mesh to the solver. The lightning icon near **Mesh** should change into a green tick, whereas the question mark near the **Setup** cell should convert into double arrows *(3)* as in Figure 20.28. Start the solver by clicking *2× LMB* on cell *A4* and, in the popup window, run the **Parallel** processing mode with the number of cores available on your computer.

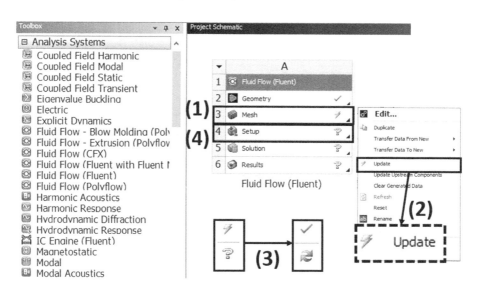

FIGURE 20.28

In the solver **Setup**, find the **General** option *(1)* (Figure 20.29). Firstly, set the **Solver Type** as *pressure-based (2)* to analyze fluid flows with a low velocity (<0.6 Ma). To provide a steady state simulation, choose the **Steady *(3)* Time** option. In a borehole operation, gravity plays a significant role, so set its value to *−9.81* along the *Y*-axis. The minus before the value means that the acceleration vector is opposite to the *Y*-axis.

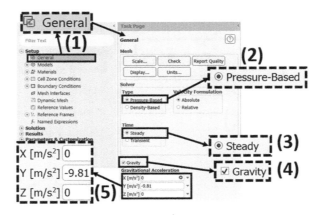

FIGURE 20.29

Move to the **Models** option (1) to select appropriate mathematical models of physical phenomena occurring in the analyzed case (Figure 20.30). Firstly, activate **Energy Equation *(2)*** for heat transfer. Do this by ticking the available option in the popup window *(3)* and confirm with **OK**. Secondly, select the **Viscous *(4)*** option, to choose a turbulence model for the fluid flow. In the analyzed case, the **k-epsilon *(5)*** turbulence model in the **Realizable *(6)*** mode is suitable. Set the **Near-Wall Treatment** as **Enhanced *(7)***, which gives good results in modeling the near-wall layer for a coarser mesh resolution in this zone.

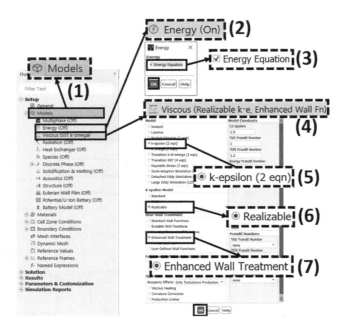

FIGURE 20.30

Tutorial 10 – Borehole Heat Exchanger   427

Now let's focus on *Materials (1)*. By default, the *air is* set as fluid and *aluminum* as solid as in Figure 20.31. In the studied case, the working *fluid* is air and the *pipe* is made of aluminum. There are also two materials: Ground and grout, not specified in *Fluent Database*, so they should be defined manually. Start by clicking *2× LMB* on the *aluminum (2)* material and open *Fluent Database (3)*. Make sure that *Solid (4) Material Type* is set and then select two random materials *(5)* to copy them for further use.

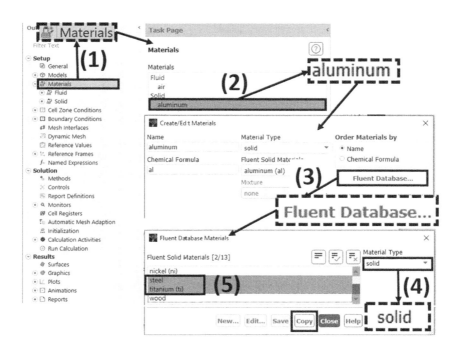

FIGURE 20.31

Being still in *Materials (1)*, click *2× LMB* on the name of any solid material *(2)* added in the previous step. Change its name to ground *(3)* and describe its properties as *Constant* values (Figure 20.32). *Density (4)* is equal to $d = 2000 \left[\frac{kg}{m^3} \right]$, *specific heat (5)* $c_p = 1600 \left[\frac{J}{kg\,K} \right]$, and *thermal conductivity (6)* $\lambda = 2.82 \left[\frac{W}{m\,K} \right]$. Confirm all the changes by clicking *Change/Create*. Repeat this step with the second solid material (not aluminum). Name it to grout and set its properties as: *Density* $d = 2700 \left[\frac{kg}{m^3} \right]$, *specific heat* $c_p = 1400 \left[\frac{J}{kg\,K} \right]$, and *thermal conductivity* $\lambda = 0.73 \left[\frac{W}{m\,K} \right]$.

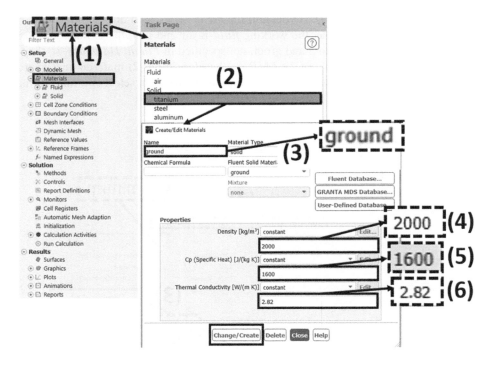

FIGURE 20.32

Find the ***Cell Zone Conditions*** option *(1)* and check if the fluid body type is defined as ***fluid*** *(2)* as in Figure 20.33. Then, click ***Edit*** *(3)* and in the popup window, set the ***fluid*** material as ***air*** *(4)*. Confirm the changes by clicking ***Apply*** *(5)*. Repeat these steps with other ***Solid*** domains: The ***pipe*** material is set as ***aluminum***, and for ***ground*** and ***grout*** select materials with the same name as the domain.

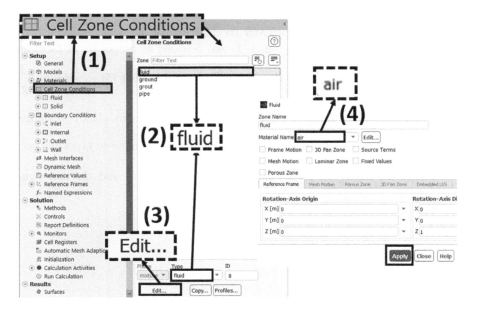

FIGURE 20.33

Tutorial 10 – Borehole Heat Exchanger

It's time to define the first boundary condition – Inlet (Figure 20.34). Click **2× LMB** on the **Boundary Conditions (1)** option and select **Inlet (2)**. Change its **Type** from velocity-inlet to **mass-flow-inlet (3)**. Open the details window by clicking **2× LMB** on the zone name **(2)**. In the **Momentum** tab, find the **Mass Flow Rate** row and click on the dropdown list **(4)** to select the **New Input Parameter** option **(5)**.

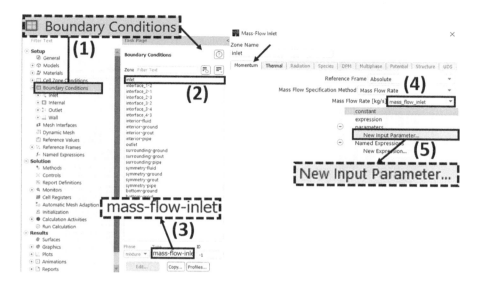

FIGURE 20.34

In the **Parameter Expression** window (as in Figure 20.35), change the default **Name** to mass_flow_inlet and set its base value to *0.02 kg/s* **(2)**. Activate the **Use as Input Parameter** option **(3)** and confirm all the changes with the OK button. Then continue inserting other values in the Momentum tab – set **Direction Specification Method** as **Normal to Boundary (4)** and change the turbulence **Specification Method** as **Intensity and Hydraulic Diameter (5)**. The default value of **Turbulent Intensity** equal to 5% is OK, but **Hydraulic Diameter** is 16 cm **(6)**. In the **Thermal** tab, set **Total Temperature** to *313* K *(7)*.

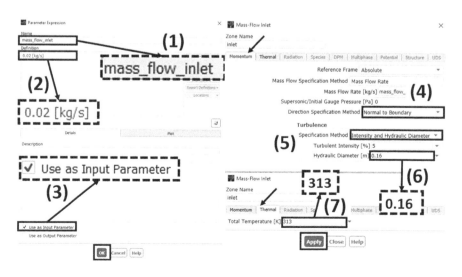

FIGURE 20.35

Being still in the **Boundary Conditions (1)**, move to the **Outlet (2)** definition (Figure 20.36). Its default type is *pressure-outlet (3)*, which describes gauge pressure instead of outlet velocity. In the **Momentum** tab, set **Gauge Pressure** to *0 Pa (4)* and define **Turbulence (5, 6)** as in the case of the Inlet boundary. In the **Thermal** tab, it is required to provide information about **Backflow Total Temperature**, if expected. In this case, leave the default value *(7)* and **Apply** all the changes.

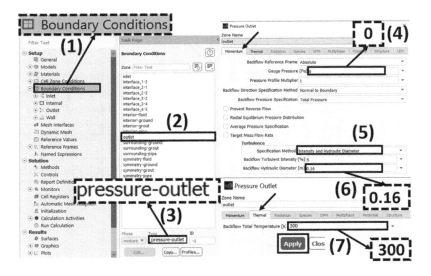

FIGURE 20.36

Now, move to the surrounding-… boundary conditions *(1)*. They are three groups based on the body name where they are located: Ground, grout, and pipe (Figure 20.37). Their type is set by default as **wall (2)** with **Heat Flux** equal to *0*. For each of them change **Thermal Conditions** to **Temperature (3)** with a constant value of *280 K (4)*. Remember to specify the **Material (5)** from which each wall is made. Confirm these settings by clicking **Apply**.

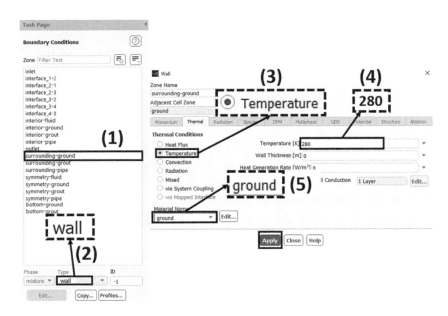

FIGURE 20.37

Tutorial 10 – Borehole Heat Exchanger

The analyzed borehole is relatively shallow (12 m) so the increase in the ground temperature due to the depth may be neglected. However, in Fluent, it is possible to describe the temperature profile with *Named Expression* as in Figure 20.38. In the *Temperature* specification method, select *New Expression (1)* and insert a mathematical formula describing the temperature increase *(2)*. It should contain a reference temperature (usually for the top layer of the ground) and a geothermal gradient (the global average value is 3 K per 100 m). To indicate the local depth, use *Variables*, and from the dropdown list choose *Mesh → Coordinate → Coordinate y (4)*. Pay attention to the location of the geometry about the coordinate system!

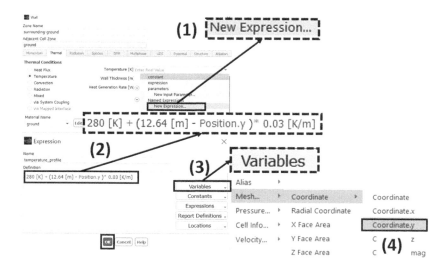

FIGURE 20.38

Among wall-typed boundary conditions, there are also bottom-ground and bottom-grout *(1)* ones. Leave for them the default *Thermal Conditions*, that is, *Heat Flux (3)* equal *0 (4)* as in Figure 20.39. Remember to choose the proper material for each surface *(5)*.

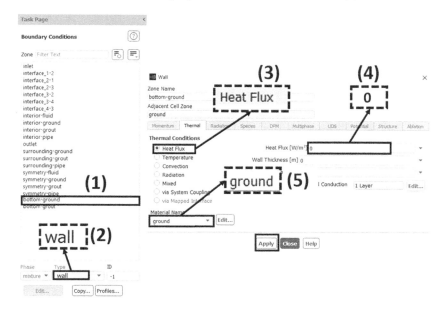

FIGURE 20.39

To provide data transfer between different parts of the geometry, use the ***Mesh Interfaces*** tab *(1)*. Click **2× *LMB*** on this option and in the popup window, create a new interface. Firstly, select Interfaces 1-2 and 2-1 *(2)* and change the ***Name Prefix*** of this pair to intf12 *(3)*. ***Create (4)*** mesh interface, which will be displayed on the list *(5)* as in Figure 20.40. Repeat this step firstly with ***interfaces 2-3*** and ***3-2 creating intf 23***, secondly, move to ***interfaces 3-4*** and ***4-3*** creating ***intf34***. Finally, ***Close*** this window.

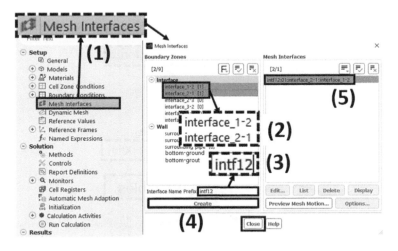

FIGURE 20.40

When the physics and boundary conditions for the model are already determined, define numeric ***Methods (1)*** for the solution process (Figure 20.41). Set the algorithm of pressure-velocity coupling as ***Coupled (2)***, which provides a stable computation process. Check whether ***Spatial Discretization*** for each parameter is based on the ***Second Order*** or ***Second Order Upwind (3)*** schemes, which provide a more accurate solution and reduce the numerical diffusion.

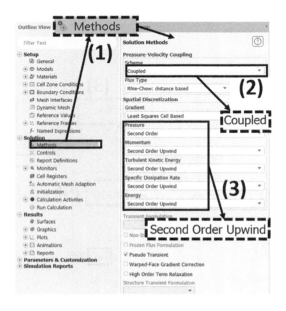

FIGURE 20.41

Tutorial 10 – Borehole Heat Exchanger

To supervise the computation process, create at least one *Report definition (1)*, as in Figure 20.42. Click 2× *LMB* on this option and a popup window will open. Create a *New* one *(2)* by choosing from the dropdown list *Surface Report* with *Area-Weighted Average* type *(3)*.

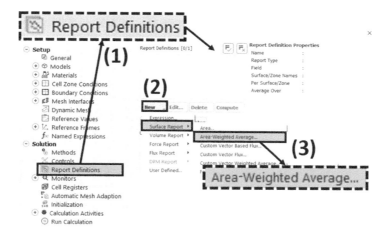

FIGURE 20.42

Change the Report Definition *Name* to temperature_outlet *(1)* and select *Temperature* as *Field Variable (2)* as in Figure 20.43. In this case, there is only one option provided, that is, *Static Temperature (3)*. As the name implies, this parameter should be calculated at the *outlet (4)*. Moreover, activate the *Print to Console* option *(5)* to monitor the changes in the outlet temperature in real-time during calculations. Choose the *Create Output Parameter* option *(6)* and confirm settings with *OK*.

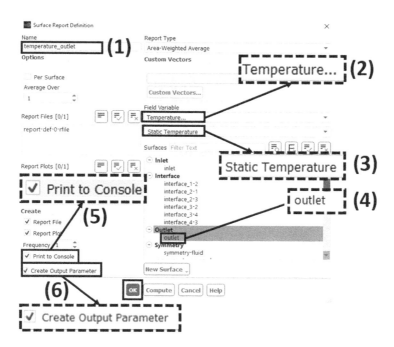

FIGURE 20.43

Expand the *Monitors* tab and select *Residual (1)* by clicking 2× *LMB* on it. In the newly opened window, all the parameters calculated during the solution are listed: *Continuity, x, y, z-velocity, energy, k* and *epsilon* as in Figure 20.44. These parameters are calculated by the solver until the divergence between two subsequent iterations is lower than the defined *Absolute Criteria (2)*. It is possible to lower these values to obtain a more accurate solution. In this analysis, *Show Advanced Options (3)* and set the *Convergence Criteria* as *none (4)*. Now your solution will be calculated with a certain number of iterations without taking into account the limits for divergence.

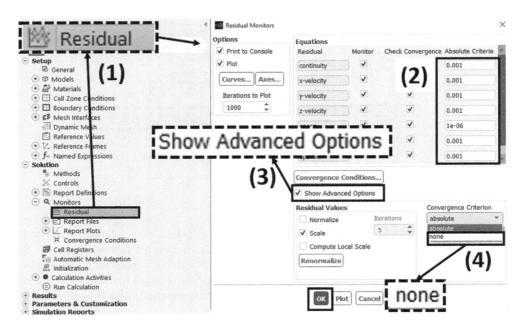

FIGURE 20.44

Initialize the solution process by finding the *Initialization* option *(1)* and selecting the *Hybrid Initialization (2)* mode. Click the *Initialize* button *(3)* and watch carefully the messages displayed in *Console (4)*. There will be ten iterations calculated and a final communicate: *Hybrid Initialization is done (5)*, as in Figure 20.45. Then, move to the next step.

Tutorial 10 – Borehole Heat Exchanger 435

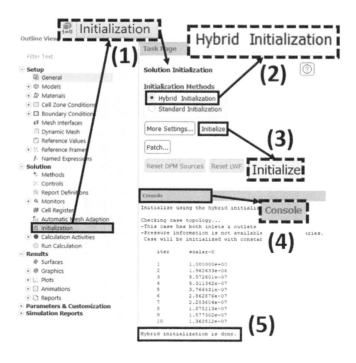

FIGURE 20.45

It is time to **Run Calculation (1)** as in Figure 20.46. Firstly, set **Number of Iterations** to **500 (2)**. Click the **Calculate** button **(3)** and observe the level of the calculation's accuracy and the value of the outlet temperature in the main window. The curves which describe the divergence in the solution for each parameter separately should be heading down. The information that calculation is complete will appear in the solver console and in a popup window.

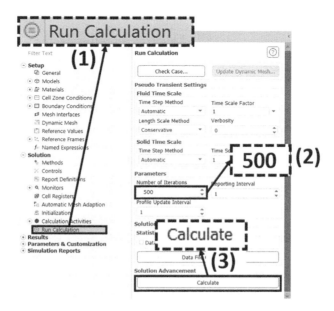

FIGURE 20.46

After 500 iterations you can observe that the outlet fluid temperature is stabilized *(1)* as in Figure 20.47. Moreover, all residuals are below the default **Absolute Criteria** *(2)*. Balancing the energy, using the option **Reports → Fluxes** (see Figure 20.42), is worth consideration. Exit the **Fluent** solver.

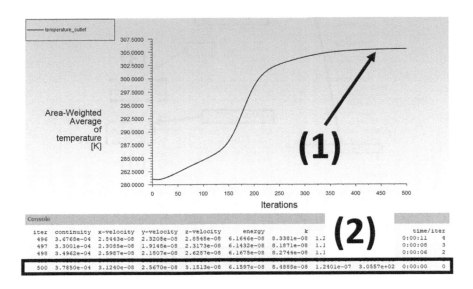

FIGURE 20.47

20.5 VARIANT ANALYSIS

A new cell, called **Parameter Set** *(1)*, is now visible in the **Project Schematic** window as in Figure 20.48. It is connected with the **Fluid Flow** system with two red arrows, which are representing the input and output parameters set in the solver. To insert new values of the input parameter, and therefore to create new design points, click 2× **LMB** on the **Parameter Set** cell.

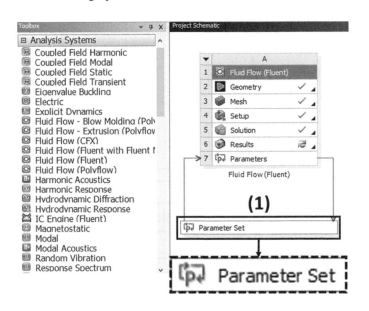

FIGURE 20.48

Tutorial 10 – Borehole Heat Exchanger

In the *Parameter Set* tab *(1)*, enter two new design points by adding new values of mass flow at the *inlet: 0.04 kg/s* and *0.01 kg/s* (column *B*) as in Figure 20.49 *(2)*. These values are twice as high and twice as low as the initial value. To save all the data from the calculation, activate the *Retain (3)* option from column *D*. Then, click *1× RMB* on the name of any design point *(4)* in column *A* and choose from the dropdown list the *Optimize Update Order* option *(5)*, to speed up the calculations. Finally, choose *Update All Design Points (6)* and wait till the end of the calculation. When the specified design point is calculated, the outlet temperature will be displayed in column C. Answer the question: How does initial mass flow/velocity of air influence the outlet temperature? Why?

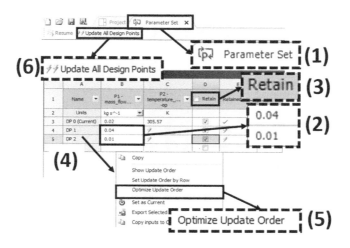

FIGURE 20.49

20.6 POST-PROCESSING

To analyze the obtained results in a more detailed way, come back to the *Project* tab *(1)* and click 2× *LMB* on the *Result* cell (A6 as in Figure 20.50), to launch the *Results* module *(2)*.

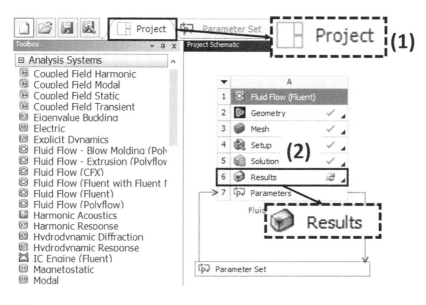

FIGURE 20.50

Select the *Contour (1)* tool and insert *Name (2)*. In the *Details of Contour* window, select *All Domains (3)* and specify *Locations* as all external surfaces. To do this, click on the three dots icon *(4)* and select all the surface names except interfaces (which are inside the geometry). Confirm your choice by clicking *OK* or Enter. Then, choose *Variable* which will be displayed: *Temperature (5)* and its *Range: Global (6)*. In the *# of Contours* cells increase the number of temperature isolines to *50 (7)*. Confirm all settings by clicking *Apply*. Create the second contour – visualize the pressure in the fluid body (Figure 20.51).

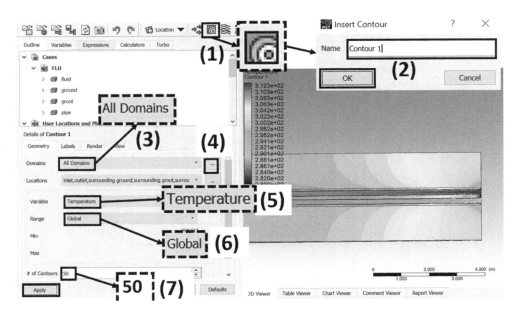

FIGURE 20.51

Deactivate the generated temperature distribution by clicking the tick in the project tree window *(1)*. Now generate a line (Figure 20.52) that will allow you to visualize temperature data as a chart. Open *Location (2)* and select *Line (3)* from the dropdown list. *Name* your location *(4)* and define its details: Line located in *All Domains (5)* with Definition Method *Two Points (6)*. Specify the first point as *(0.00, 10.64, 0.00)* and the second one as *(0.00, 10.64, 3.20)*. These coordinates represent a horizontal line located on the *symmetry* plane at the height of 10.64 m (that is, borehole depth 2 m). By changing the *Y* coordinate, you can easily move the line up and down. Check if *Line Type* is *Cut (8)* and confirm your selections. Using the procedure described in this step, add two other lines and locate them at the heights of *7* and *3 m*.

Tutorial 10 – Borehole Heat Exchanger

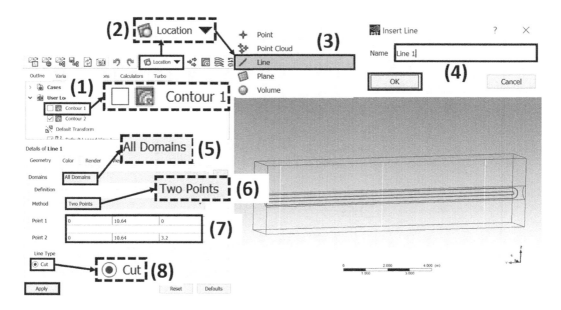

FIGURE 20.52

Find the *Chart* icon *(1)* and add a new chart. You can change its *Name* if necessary *(2)*. In the *Details of Chart* window, find the *Data Series* tab *(3)* and insert *Name* of the new data series: *Temperature1* *(4)* and set its *Location* on *Line1* *(5)* as in Figure 20.53. To add new data series, click on the *New* icon *(6)* and repeat the above steps creating: *Temperature2* on *Line2* and *Temperature3* on *Line3*.

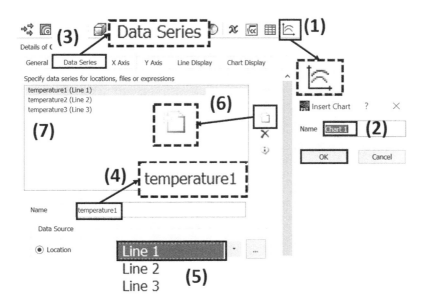

FIGURE 20.53

In the *General* tab, it is possible to choose *Type* of the chart – in this case choose *XY-scatter*, as in Figure 20.54 *(1)*. In the *X-Axis* tab, define the *Variable* as a Z coordinate and set the *Y-Axis Variable* as *Temperature (3)*. In the **Line Display** tab, it is possible to select which data series will be printed. Select all of them *(4)* and confirm your settings with *Apply*. Based on the chart, discuss the changes in the temperature profile in the ground and in the fluid with increasing borehole depth.

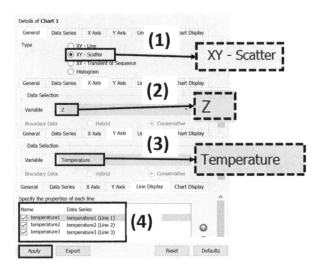

FIGURE 20.54

Now switch the view from *Chart Viewer* to *3D Viewer (1)* and create *Streamline (2)*. Leave the default *Name (3)* and specify the *Domain* as fluid *(4)*. Set the *inlet (5)* as the face from which the streamlines start and *# of Points* as 10 *(6)*. *Apply* the changes to create a new visualization (Figure 20.55).

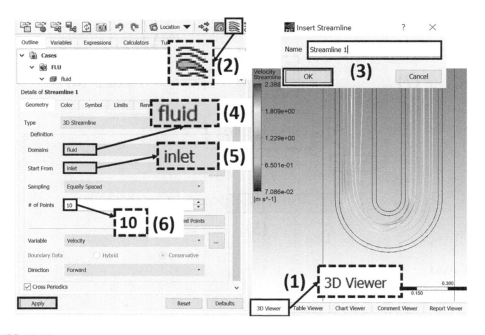

FIGURE 20.55

Tutorial 10 – Borehole Heat Exchanger

Now it is time to import data on two other design points. To do this, open the *File* dropdown list and choose the *Load Results (1)* option. Then, find the location of the folder with your *project_files* and select the following path: *dp1\FLU\Fluent\.....dat.h5* to import the data on design point 1. After successful import, all the created contours, streamlines, etc. should be visible for both cases as in Figure 20.56 *(2)*. Repeat this procedure with design point 2.

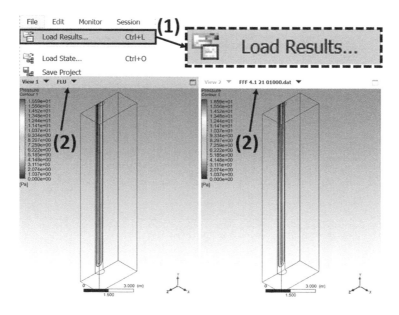

FIGURE 20.56

Index

A

Absorption 61, 64–66, 166, 333
Absorption 124
Animation 261, 370, 371
Animation 261, 287
Autosave 286
Axisymmetric 325, 326, 330, 331
Axisymmetric 331

B

Backflow temperature 99, 336
Betz limit 218
Biomass energy 305, 308
Body Force 26, 43
Body Force Weighted 127
Body Sizing 145, 187, 240, 420
Boolean 291
Boundary condition: 10, 13–15, 97, 151–153, 181, 196–198, 248, 255, 334–337, 389, 390, 407, 408, 429; Dirichlet boundary condition 14; Neumann boundary condition 14, 15; Robin boundary condition 15
Boundary layer 11, 26, 32, 35, 36, 37, 39, 120, 144, 189, 275, 278, 281, 331, 357
Boundary layer 117
Boussinesq hypothesis 31
Buoyancy 95, 96

C

Characteristic length 44, 184, 270
Chemistry: Fast 44, 45; Slow 43–45
Combine 82–86, 140, 232, 324, 377, 379
Combustion 5, 6, 29, 43, 44–49, 63, 297, 305, 308–312, 314, 315, 318–322, 325, 332, 334, 338, 339, 341; Heterogeneous combustion 47, 48; Homogenous combustion 47; Non-premixed combustion 44, 46, 47; Premixed combustion 44, 46, 47
Computational Cells 3, 5
Connections 237, 294, 425
Conservation Laws 4
Contour 105, 130, 157, 202, 203, 257, 259, 286, 288, 299, 300, 393, 438
Convergence 11, 12, 17, 21–23, 101, 127, 156, 190, 202, 250, 251, 255, 340, 365, 370, 391, 434
Coupled 127, 128, 154, 197, 199, 201, 285, 432
Coupled algorithm 23
Cyclones 311, 312, 347

D

Discrete Ordinates 111, 123, 133
Discrete Phase Model 52, 335

Discrete Phase Model 347, 360
Discretization 4–6, 9, 11, 14–17, 20–23, 123, 127, 128, 154, 201, 250, 285, 338, 364, 432
Domain: Computational Domain 17, 19, 23, 45, 77, 88, 122, 278; Continuous domain 4; Discrete domain 4; Physical domain 9
Drag coefficient 218
Dynamic mesh 221
Dynamic Mesh 249, 250

E

Eddy: Eddy dissipation concept (EDC) 44, 46; Eddy dissipation model (EDM) 44, 45; *Eddy dissipation model (EDM)* 325, 333; Eddy viscosity 31, 32, 36–39; *Edge sizing* 274, 293, 421
Element size 87, 88, 143, 145, 237, 240, 241, 273, 274, 293, 328, 380, 418, 420
Emissivity 71, 73, 126, 153
Enclosure 181, 184, 185, 187, 189, 195, 270
Energy cascade 33, 34
Enhanced Wall Treatment 37
Enhanced Wall Treatment 194, 246
Equation: Continuity Equation 13, 17, 23, 25, 26; Energy Equation 27, 37, 96, 122, 133; *Energy Equation* 147, 191, 322, 372, 411; Governing Equations 10, 43, 101–103, 128; Momentum Equation 5, 17, 23, 26, 30, 134, 364; Equation of State 13; Navier-Stokes Equations 3, 4, 5, 15, 30; Transport Equation 16, 18–22, 25, 27, 30, 33, 35–37, 47, 49, 52, 122, 123, 131, 251, 255, 288, 360, 364

F

Face sizing 240, 273, 274
Fill Factor 173, 174
Finite: Finite Difference Method 4, 15; Finite Element Method 4, 15, 16; Finite Rate Model (FRM) 44, 45; Finite rate/Eddy dissipation model (FREDM) 44; Finite Volume 4, 10, 15–17, 25; Finite Volume Method 4, 15–17, 25, 27
Flame 44, 46–48
Flows Supersonic 10
Fluidized bed 49, 50, 308–310, 315–317

G

Gas cleaning 311, 312
Gasification 45, 48, 50, 52, 63, 308, 313, 314, 316–318, 320–322, 325, 332
Geometry design 9
Geothermal energy 397, 399–401, 403, 406
Gravity 26, 62, 96, 122, 148, 166, 312, 331, 347, 359, 385, 426

443

H

Heat: Heat exchanger 29, 49, 50, 62, 169, 309, 310, 317, 318, 321, 373–375, 385, 386, 392, 395, 396, 401, 403, 406, 409, 411; Heat transfer 3, 5, 6, 17, 18, 48, 50–52, 60, 64, 66, 69, 71, 96, 125, 126, 134, 149, 153, 169, 193, 197, 199, 247, 309, 316, 331, 361, 385, 403, 404, 426; *Heat Transfer* 99, 126, 389, 390; Heat transfer Coefficient 50, 69, 71, 99, 126, 390; Heat receiver 63, 135, 148, 158

Hydraulic Diameter 110
Hydraulic Diameter 151, 248, 336, 363, 429

I

Ideal gas: *Air Ideal Gas* 385; Incompressible Ideal Gas 27
Imbalance 17, 21, 22, 102, 110, 255, 288
Incidence Angle Modifier 73
Inflation 144, 149, 189, 191, 241, 263, 275, 331, 357, 285
Inflation 144, 241, 275, 276
Injection 49, 51, 363, 366, 402
Injections 360–362, 365, 367
Isolines 157, 342, 393, 438
Isothermal 15, 17, 25, 122, 247, 282, 314
Isotropic 30, 36
Isotropic Phase Function 124

K

k-epsilon 23
k-epsilon 194, 246, 331, 360, 385, 426
Kolmogorov length scale 33
k-omega 122, 149, 246, 281

L

Lagrangian Particle Tracking 49
Laminar flow 45, 110
Lax's theorem 22
Lift force 6, 212, 218, 219, 223

M

Material properties 14, 93, 121, 173, 193, 244
Material Properties 94, 386
Maximum Power Point 173
Mesh: Mesh independence 11; Mesh interface 248, 432; *Mesh metrics* 146, 325, 329, 382, 422; Mesh motion 247, 281; Mesh quality 11, 12, 35, 89, 122, 142, 145, 146, 190, 192, 280, 328, 355, 359, 371, 381, 382, 422; Mesh refinement 11, 328; Mesh structured 2, 186, 373; Mesh unstructured 12
Mixing length 32
Model: Discrete Phase Model 52, 347, 360; Eddy Dissipation Model 44, 45, 325; *Eddy Dissipation Model* 325, 333; Eulerian Model 51; Finite rate model (FRM) 44, 45; Finite rate/Eddy dissipation model (FREDM) 44; Lagrangian Model 49, 50, 51; Zero-equation Model 43, 44
Monitor Points 101, 107, 392, 393
Multiphase flow 48, 49–52
MultiZone 381

N

Named Expression 431
Named Selection 90, 91, 95, 97, 98, 115, 147, 185, 239–242, 276, 277, 294, 330, 352, 354, 383, 384, 423, 424
Natural Convection 26, 63, 122, 124, 127, 132
Number: Damköhler number 43, 44, 45; Reynolds number 4, 5, 29–32, 34, 37, 43, 96, 110, 221

O

Operation pressure 27
Optical efficiency 71
Orthogonal Quality 12, 88, 146, 191, 280, 329

P

Parameter set 436, 437
Particle tracks 365
Particulate matter 29, 311–313, 347, 360
Pattern 12, 223, 232, 419
Pattern 232, 291, 292
Phase: Continuous phase 49, 51, 52, 365, 367; *Continuous phase* 360; Dispersed phase 48, 49, 51; Photovoltaics 163, 165, 166, 170, 174; Polyhedral Mesh 11, 111, 114, 118, 120, 181, 353; Postprocessing 9, 23, 77, 104, 109, 129, 157, 202, 245, 256, 297, 341, 365
Power: Power coefficient 217, 218; Solar Power 57, 58, 60, 63; Pressure-Velocity coupling 22, 23, 154, 338, 432
Pressure-Velocity coupling 201, 250, 285, 364
Probability density function 47
Projection 141, 231, 232
Pseudo transient 17
Pseudo Transient 201, 338
Pyrolysis 45, 51, 308, 315, 316, 318–320, 325

R

Reference values 284
Report definition 154, 155, 200, 252, 253, 285, 338, 339, 433
Residuals 9, 103, 128, 202, 251, 252, 288, 340, 436
Reynolds: Reynolds Decomposition 29, 30; Reynolds Stress Tensor 5, 31
Rotor Solidity 221

S

Scalable wall function 331, 385
Scheme: Central Differencing Scheme 21; *Upwind Scheme* 338
Share topology 84, 115, 118, 185, 236, 271, 356
Shockley-Queisser limit 172
Siegert method 321
SIMPLE 250, 285, 364
SIMPLEC 285

Index

Simulation: Direct Numerical Simulation (DNS) 4; Large Eddy Simulation (LES) 5; Reynolds Averaged Numerical Simulation (RANS) 5
Sketching 78–82, 112, 137–140, 182, 224, 225, 227, 233, 235, 264–266, 269, 290, 326, 348, 349, 351, 375, 376, 412, 416
Skewness 11, 12, 142, 146, 243, 244, 328, 355, 381, 422
Solar: Solar calculator 123, 194, 205; Solar collector 61–63, 66, 70, 72, 73, 77, 108, 109, 111, 163, 166, 169, 176
Solar Power 57, 58, 60, 63
Solver: *Solver Control* 101, 103, 391; Density Based Solver 13; Pressure Based Solver 13; Solving 9, 15–17, 35, 52, 103, 117, 288
Spatial Discretization 17
Spatial Discretization 127, 154, 201, 250, 285, 338, 364, 432
Split body 84, 86, 114, 378
Steady state 13, 17, 148, 193, 201, 245, 331, 405, 426
Streamlines 9, 23, 49, 181, 395, 440
Surface: *Surface Integrals* 131, 134, 367; *Surface-to-Surface* 181, 194; Surfaces from sketches 268
Sweep 79, 81, 88, 135, 138, 139, 227, 376, 411, 417, 419, 421

T

Time Scale Factor 128
Timestep 261, 287, 288, 299

Transient 13, 17, 201, 245, 251, 255, 263, 299, 407
Turbulences 30, 34, 43, 46, 311
Turbulence dissipation 33, 35–37
Turbulence intensity 212
Turbulence Intensity 214, 248
Turbulent kinetic energy 21, 33–35, 37
Turbulent Kinetic Energy 285

U

Under-Relaxation Factors 250

V

Vacuum-tube 111, 131
Vectors 204, 259, 394, 294, 379
Vectors 127, 130, 202, 379
View Factors 198, 199
Volume Rendering 105, 107

W

Wind turbine: Horizontal Axis Wind Turbine 209, 214; Vertical Axis Wind Turbine 209, 263

Y

Y-plus 36